T0297556

Plant Functional Diversity

Plant Functional Diversity

Organism Traits, Community Structure, and Ecosystem Properties

Eric Garnier

Director of Research at the Centre National de la Recherche Scientifique
Centre d'Ecologie Fonctionnelle et Evolutive (UMR 5175), Montpellier, France

Marie-Laure Navas

Professor of Ecology
Montpellier SupAgro, Montpellier, France

Karl Grigulis

Research Engineer at the Centre National de la Recherche Scientifique
Laboratoire d'Ecologie Alpine (UMR 5553), Université Joseph Fourier, Grenoble, France

OXFORD
UNIVERSITY PRESS

Great Clarendon Street, Oxford, OX2 6DP,
United Kingdom

Oxford University Press is a department of the University of Oxford.
It furthers the University's objective of excellence in research, scholarship,
and education by publishing worldwide. Oxford is a registered trade mark of
Oxford University Press in the UK and in certain other countries

First Edition published in 2016

Impression: 1

Published in the United States of America by Oxford University Press
198 Madison Avenue, New York, NY 10016, United States of America

British Library Cataloguing in Publication Data

Data available

Library of Congress Control Number: 2015944354

ISBN 978–0–19–875736–8 (hbk.)
ISBN 978–0–19–875737–5 (pbk.)

Printed and bound by
CPI Group (UK) Ltd, Croydon, CR0 4YY

To Benoît, Coline and Violette who have been hearing about this book for (too) long

To our parents, for whom (scientific) ecology remains a mysterious planet

As it happens that classifying objects
into categories clarifies their study,
it is useful to do so for those that lend themselves to this exercise.
Here are the fundamental and essential categories
which comprise, as it seems, most if not all plants:
trees, shrubs, subshrubs, and herbs plants.

(Theophrastus, c. 315 BC)

To the detriment of ecology as a predictive science,
the evidence of a consistent predisposition
in the fundamental architecture of evolution
has been obscured by a focus on micro-evolution and,
until recently, a tendency to ignore widespread evidence
of repeated patterns in macro-evolution.

(Grime & Pierce, 2012)

This book is based on 'Diversité Fonctionnelle des Plantes – Traits des Organismes, Structure des Communautés, Propriétés des Ecosystèmes' authored by Eric Garnier and Marie-Laure Navas, and published in 2013 by De Boeck. It has been substantially enriched compared to this French version, and some chapters have been extensively revised and completed.

Foreword

As society seeks to meet the needs of a growing human population and rising aspirations for economic consumption, there has been a corresponding global decline in biodiversity and other benefits that society receives from ecosystems. These changes have accelerated over the last sixty years and may be approaching or exceeding the limits of tolerable environmental change.

Perhaps the most serious of these changes is the loss of biodiversity because, when species extinction occurs, it is irreversible. Species extinction is occurring orders of magnitude more rapidly than the evolution of new species. We are in the midst of the sixth major extinction event in the history of life on Earth—the first that has been biologically mediated—specifically by human activities that alter habitats and environment at globally significant scales.

Given the rising public concern over biodiversity loss, there has been considerable progress in specifying where biodiversity is declining most rapidly and the major factors that are driving this decline. Other research has documented the groups of organisms that have suffered the greatest decline and the impact of these losses on the overall diversity in the tree of life. Together this research provides critical information for designing conservation strategies that prioritize efforts to reduce the rates of biodiversity loss.

A critical gap in developing targeted biodiversity-conservation strategies is our incomplete understanding of the mechanisms by which biodiversity influences community and ecosystem processes. Sadly, we therefore do not fully understand the functional consequences of biodiversity loss, making it difficult to provide a compelling rationale for biodiversity protection to policy makers and managers who are tasked with deciding conservation priorities among many competing demands that are intended to benefit society.

This book takes a major step toward filling this conceptual and practical gap by providing a framework for describing plant functional diversity and its impacts on the biotic composition and functioning of managed and natural ecosystems. It draws on a broad spectrum of research focused on functional approaches to plant physiology, community dynamics, biogeography, and the dynamics of managed and unmanaged ecosystems to identify the linkages between plant traits and the functional consequences that are important to ecological dynamics and to society. Plant traits are critical to any feedback involving vegetation because they influence both plant *responses* to the biophysical environment and plant *effects* on the environment and on other species. The book incrementally assembles the complexity of trait-based interactions in ecosystems, beginning with the functional characterization of individual plants, then moving to the distribution of traits in entire communities. Finally, this trait distribution of communities influences pools and fluxes of materials in ecosystems in ways that govern ecosystem properties and the benefits (ecosystem services) that ecosystems provide to society.

The functional-traits framework developed in this book provides a valuable conceptual framework and pragmatic tools to address a wide range of ecological issues, including the distribution of plants along environmental gradients, the responses of ecosystems to disturbances and other environmental changes, and the management of agricultural and other ecosystems to meet the needs of a growing human population. This book will therefore be valuable to a wide range of ecologists, managers, and policy makers who seek to understand why vegetation is changing and how this affects biodiversity, ecosystems, and society as a whole.

F Stuart Chapin III
University of Alaska Fairbanks

Acknowledgements

This book is the result of many interactions and exchanges with a large number of people whom we would like to thank here:

- The members of the ECOPAR research group (Ecologie Comparative des Organismes, Communautés et Ecosystèmes) at the Centre d'Ecologie Fonctionnelle et Evolutive (UMR 5175), with which a large part of the research presented in this book has been carried out. Catherine Roumet and Cyrille Violle have played a major role in the development of a shared vision of the concepts and experimental approaches presented in the text, together with Denis Vile and Eléna Kazakou, and more recently, Maud Bernard-Verdier, Marie-Angélique Laporte, and Karim Barkaoui.
- Sandra Lavorel who, with her vision and energy, has greatly contributed to the development and spread of the trait-based approach to ecology, and more recently to its application in the context of ecosystem services.
- Hendrik Poorter, for his rigour and difficult questions, despite these sometimes being somewhat rough and destabilizing.
- The members of the 'TRAITS' Research Group (GDR 2574 of CNRS), and in particular Pablo Cruz, Frédérique Louault, Pascal Carrère, Stéphanie Gaucherand, Anne Bonis, and Bernard Amiaud. This Research Group has served as an exchange forum for a large number of the ideas presented in the text, and allowed for the establishment of large-scale coordinated experiments on French grasslands.
- The members of the Montpellier-Sherbrooke International Research Group, in the context of which productive exchanges were established with Bill Shipley.
- The members of the Glopnet network and the consortium of the VISTA European project, which have both contributed to considerable advances in the field of plant traits, in particular via the development and sharing of large trait databases.
- Alison Munson, whose foresight and persistence led to the establishment of a series of ongoing international courses on the traits of organisms.
- Gaëlle Damour and Sabrina Gaba, who helped us test the approach presented in this book in cultivated systems, with enthusiasm and dedication.

We would also like to thank the CEntre for the Synthesis and Analysis of Biodiversity (CESAB) from the French Foundation for Research on Biodiversity (FRB), who graciously provided us with working space in its premises during decisive moments in the writing of this book. Many thanks to Magali Grana and Baptiste Laporte for their warm welcome on these occasions.

Special thanks to our colleagues who have provided us with feedback on one or several of the chapters: Terry Chapin, Isabelle Chuine, Hans Cornelissen, Francesco de Bello, Michel Duru, Bruno Fady, Grégoire Freschet, Jens Kattge, Michael Kleyer, Sandra Lavorel, Alison Munson, Robin Pakeman, Hendrik Poorter, Bill Shipley, Evan Weiher, and Ian Wright.

Acknowledgements

Contents

Glossary

This glossary groups those terms for which a definition is given in the text, excluding general concepts of ecology and traits. For the latter, definitions for a small set of traits are given in Table 9.5 in Chapter 9 and for a broader range of terms see the online thesaurus of plant traits (www.top-thesaurus.org).

Term	Definition
Assembly rules	Rules defining the composition of communities and determining which subset of species existing in a given geographical region can be found existing together in a given habitat
Attribute	The value or state of a trait determined in a given location at a given time
Autecology	An ecological approach whose objective is to understand as precisely as possible the relationships between a single species and the biotic and abiotic factors of the environment
Biological diversity	The living tissue of the planet (a synonym of biodiversity)
Biome	A grouping of terrestrial ecosystems that are similar in vegetation structure and physiognomy, in the major features of the environment to which this structure is a response, and in some characteristics of their animal communities
Comparative ecology	An ecological approach where the objective is to determine patterns or general laws from comparisons between entities corresponding to different levels of organization (organisms, higher-level taxa, communities, ecosystems, etc.); applied at the level of species, it involves particular measurements taken from a large number of entities using screening methods
Competitive exclusion	The principle according to which two species using the same limiting resource cannot coexist, with one species eventually excluding the other based on their competitive hierarchy
Complex gradient	An environmental gradient induced by the variation of a number of factors. This is notably the case for indirect gradients
Controlled vocabulary	A list of key terms and their definitions, established by a community of users; a controlled vocabulary can be seen as a language spoken by people within a field of knowledge so as to be mutually understood
Convergence	Characterizes a trait distribution in which there exists a greater similarity between species/individuals within a selected community than that which would be predicted by a neutral model. We can also term this an aggregated structure
Crop weed	Plant growing in a cultivated field without having been intentionally established there
Data integration	The entirety of the procedures aimed at matching and combining information originating from different sources in a pertinent and useful manner
Direct gradient	An environmental gradient in which the factor which varies has a direct impact on plants
Disturbance	An event which is discontinuous in time and which provokes a modification in the structure of an ecosystem, community, or population, and which leads to changes in the availability of resources or substrates, or in the physical environment. Sometimes defined in a more restrictive manner as an event leading to a loss of biomass
Divergence	Characterizes a trait distribution for which there exists a greater difference between species/individuals within a selected community than that which would be predicted by a neutral model. We can also term this an over-dispersed structure

Term	Definition
Dominance hypothesis	States that it is the traits of the dominant species which have the greatest influence on ecosystem properties (also termed 'mass ratio hypothesis')
Ecoinformatics	A field of research and development focused on the interface between ecology, computer science, and information technology
Ecological niche	A hypervolume of n dimensions (relating to n environmental gradients) within which an organism can grow, survive, and reproduce
Ecological performance	The response curve of the performance traits of an organism to an environmental factor. Related to 'habitat preference' and 'ecological preference'
Ecological strategy	A group of genetically similar or analogous genetic characteristics which are frequently found together within species or populations, and which induce ecological similarities between these species or populations (see Chapter 4 for a slightly different definition)
Ecosystem	A bordered ecological system consisting of all of the organisms found within these borders together with the physical environment with which they interact
Ecosystem properties	Pools (quantities) and fluxes (flow) of materials and energy in ecological systems
Ecosystem services	Benefits that human populations obtain either directly or indirectly from the functioning of ecosystems
Effect trait	A trait which influences the properties of an ecosystem (structure and/or functioning)
Environmental gradient	A gradual change in a given biotic or abiotic environmental factor through space or time
Filtering effect (of an environmental factor)	A restriction by an environmental factor of the range of trait values present in a community in comparison to those found in the totality of the species pool present in a given geographical region
Function	A specific activity carried out by part of a whole or by the whole itself
Functional divergence	Describes the unevenness in the distribution of trait values in the volume of the functional space occupied by the species/individuals of a community
Functional diversity	The variation in the degree of expression of functions between the different individuals of a population, between populations of the same species, between species, or between ecosystems. It was initially defined as being the value, the range, and the relative abundance of the traits measured in a given community; however, when it refers to the distribution of trait values the term 'functional structure' should be preferred, as the term 'diversity' is often used as being synonymous with 'divergence'
Functional evenness	Describes the regularity of the distribution of trait values in the volume of functional space occupied by the species/individuals or a community
Functional group (or type)	A non-phylogenetic group of species which respond in a similar manner to variation in environmental factors (functional response group) or which have similar effects on ecosystem properties (functional effect group)
Functional marker	A trait linked to a plant function, which is relatively easy to observe and rapid to quantify (also called a 'soft trait', a term which is best not used due to possible confusion: cf. Chapter 2)
Functional richness	Volume of the functional space occupied by the species/individuals of a community
Functional structure	The distribution of trait values measured in a given community; this can be described by quantifying the different associated moments (previously called 'functional diversity': cf. definition above)
Functional trade-off	Corresponds to the fact that there cannot be an optimization of the values of all trait values in a given environment: a trait value favourable in one environment may be unfavourable in another
Functional trait	A trait indirectly influencing the fitness of an individual via its effects on growth, reproduction, or survival
Fundamental niche	The part of the niche hypervolume occupied by an organism in the absence of interactions with other organisms
Genetic variability	The phenotypic variability between individual genotypes
Indirect gradient	A gradient for which the environmental factor used to organize the observations does not in itself have a physiological impact on plant functioning
Leaf economics spectrum	Characterizes a suite of leaf traits which opposes plants with a high rate of resource acquisition and poor resource conservation ('fast and leaky') with others having opposite characteristics ('slow and tight')

Term	Definition
Limiting similarity hypothesis	Hypothesis stating that species cannot coexist unless they occupy ecologically different niches, that is to say that they are differentiated along at least one niche axis, thus limiting the negative effects of biotic interactions on their performance
Meta-analysis	The statistical synthesis of results obtained in different studies
Metadata	Information about a data set which is necessary to understand and interpret the data, such as the contents of the data set, their experimental context, their structure, their accessibility, etc.
Meta-phenomics	A methodology derived from meta-analysis which determines in a systematic manner the intraspecific response curves of different traits to environmental factors ('dose-response curves')
Neutral theoretical model	A model hypothesizing the absence of structuring by ecological factors; it can be tested by the use of null models
Niche complementarity hypothesis	States that it is the presence of species using environmental resources in a complementary manner that has the greatest influence on ecosystem properties
Null model	A model based on the randomization of ecological data or of samples drawn from a statistical distribution; it is used to obtain the expected pattern in the absence of any influence from ecological mechanisms
Ontology	A formal representation of the concepts of a field of knowledge and the relationships between these concepts
Performance trait	A trait directly linked with fitness. There are three of these: plant biomass, reproductive success (e.g. number and biomass of seeds), and survival
Phenotypic plasticity	The potential of each genotype to produce different phenotypes under differing environmental conditions
Phylogenetic conservatism	The propensity of species to conserve ancestral ecological characteristics, which can include trait values. Traits which are poorly conserved within a phylogeny are termed labile
Phylogenetic diversity	The diversity of living organisms established on the basis of the degree of evolutionary relatedness between them
Plant community	The individuals of all species that potentially interact within a single patch or local area of habitat
Realized niche	The part of the niche hypervolume occupied by an organism when it is in interaction with other organisms present
Resources	An environmental factor consumed by a plant, the quantity and/or quality of which is modified by this activity, and which allows a plant to grow, survive, and reproduce
Response trait	A trait whose values respond to variation in environmental conditions, whether these be biotic or abiotic in nature; this response is hypothesized to be adaptive
Screening	A method which consists of measuring one or multiple traits simultaneously on a large number of species or populations
Secondary succession	The phenomenon of the colonization of a biotype by living organisms and the changes over time in the floristic and faunistic compositions of a site after a disturbance has partially or totally destroyed the pre-existing ecosystem
Simple gradient	An environmental gradient induced by variation in one single factor
Standard	A published reference, the use and diffusion of which is widespread and accepted by a large proportion of a given community. A standard is different from a norm, which is a reference established by consensus and approved by a recognized organization
Suite of traits	A group of traits correlated within an ensemble of species, genotypes, or individuals
Taxonomic diversity	The diversity of living organisms established on the basis of the evaluation of the numbers of taxonomic units
Thesaurus	Permits the organization and structuring of a controlled vocabulary on the basis of the semantic relationships of hierarchy, association, or equivalence
Trait	Any morphological, physiological, or phenological *heritable* feature measurable at the individual level, from the cell to the whole organism, without reference to the environment or any other level of organization
α niche	A component of the niche at a small spatial scale corresponding to the differences in ecological preferences between species coexisting within a single community
β niche	A component of the niche corresponding to the differences in ecological preferences between species expressed at the spatial scale of macrohabitat factors or of climate

Table of abbreviations

This table regroups the abbreviations frequently used in this book, and their meanings.

Abbreviation	Signification
$[m]_T$	Mineral element concentration in green biomass
$[m]_{Lit}$	Mineral element concentration in litter
AGB	Above-ground standing biomass
AGB_0	Initial above-ground standing biomass
AGB_{max}	Maximum above-ground standing biomass over the growing season
AGR	Absolute Growth Rate
A_{max}	Mass-based maximum leaf photosynthetic rate
ANPP	Above-ground Net Primary Productivity
B_{Litter}	Litter biomass of a plant community
B_{Living}	Living biomass of a plant community
$B_{root\ (i,j)}$	Root biomass of species j in stratum i
CArM	Community arithmetic mean (unweighted) of trait values
CFP	Community Functional Parameter
CSR	Competitor (C), Stress tolerator (S) and Ruderal (R) Strategies
CWM	Community Weighted Mean of trait values
Decomp	Decomposition rate
d_{root}	Root diameter
e	Relative rate of biomass loss
EP	Ecosystem Properties
$E_{\Delta t}$	Soil water evaporation over a period of time Δt
FDv	Functional Divergence
GCTE	Global Change and Terrestrial Ecosystems
Glopnet	Global plant trait network
INRA	Institut National de la Recherche Agronomique (National Agronomic Research Institute)
$k_{i,j}$	Hydraulic conductivity per unit of root area of species j in soil layer i
LA	Area of a single leaf (one side)
LA_T	Total plant leaf area

Abbreviation	Signification
LAR	Leaf Area Ratio
LCC	Leaf Carbon Concentration
LDMC	Leaf Dry Matter Content
LES	Leaf Economics Spectrum
LLS	Leaf Life Span
LMA	Leaf Mass per Area
LMF	Leaf Mass Fraction
LM_{WP}	Leaf mass of the whole plant
LNC	Leaf Nitrogen Concentration
LPC	Leaf Phosphorus Concentration
LRes	Structural resistance of leaves
L_{root}	Root length
L_{Th}	Leaf Thickness
M_{cci}	Mass of the biochemical fraction i
MEA	Millennium Ecosystem Assessment
M_i	Mass of tissue i
M_{Lit}	Mass of litter
$M_{L(D)}$	Leaf dry mass
$M_{L(F)}$	Leaf fresh mass
MLCF	Mass Loss Correction Factor
MRT	Mean Residence Time of a nutrient
M_{Tkt1}	Total plant biomass of species k at time t_1
NAR or ULR	Net Assimilation Rate or Unit Leaf Rate
N_k	Number of individuals of species k
NNI	Nitrogen Nutrition Index
NPP	Net Primary Productivity
OFL	Onset of flowering
p_k	Proportion of species k in community
PNI	Phosphorus Nutrition Index
RA	Root area
r_D	Rate of litter decomposition
R_{eff}	Resorption efficiency
RGR	Relative Growth Rate
RGR_{max}	Maximum Relative Growth Rate
R_m	Mass-based dark respiration rate
RMF	Root Mass Fraction
RNC	Root Nitrogen Concentration
RPH	Reproductive Height

Abbreviation	Signification
RTD	Root Tissue Density
$R_{\Delta t}$	Rainfall over a period of time Δt
SLA	Specific Leaf Area
SM	Seed Mass
SNC	Seed Nitrogen Concentration
SNPP	Specific Net Primary Productivity
SOM	Soil Organic Matter
SR	Species richness
$SRA_{i,j}$	Specific Root Area of species *i* in stratum *j*
SRL	Specific Root Length
StMF	Stem Mass Fraction
SWC_{FC}	Soil water content at field capacity
SWC_t	Soil water content at time t
TEEB	The Economics of Ecosystems and Biodiversity
V_i	Tissue volume
V_L	Leaf volume
VPH	Vegetative Height
$W_{ab\Delta t}$	Water absorption by plants over a period of time Δt
$\delta^{13}C$	^{13}Carbon isotope ratio
ΔB_{Litter}	Litter mass variation over a period of time (ΔT)
ΔB_{Living}	Living biomass variation over a period of time (ΔT)
$\Delta B_{Standing}$	Standing biomass variation over a period of time (ΔT)
ρ_F	Fresh leaf density
ψ_j	Water potential of species *j*
$\psi_{soil,i}$	Soil water potential in layer *i*

CHAPTER 1

A functional approach to biological diversity

1.1 Diversity or diversities?

Biological diversity[1] consists of the living world which surrounds us and to which we belong; it represents the living tissue of our planet. Its origin is ancient, as the first forms of life appeared on Earth some three and a half billion years ago. Since then, living organisms have evolved and have transformed to make up the biological diversity that we see today, with all of its multiplicity of forms, functions, and organization. Initially limited to aquatic environments, organisms began to colonize the land slightly less than 400 million years ago. Most of the Earth's environments have been progressively colonized, including some of the harshest areas on the planet.

The quantification of this diversity remains a major challenge, however. One reason for this is related to its plural nature, which poses the question of which entity to take into account. The majority of current quantifications deal with the number of species, reflecting primarily the 'taxonomic facet' of diversity. There are probably some nine million species on Earth[2] today, of which only some 1.2 million have been described (Mora et al., 2011). Beyond these somewhat imprecise estimations, the work of palaeontologists, geneticists, and evolutionists has allowed for the reconstruction of the origins of life and the evolution of organisms. Based on the theory of evolution and the study of fossils, they have been able to propose a classification of life, the tree of life, whose branches are linked on the basis of the degree of relatedness between organisms; each organism being situated in the context of its evolutionary history, and the relationships between organisms representing the 'phylogenetic facet' of diversity. However, the diversity of organisms is also related to other differences which transcend taxonomic boundaries or the degrees of relatedness between them. One of these relates to the diversity of functions carried out by this multitude of organisms in relation to the environments which they occupy: this is the 'functional facet' of diversity.

If we wish to understand the functioning of ecological systems, and ultimately that of our planet, taking into account the functional components of organisms is essential. To illustrate this, it is only necessary to look at the comparison carried out by Lovelock (1979) between the atmospheres of the two planets from our solar system closest to ours, Venus and Mars, with those of the Earth today and of a hypothetical Earth on which life would not exist (Table 1.1).

The message is clear: the atmosphere of our planet, with its exceptionally high concentration of oxygen and its exceptionally low concentration of carbon dioxide, is a product of living organisms. Oxygen, necessary for aerobic respiration in the large majority of known organisms, is a product derived from photosynthesis, the process by which autotrophic organisms synthesize organic matter

[1] The expression 'biological diversity' can be considered as synonymous with 'biodiversity', a term widely used since this neologism was coined during the National Forum on Biodiversity held in Washington (USA) in September 1986 (see for example Le Roux et al., 2009 for the various nuances of this concept).

[2] These estimates should be treated with caution as they aggregate entities under the category of species which are defined in very different manners, for example by botanists, zoologists, and microbiologists.

Table 1.1 Atmospheric composition and mean temperatures at the surface of Mars, today's Earth, Venus, and a hypothetical Earth without life (taken from Lovelock, 1979). Reproduced with permission from Oxford University Press.

Gas/Temperature	Planet			
	Venus	*Earth without life*	*Mars*	*Today's Earth*
Carbon dioxide	98%	98%	95%	0.03%
Nitrogen	1.9%	1.9%	2.7%	79%
Oxygen	negligible	negligible	0.13%	21%
Surface temperature (°C)	477	290 ± 50	−53	13

from solar energy, water, carbon dioxide, and mineral elements. The evolution of the concentration of oxygen in the atmosphere over time has been marked by major changes (Figure 1.1; Holland, 2006), often accompanied by the appearance of new forms of life (cf. Canfield, 2005; Knoll, 2003): initially close to zero, this concentration begins to rise with the appearance of the first oxygen-producing cyanobacteria approximately 2.5 billion years ago, then, after a stable phase, it rises again in parallel with the diversification of terrestrial plants and animals, reaching a maximum during the Carboniferous period before decreasing to its current value. This 'oxygen revolution' allowed for both an extraordinary increase in the primary production of the biosphere and an explosive diversification of different forms of life (Canfield, 2005). In this example it is therefore less the origin and history of organisms that is at stake than their function as oxygen producers.

The importance of functions carried out by organisms and their consequences for ecological systems can be evaluated at different scales of space and time: from the micron to the planet, and from the second to millions of years (Figure 1.2; cf. Osmond et al., 1980). The aim of this book is to present this functional aspect of diversity using terrestrial vegetation as a case study.

1.2 Plants and functions

Before going further, it is necessary to define what is meant by 'function'. A biological definition directly applicable to organisms is 'a mode of action by which a thing fulfils its purpose' (based on the Oxford Dictionary of English). However, as discussed by Calow (1987), 'purpose' raises a philosophical question long debated by biologists: that of the argument from design, similar to the teleological argument of Aristotle, according to which there exists in nature rational design or purpose. Although we acknowledge that this is an important debate, we will not discuss this subject in this book, but, following Calow (1987), have rather opted to use a neutral definition according to which

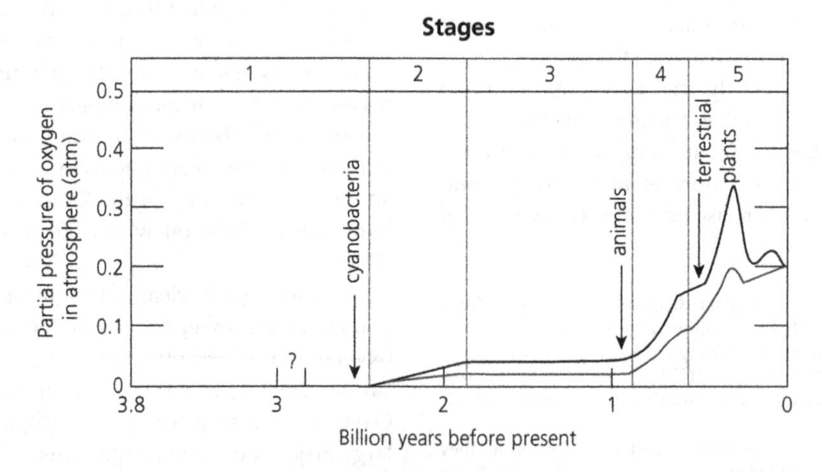

Figure 1.1 Estimated change in oxygen partial pressure in the atmosphere over the last 3.8 billion years. Adapted from Holland (2006). Reproduced with permission from The Royal Society.

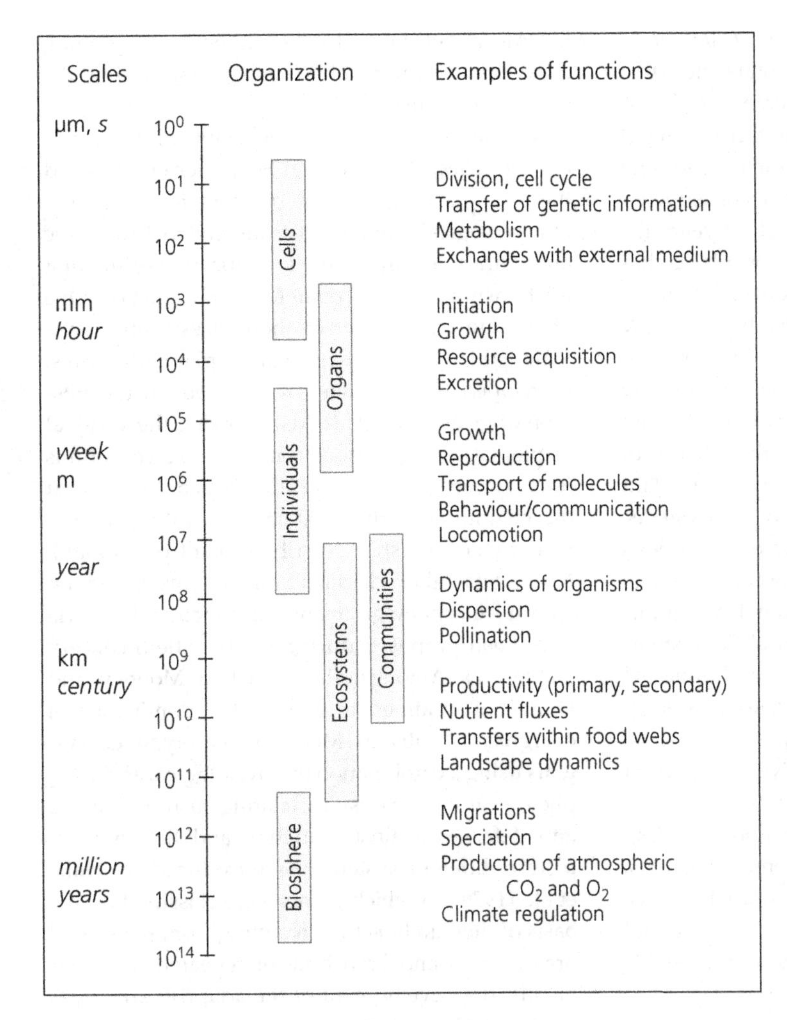

Figure 1.2 Examples of functions, and the scales of space and time at which they are generally studied in terrestrial systems. The envelopes of values given for the scales of space and time (on a logarithmic scale) should be considered as approximate orders of magnitude (adapted from Osmond et al., 1980). Key: µm, micrometre; s, second; mm, millimetre; m, metre; km, kilometre.

a function is 'a specific activity carried out by part of a whole or by the whole itself' (translated from CNRTL: http://www.cnrtl.fr/definition/fonction). It should be noted that this definition is applicable to different levels of organization of the living world, from the cell to that of the planet, and not only to organisms. Figure 1.2 presents some examples of functions at different levels of organization, and indicates the temporal and spatial scales at which they act. It illustrates the great diversity of functions that can be identified depending on the question being posed. The understanding of these functions and of how they are controlled, and the manner in which they are organized within communities and ecosystems, or even at the planetary scale, are major themes of ecological research which will be addressed in the following chapters.

However, as this general scheme can be applied to all living organisms, why only deal in this book with plants? There are two major reasons for this. The first is that plants make up more than 99% of living material in terrestrial ecosystems. As a consequence, their functioning, through the acquisition and transformation of resources, largely determines the quantity and quality of habitats and substrates utilized by other organisms, as well as playing an important role in controlling the Earth's atmospheric composition and climate (Grime, 2003; Keddy, 2007). This leads to the identification of two questions essential for the understanding and prediction of the functioning of terrestrial ecosystems: what are the environmental factors that control the distribution and functioning of plants? What is the impact of plants on the functioning of ecological systems? The second reason for choosing to concentrate on plants is a direct consequence of the first, and is that over the last 20 years the functional approach to diversity for plants has seen major progress due to the development of major conceptual advances (reviewed by Garnier & Navas, 2012; Lavorel et al., 2007). This development is partly explained by the need to simplify the complexity of the plant world[3], with the aim of understanding and predicting the effects of global change or management practices on plant distributions and the properties of ecosystems, both locally as well as at global scales (references in Lavorel et al., 2007). The momentum given to this approach by large international programs such as Global Change and Terrestrial Ecosystems (GCTE: http://www.igbp.net/researchprojects/astprojects/globalchangeandterrestrialecosystems) or DIVERSITAS (http://www.diversitas-international.org/) has been decisively important (Canadell et al., 2007; Smith et al., 1997). It has also shown the usefulness of this approach in providing answers to today's major environmental questions. In the following part of this introduction, we provide a brief historical outline of the functional approach before presenting an outline of this book's structure.

1.3 Functional diversity and comparative plant ecology

Non-taxonomic approaches to plant diversity are very ancient. In fact, the first classification of plants into 'trees', 'shrubs', 'subshrubs', and 'herbs', proposed by the Greek philosopher Theophrastus (372–287 BC), the founder of botany (see Amigues, 2010), contains the germ of the functional approach to the diversity of plants (cf. Weiher et al., 1999). This approach was further developed during the second half of the nineteenth century, following the founding work in biogeography carried out by Alexander von Humboldt, before beginning a period of remarkable development during the twentieth century (Table 1.2).

At the turn of the nineteenth and twentieth centuries, Warming and Schimper emphasized the ecological significance of plant forms, noting that ecological similarities transcended taxonomic boundaries, leading to the identification of the first relationships between plant form and function. One of the most important of these classification systems, still used today, is that of plant life-forms. Developed by Raunkiaer, it is based on the position of growth-points (buds) allowing the survival of plants during seasons with adverse conditions (this system was further developed by Box at the beginning of the 1980s, incorporating in particular characteristics describing plant phenology). Numerous other classification systems based on various morphological and/or functional criteria have been proposed during the twentieth century (Table 1.2). Amongst these, that of Mooney and Dunn (1970) aimed to explain the dominance of evergreen shrubs in Mediterranean-type ecosystems using a combination of physiological and morphological characteristics, relating to resource use and defences against predation and fire. A completely different system was introduced by Hallé et al., (1978), in which plants were classified on the basis of their architecture, resulting from the spatial organization and branching of repeated modules. Finally, the development of the adaptive strategies model of Grime based on a combination of morphological, physiological, and reproductive characteristics was, and continues to be, remarkably influential in plant ecology (see Chapter 4). The reader can find more details on the history of different functional approaches to plant classification in Duckworth et al., (2000), Grime (2001), and Grime and Pierce (2012).

The different characteristics used to describe plants, such as those mentioned above (life-form, bud position, resource use, etc.) are called 'traits'. These traits, which are characters of organisms related to their functioning and influencing their fitness (see McGill et al., 2006; Violle et al., 2007; and Chapter 2 for a definition and description of the concept), are the tools used to implement the functional approach to diversity. They allow us

[3] The number of plants is estimated at some 315 000 species, some 3.5% of the total number of species on the planet (Mora et al., 2011).

Table 1.2 Some important stages in the development of the functional approach to plant diversity from the beginning of the nineteenth century (based on Duckworth *et al.*, 2000, and further completed by Grime & Pierce, 2012). Reproduced with permission from Wiley.

Authors	Comments
von Humboldt (1806)	First recognized relationship between plant form and function. Developed classification based on growth-forms
Grisebach (1872)	Classification of 60 vegetative forms correlated with climate
Warming (1884, 1909)	Classification based on simple life-history features (e.g. lifespan and vegetative expansion power)
Schimper (1903)	Recognised convergences between plant form and function, despite taxonomic differences, between vegetation types from geographically different, but climatically similar, areas
Raunkiaer (1907, 1934)	Life-forms system
Kearney and Schantz (1912)	Proposed four basic strategies of plants in arid regions in response to drought
Braun-Blanquet (1928)	Added further detail to the life-form system
Ramenskii (1938)	Used systematic observations on vegetation in fixed quadrats to record the growth and development of species and to propose primary functional types
Gimingham (1951)	Growth-forms system which also considered branching of stems
Dansereau (1951)	Classification system based on life-form, morphology, deciduousness, and cover
Curtis (1959)	Best known for surveying and classifying the vegetation of Wisconsin, but recently documents have come to light showing that just before his premature death he had received NSF funding to use morphological traits in a functional approach
Küchler (1967)	Hierarchical classification, with an initial division of plants into woody or herbaceous. Lower-order groups are based on life-forms, leaf characteristics, and cover
Mooney and Dunn (1970); Mooney (1974)	Investigation of form–environment relationships in the context of convergent evolution
Hallé *et al.* (1978)	Models of tree architecture based on the underlying 'blueprint' for development rather than morphology at any given moment
Box (1981)	Developed a global classification based on structural and phenological attributes in relation to climate
Grime (1974, 1979)	Plant strategy theory and CSR system ('competitive', 'stress-tolerant' and 'ruderal') of plant functional types
Noble and Slatyer (1980)	'Vital attributes' classification of plants on basis of life history factors in relation to response to disturbance

to understand the interactions between organisms and the components of their environment, whether these be biotic or abiotic, and to establish links between the different levels of organization present in ecological systems. Their use will be omnipresent in this book, reflecting the extraordinary expansion over the last three decades of the field of trait-based ecology, an approach in ecology based on the utilization of traits (see in particular the reviews in Garnier & Navas, 2012; Lavorel et al., 2007; Naeem & Bunker, 2009).

1.4 Questions addressed by a functional approach to diversity

The development of a functional approach to diversity requires the reformulation of certain questions relative to ecological systems and their components. While taking into account the taxonomic aspect of biodiversity remains pertinent to answering many of these questions, the functional approach to diversity has great potential to answer questions relating to, for example, the distribution of organisms along environmental gradients (e.g. Chapin et al., 1993; Grime, 1979; and Westoby et al., 2002), the identification of rules governing community assembly (e.g. McGill et al., 2006; Shipley, 2010; and Suding et al., 2008), and understanding how the functioning of organisms translates to effects at the level of the ecosystem (e.g. Chapin, 1993; Lavorel & Garnier, 2002; Reich et al., 1992) and how it controls certain services that ecosystems provide to human society (Díaz et al., 2006, 2007).

The aim of this book is to show how taking into account the functional facet of biodiversity improves our understanding of these important

Table 1.3 The main questions addressed in this book (inspired by Keddy, 1990, and by Bill Shipley, personal communication), and the chapters in which they are discussed.

Questions addressed by a functional approach to diversity	Chapter
What is a trait, and in what ecological context are traits used?	*2: Traits and comparative ecology: definitions, methods, and conceptual framework*
What are the fundamental traits?	*3: Plant functional characterization*
How do traits vary along environmental gradients?	*4: Gradients, response traits, and ecological strategies* *8: Functional diversity in agriculture*
What are the determinants of trait variability in a community?	*5: Community functional structure* *8: Functional diversity in agriculture*
What are the rules of community assembly in terms of traits?	*5: Community functional structure*
How do the values and variability of traits affect the functioning of ecosystems and the delivery of ecosystem services to human societies?	*6: Traits and ecosystem properties* *7: Functional diversity and ecosystem services* *8: Functional diversity in agriculture*
How are data on traits stored and accessed?	*9: Managing functional diversity data*

questions, which are at the core of current ecological research (Table 1.3). After, in Chapter 2, outlining several key definitions and describing the conceptual outline which has guided the organization of this book, Chapter 3 presents how traits can be used to characterize plants from the functional perspective. The relationships between the values of traits and environmental factors will be first presented in Chapter 4, and then revisited in Chapter 8 in an agronomic context. The distribution of trait values in communities, and the biotic and abiotic determinants of these distributions, will be discussed in Chapters 5 and 8, while the relationships between the components of functional diversity and the functioning and services of ecosystems will be presented in Chapters 6, 7, and 8. Recent progress in methods to improve the storage and accessibility of trait data will be discussed in Chapter 9, before the book concludes by proposing some future perspectives for the further development of the functional approach to diversity (Chapter 10).

1.5 Key points

1. Biological diversity, which constitutes the living tissue of our planet, presents a number of facets: taxonomic, phylogenetic, and functional. This work deals with the last of these facets, which can be defined as the variation in the degree of expression of functions at the different levels of organization of the living world. A 'function' is defined as 'an activity carried out by part of a whole or by the whole itself', and can be applied at different levels of organization, from that of the cell to the planet. The understanding of these functions, how they are controlled, and the manner in which they are organized within communities and ecosystems forms the basis of the subjects presented in the different chapters of this book.
2. This book is focused on plants, as these constitute more than 99% of the living matter of terrestrial ecosystems. Consequently, their functioning largely determines the quantity and quality of habitats and substrates used by other organisms, as well as playing an important role in controlling the Earth's atmospheric composition and climate. It is also for plants that the functional approach to diversity has experienced its strongest development over the last 20 years. While being very ancient, non-taxonomic approaches to the diversity of plants experienced an active phase of development during the second half of the nineteenth century, and again even more so during the twentieth century.
3. The characteristics used to describe plants from a functional perspective are called 'traits'. Related to the functioning of organisms, traits modulate their fitness, allow for an understanding of the interactions between organisms and the

components of their environment, and establish links between the different levels of organization of ecological systems.

4. The aim of this book is to show how taking into account the functional facet of diversity improves our understanding of important questions that are at the core of current ecological research, such as, for example, the distribution of organisms along environmental gradients, the identification of rules governing community assembly, and understanding how the functioning of organisms translates to effects at the level of the ecosystem and how it controls certain services that ecosystems provide to human society.

1.6 References

Amigues, S. (2010). *Théophraste: Recherches sur les plantes. A l'origine de la botanique*. Paris: Belin.

Box, E. O. (1981). *Macroclimate and Plant Forms: An Introduction to Predictive Modeling in Phytogeography* (Vol. 1). The Hague: Dr. W. Junk Publishers.

Braun-Blanquet, J. (1928). *Pflanzensoziologie. Grundzüge der Vegetationskunde*. Berlin: Springer.

Calow, P. (1987). Towards a definition of functional ecology. *Functional Ecology, 1*, 57–61.

Canadell, J., Pataki, D., & Pitelka, L. (eds) (2007). *Terrestrial Ecosystems in a Changing World*. Berlin: Springer-Verlag.

Canfield, D. E. (2005). The early history of atmospheric oxygen: Homage to Robert A. Garrels. *Annual Review of Earth and Planetary Sciences, 33*, 1–36.

Chapin, F. S., III. (1993). Functional role of growth forms in ecosystem and global processes. In J. R. Ehleringer & C. B. Field (eds), *Scaling Physiological Processes. Leaf to Globe* (pp. 287–312). San Diego: Academic Press, Inc.

Chapin, F. S., III, Autumn, K., & Pugnaire, F. (1993). Evolution of suites of traits in response to environmental stress. *The American Naturalist, 142*, S78–92.

Curtis, J. T. (1959). *The Vegetation of Wisconsin: An Ordination of Plant Communities*. Madison, WI: The University of Wisconsin Press.

Dansereau, P. (1951). Description and recording of vegetation upon a structural basis. *Ecology, 32*(2), 172–229. doi:10.2307/1930415

Díaz, S., Fargione, J., Chapin, F. S., III, & Tilman, D. (2006). Biodiversity loss threatens human well-being. *PLoS Biology 4(8)*, e277. doi: 10.1371/journal.pbio.0040277.

Díaz, S., Lavorel, S., de Bello, F., Quétier, F., Grigulis, K., & Robson, M. (2007). Incorporating plant functional diversity effects in ecosystem service assessments. *Proc. Natl. Acad. Sci. USA, 104*, 20684–20689.

Duckworth, J. C., Kent, M., & Ramsay, P. M. (2000). Plant functional types: an alternative to taxonomic plant community description in biogeography? *Progress in Physical Geography, 24*(4), 515–542. doi:10.1177/030913330002400403

Garnier, E., &Navas, M.-L. (2012). A trait-based approach to comparative functional plant ecology: concepts, methods and applications for agroecology. A review. *Agronomy for Sustainable Development, 32*, 365–399. doi:10.1007/s13593-011-0036-y

Gimingham, C. H. (1951). The use of life form and growth form in the analysis of community structure as illustrated by a comparison of two dune communities. *Journal of Ecology, 39*(2), 396–406. doi:10.2307/2257920

Grime, J. P. (1974). Vegetation classification by reference to strategies. *Nature* (London), *250*, 26–31.

Grime, J. P. (1979). *Plant Strategies and Vegetation Processes*. Chichester: John Wiley & Sons.

Grime, J. P. (2001). *Plant Strategies, Vegetation Processes, and Ecosystem Properties* (2nd ed.). Chichester: John Wiley & Sons.

Grime, J. P. (2003). Plants hold the key. Ecosystems in a changing world. *Biologist, 50*(2), 87–91.

Grime, J. P., & Pierce, S. (2012). *The Evolutionary Strategies that Shape Ecosystems*. Oxford: Wiley-Blackwell.

Grisebach, A. (1872). *Die Vegetation der Erde nach ihrer Klimatischen Anordnung: Ein Abriss der Vergleichende Geographie der Pflanzen*. Leipzig: Wilhelm Engelmann.

Hallé, F., Oldeman, R. A. A., & Tomlinson, P. B. (1978). *Tropical Trees and Forests. An Architectural Analysis*. Berlin: Springer-Verlag.

Holland, H. D. (2006). The oxygenation of the atmosphere and oceans. *Philosophical Transactions of the Royal Society B: Biological Sciences, 361*(1470), 903–915. doi:10.1098/rstb.2006.1838

Kearney, T. H., & Shantz, H. L. (1912). The water economy of dry-land crops. *Yearbook of the United States Department of Agriculture, 1911* (pp. 351–362). Washington, DC: USA.

Keddy, P. A. (1990). The use of functional as opposed to phylogenetic systematics: a first step in predictive community ecology. In S. Kawano (ed.), *Biological approaches and evolutionary trends in plants* (pp. 387–405). London: Academic Press Limited.

Keddy, P. A. (2007). *Plants and Vegetation – Origins, Processes, Consequences*. Cambridge: Cambridge University Press.

Knoll, A. H. (2003). The geological consequences of evolution. *Geobiology, 1*(1), 3–14. doi:10.1046/j.1472-4669.2003.00002.x

Küchler, A. W. (1967). *Vegetation mapping*. New-York: Ronald Press.

Lavorel, S., Díaz, S., Cornelissen, J. H. C., Garnier, E., Harrison, S. P., McIntyre, S.,... Urcelay, C. (2007).

Plant functional types: Are we getting any closer to the Holy Grail? In J. Canadell, D. Pataki, & L. Pitelka (eds), *Terrestrial Ecosystems in a Changing World* (pp. 149–164). Berlin: Springer-Verlag.

Lavorel, S., & Garnier, E. (2002). Predicting changes in community composition and ecosystem functioning from plant traits: revisiting the Holy Grail. *Functional Ecology, 16*, 545–556.

Le Roux, X., Barbault, R., Baudry, J., Burel, F., Doussan, I., Garnier, E., ... Trommetter, M. (eds) (2009). *Agriculture et biodiversité. Valoriser les synergies*. Versailles: Quae.

Lovelock, J. E. (1979). *Gaia – A new look at life on Earth*. Oxford: Oxford University Press.

McGill, B. J., Enquist, B. J., Weiher, E., & Westoby, M. (2006). Rebuilding community ecology from functional traits. *Trends in Ecology and Evolution, 21*, 178–185.

Mooney, H. A. (1974). Plant forms in relation to environment. In B. R. Strain & W. D. Billings (eds), *Handbook of vegetation science. Part IV. Vegetation and environment* (pp. 111–122). The Hague: Junk.

Mooney, H. A., & Dunn, E. L. (1970). Convergent evolution of Mediterranean-climate evergreen sclerophyll shrubs. *Evolution, 24*, 292–303.

Mora, C., Tittensor, D. P., Adl, S., Simpson, A. G. B., & Worm, B. (2011). How many species are there on Earth and in the Ocean? *PLoS Biology, 9*(8), e1001127. doi:10.1371/journal.pbio.1001127

Naeem, S., & Bunker, D. A. (2009). TraitNet: furthering biodiversity research through the curation, discovery, and sharing of species trait data. In S. Naeem, D. A. Bunker, A. Hector, M. Loreau & C. Perrings (eds), *Biodiversity, Ecosystem Functioning, and Human Wellbeing – An Ecological and Economic Perspective* (pp. 281–289). New York City: Oxford University Press.

Noble, I. R., & Slatyer, R. O. (1980). The use of vital attributes to predict successional changes in plant communities subject to recurrent disturbances. *Vegetatio, 43*, 5–21.

Osmond, C. B., Björkman, O., & Anderson, D. J. (1980). *Physiological Processes in Plant Ecology. Toward a synthesis with Atriplex*. Berlin: Springer-Verlag.

Ramenskii, L. G. (1938). *Introduction to Comprehensive Soil-Plant Studies of Landscapes* (in Russian). Moscow: Sel'khozgiz.

Raunkiaer, C. (1907). *Planterigets Livsformer og Deres Betydning for Geografien*. Copenhagen: Munksgaard.

Raunkiaer, C. (1934). *The Life Forms of Plants and Statistical Plant Geography* (English ed., H. Gilbert-Carter & A. Fausbøll, trans.). Oxford: Oxford University Press.

Reich, P. B., Walters, M. B., & Ellsworth, D. S. (1992). Leaf life-span in relation to leaf, plant, and stand characteristics among diverse ecosystems. *Ecological Monographs, 62*, 365–392.

Schimper, A. F. W. (1903). *Plant Geography upon a Physiological Basis*. Oxford: Oxford University Press.

Shipley, B. (2010). *From Plant Traits to Vegetation Structure. Chance and Selection in the Assembly of Ecological Communities*. Cambridge: Cambridge University Press.

Smith, T. M., Shugart, H. H., & Woodward, F. I. (1997). *Plant Functional Types: Their Relevance to Ecosystem Properties and Global Change*. Cambridge: Cambridge University Press.

Suding, K. N., Lavorel, S., Chapin, F. S., III, Cornelissen, J. H. C., Díaz, S., Garnier, E., ... Navas, M.-L. (2008). Scaling environmental change through the community-level: a trait-based response-and-effect framework for plants. *Global Change Biology, 14*, 1125–1140.

von Humbolt, A. (1806). *Ideen zu einer Physiognomik der Gewächse*. Tübingen.

Violle, C., Navas, M.-L.,Vile, D., Kazakou, E., Fortunel, C., Hummel, I., & Garnier, E. (2007). Let the concept of trait be functional! *Oikos, 116*, 882–892.

Warming, E. (1884). *Om skudbygning, overvintring og foryngelse*. Copenhagen: Festskr. Naturh. Foren.

Warming, E. (1909). *Oecology of Plants: An Introduction to the Study of Plant Communities*. Oxford: Oxford University Press.

Weiher, E., van der Werf, A., Thompson, K., Roderick, M., Garnier, E., & Eriksson, O. (1999). Challenging Theophrastus: A common core list of plant traits for functional ecology. *Journal of Vegetation Science, 10*, 609–620.

Westoby, M., Falster, D. S., Moles, A. T., Vesk, P. A., & Wright, I. J. (2002). Plant ecological strategies: Some leading dimensions of variation between species. *Annual Review of Ecology and Systematics, 33*, 125–159.

CHAPTER 2

Trait-based ecology: definitions, methods, and a conceptual framework

2.1 Introduction

As presented in the previous chapter, the diversity of functions can be quantified at different levels of the organization of biological systems (cf. Figure 1.2). However, in this book, the individual will be considered as the primary entry point, as individual organisms lie at the heart of the ecological and evolutionary nexus of the living world (Fitter & Hay, 2002). It is at this level that adaptation occurs, as it is the differential survival, reproduction, and mortality of individuals that leads to modification of gene frequencies, and thus of their evolution in response to natural selection. Additionally, the responses of individuals determine those of populations and communities, and thus the dynamics and functioning of the entire ecosystem, in conjunction with the exchanges of resources between organisms and their environment. Analysing and understanding the variation in the expression of functions between individuals, and the impacts of this variation at the level of populations and ecosystems, is central to the understanding of the diversity of life at all of its levels of organization.

The left hand side of Figure 2.1 presents some examples of functions defined at the level of individual plants. Variations in the degree of expression of these functions between the different individuals of a population, between populations of the same species, or between different species constitute plant functional diversity. Some of these functions are very difficult to quantify precisely: how is it possible to measure the growth of an individual plant in a multispecies community in which root systems may be completely entangled and descend to depths of many metres (cf. Figure 3.15)? How is it possible to estimate the fecundity of a tree or a clonal plant that might live for thousands of years (Thomas, 2013)? How is it possible to understand the process of seed dispersal when some seeds might be found at the base of the plant that produced them and others many kilometres away (Fenner & Thompson, 2005)? And finally, how is it possible to measure these functions for the 315 000 species of plants found on the planet? The task is enormous ... and is not manageable without accepting a certain degree of simplification, and thus of approximation: the growth of a plant can be estimated from certain characteristics of its leaves, its fecundity by the mass of its seeds and of its vegetative size, and the distances of seed dispersal from the existence of structures which allow seeds to be carried by the wind or attract animal dispersers (see Chapter 3 for more details). These different characteristics of plants, correlated with their functions but easier to measure than the former, have been named 'functional markers'[1] (Garnier et al., 2004). The right hand side of Figure 2.1 presents examples of functional markers, next to the functions which they serve to assess.

[1] This dichotomy between functions and functional markers has been discussed in the literature using the terms 'hard trait' for the former and 'soft trait' for the latter (Hodgson et al., 1999; Weiher et al., 1999). These terms can sometimes lead to confusion (discussed in Violle et al., 2007), and we prefer not to use them in this book.

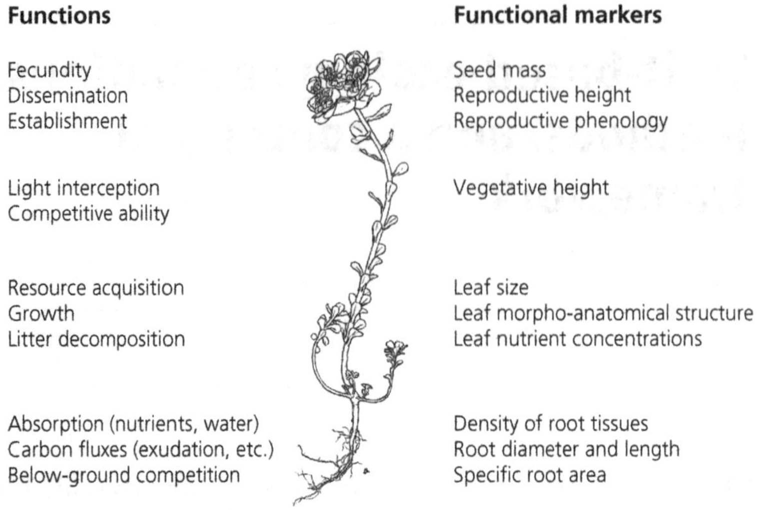

Functions	Functional markers
Fecundity Dissemination Establishment	Seed mass Reproductive height Reproductive phenology
Light interception Competitive ability	Vegetative height
Resource acquisition Growth Litter decomposition	Leaf size Leaf morpho-anatomical structure Leaf nutrient concentrations
Absorption (nutrients, water) Carbon fluxes (exudation, etc.) Below-ground competition	Density of root tissues Root diameter and length Specific root area

Figure 2.1 Examples of functions defined at the level of the individual and corresponding functional markers linked with them (see also Figure 1.2). 'Individual' might refer either to a 'genet' (a genetic individual) or a ramet (that is, a recognizable separately rooted, above-ground shoot) in cases when the genet cannot be not clearly identified. Drawing of *Euphorbia helioscopia* L. by Baptiste Testi. Adapted from Garnier & Navas (2012).

Functions and functional markers taken together constitute measurable characteristics of organisms called 'traits'[2] (see McGill et al., 2006; Violle et al., 2007, for recent reviews). The following sections describe the scope of the trait concept, and how it is used to characterize the functional diversity of organisms in an ecological context. We will then present a conceptual framework within which traits are used to analyse diversity at the levels of communities and ecosystems, and upon which this book is based.

2.2 Traits: definitions and protocols

2.1.1 What is a trait?

A detailed reading of the literature reveals that the term 'trait' has been used with a variety of relatively different meanings in ecology and population biology (Violle, et al., 2007). Here a trait will be defined at the level of the individual (Crisp & Cook, 2012; Lavorel et al., 1997; McGill et al., 2006). We propose here a definition slightly modified from that given by Violle et al. (2007), to account for the fact that a trait is heritable (e.g. Crisp & Cook, 2012): a trait is 'any morphological, physiological, or phenological *heritable* feature measurable at the individual level, from the cell to the whole organism, without reference to the environment or any other level of organization'. The trait concept is thus defined independently of environmental characteristics or any other levels of organization (population, community, or ecosystem). Some examples of traits that can be considered as functions or functional markers relating to, for example, reproduction (fecundity, dispersal, etc.) or plant growth (light interception, mineral absorption, etc.) are presented in Figure 2.1. Traits can be continuous (e.g. plant height, seed mass) or categorical when they are characterized as classes of values (e.g. annual or perennial plant, deciduous or evergreen trees).

As discussed in Violle et al. (2007), this definition requires some additional explanation: (1) measuring a trait consists of obtaining a value or a category for this trait, measured in a certain place and at a certain time; (2) a trait may have different values or categories for a given species depending on

[2] The term 'trait' is practically synonymous with 'phenotypic character', a term widely used by geneticists who generally emphasize heritability. In ecology, the term 'trait' is in common usage, and thus is the term we use in this book.

environmental conditions and/or over time (seasonal or ontogenic variations, for example); (3) for a given trait, values (or categories) are often measured at the population level, by taking the average of the values from a collection of individuals. These two last points imply that obtaining values for a trait should also require recording information describing the local environment in which the measures were taken, so that an assessment of the ecological or evolutionary significance of the values can be made (Bartholomeus et al., 2008; McGill, et al., 2006). Thus, a 'trait' such as 'vegetative height' makes no reference to the environment since a plant will have a height in every environment, but the 'trait value' (e.g. 100 cm) will usually be environmentally dependent. Furthermore, it is also necessary to know the relative importance of intraspecific versus interspecific variability of the trait (Albert et al., 2010; Cornelissen et al., 2003a; Garnier et al., 2001; Mokany & Ash, 2008; Roche et al., 2004, and see Chapter 4). If the values of a trait do not vary significantly over time or in different environments, however, a precise description of the environment in which these were measured is less critical.

Violle et al. (2007) suggested applying to plants the 'morphology, performance, and fitness' paradigm of Arnold (1983), initially developed for animals, which proposes a three-level hierarchical classification for traits: the three components of fitness, which are growth, survival, and reproduction, being characterized by three 'performance traits', these being plant vegetative biomass, reproductive output, and survival, which are themselves modified by morphological, physiological, and phenological traits, henceforth called 'functional traits' (Figure 2.2). A current limit of this framework is that while being a subject of many ecological studies, relationships between functional traits and performance traits, and thus their relationships to fitness, have rarely been demonstrated (Ackerly & Monson, 2003; Craine, et al., 2012).

Plant characteristics which require the identification of an explicit relationship with environmental factors are thus not considered as traits. This is in particular the case for, for example, resistance to drought or grazing, heavy metal or shade tolerance, or Ellenberg indicator values (Ellenberg, 1988; Ellenberg et al., 1992), which describe the performance of individuals in relation to certain environmental variables. These types of characteristics require that trait values be measured under different environmental conditions, with the range

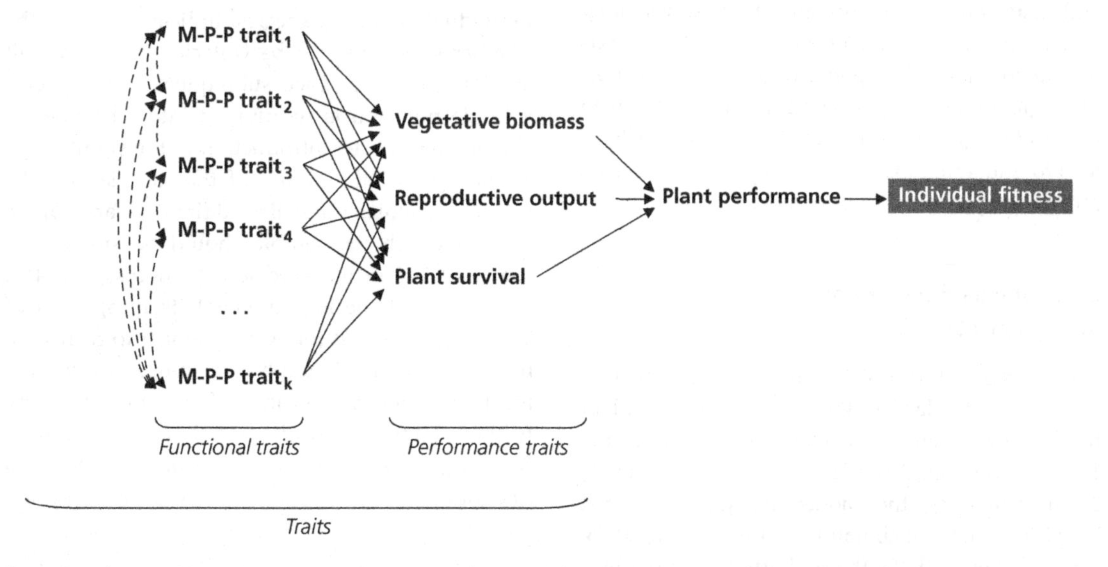

Figure 2.2 Three-level hierarchical structure of the concept of traits as defined by Arnold (1983). Morphological, physiological, or phenological traits (M-P-P) (from 1 to k) have impacts on one or several performance traits (plant vegetative biomass, reproductive output, and survival) which determine plant performance and fitness of individuals. The traits M-P-P can be inter-related (double arrowed lines). For clarity, the interaction between performance traits and their impacts on M-P-P traits are not shown. Taken from Violle et al. (2007).

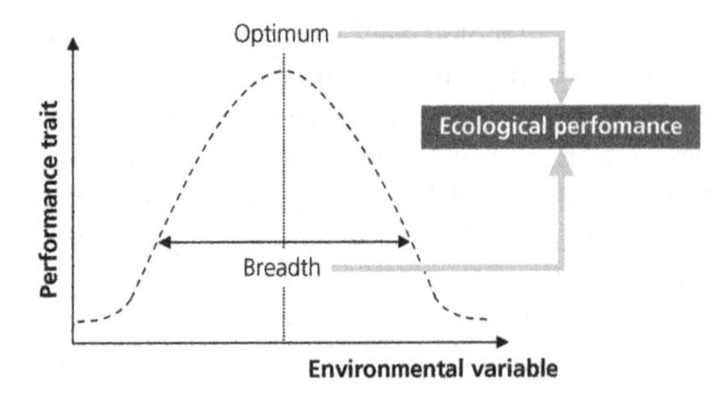

Figure 2.3 Graphical representation of ecological performance. In this example, the performance can be described by the position of the optimum and the spread of the distribution of values of one of the three performance traits (plant biomass, reproductive success, individual survival: cf. Figure 2.2) along an environmental gradient. See the text of section 2.1 for further details. Taken from Violle et al. (2007).

of values recorded corresponding to the *response* of a trait to environmental factors, rather than to the trait itself. Following proposals by animal ecologists (Irschick, 2003; Irschick & Garland, 2001), Violle et al. (2007) proposed that the response of an organism's performance traits to an environmental factor be called 'ecological performance' (referred to also as 'habitat preference' or 'ecological preference'). The latter can be defined using the optimum and distribution of performance trait values along an environmental gradient (Figure 2.3). It depends on the coordinated response of a number of morphological, physiological, and phenological traits to the gradient (e.g. Craine et al., 2012, and see Chapter 4 for numerous examples), in agreement with the scheme proposed by Violle et al. (2007).

2.1.2 Standardized protocols of trait measurements

Considerable effort has been made over the last two decades to standardize the measurement of plant traits (Cornelissen et al., 2003b; Hendry & Grime, 1993; Knevel et al., 2005; Pérez-Harguindeguy et al., 2013). Following the pioneering work initiated by Grime and collaborators (Grime et al., 1988; Grime & Hunt, 1975), these efforts have undoubtedly contributed to the rapid expansion of trait measurements for an increasing number of species, often found in very different environments (Chave et al., 2009; Díaz et al., 2004; Freschet et al., 2010; Grime et al., 1997; Moles et al., 2005; Niinemets, 2001; Wright et al., 2004, to cite only a few examples).

Four principal manuals of measurement protocols have been produced since 1993. The first (Hendry & Grime, 1993) is a laboratory manual which describes how to measure 68 traits and/or their response to a number of environmental factors under controlled conditions. The other three manuals describe trait measurements under natural uncontrolled conditions. They are constructed in a similar fashion with, for each trait, a description of the protocol to be followed to collect samples and carry out the measurement. The manual of Cornelissen et al. (2003b) describes protocols for 28 traits; that of Knevel et al. (2005), developed in conjunction with the establishment of the LEDA plant traits database (Kleyer et al., 2008), includes 31 traits, and in addition contains specific information on how to code the trait for entry into computer databases. Finally, the manual of Pérez-Harguindeguy et al. (2013) is an update of that of Cornelissen et al. (2003b), and contains information for an additional 20 new traits.

In these last two manuals[3], preliminary sections describe methods to be used in selecting species within a community, and also in selecting individuals within a species. Indications as to the number of samples to be collected for each trait are also provided. Each subsequent section treats a single trait and is subdivided into a number of headings: 1) a brief introduction to the significance of the trait; 2) the nature of the sample and the method by which it is collected (if necessary); 3) the way in which the sample should be stored and prepared before measurement (if necessary); 4) a description of the measurement itself; 5) a list of any exceptions and the way in which to deal with these; and 6) bibliographic references discussing the theory and significance of the trait, and any further methodological precisions if required. Box 2.1 gives an example of a section devoted to a categorical trait, the Raunkiaer life forms, and of a

[3] The manual of Cornelissen et al. (2003b) is widely used by the scientific community, as evidenced by the hundreds of citations every year during the period 2010–2014 (source: Web of Science).

Box 2.1 Two examples of standardized trait measurement protocols extracted from the manual of Cornelissen et al. (2003b): life forms (categorical trait) and seed mass (continuous trait).

Life form

Brief trait introduction

Life form is another classification system of plant form designed by Raunkiaer (1934) and adequately described by Whittaker (1975): 'instead of the mixture of characteristics by which growth forms are defined (. . .), a single principal characteristic is used: the relation of the perennating tissue to the ground surface. Perennating tissue refers to the embryonic (meristematic) tissue that remains inactive during a winter or dry season, and then resumes growth with return of a favourable season. Perennating tissues thus include buds, which may contain twigs with leaves that expand in the spring or rainy season. Since perennating tissue makes possible the plant's survival during an unfavourable season, the location of this tissue is an essential feature of the plant's adaptation to climate. The harsher the climate, the fewer plant species are likely to have buds far above the ground surface, fully exposed to the cold or the drying power of the atmosphere'. Furthermore, for species that may be subject to unpredictable disturbances, such as periodic grazing and fire, the position of buds or bud-forming tissues allows us to understand the likelihood of their surviving such disturbances. It is important to note that the categories below refer to the highest perennating buds for each plant.

How to record?

Life form is a categorical trait assessed from field observation, descriptions, or photos in the literature. Many floras give life forms as standard information on plant species. Five major life forms were initially recognized by Raunkiaer, but his scheme was further expanded by various authors (e.g. Ellenberg & Müller-Dombois, 1967). Here we present one of the simplest, most widely used schemes.

1. *Phanerophytes:* plants that grow taller than 0.50 m and whose shoots do not die back periodically to that height limit (e.g. many shrubs, trees, and lianas).
2. *Chamaephytes:* plants whose mature branch or shoot system remains below 0.50 m, or plants that grow taller than 0.50 m, but whose shoots die back periodically to that height limit (e.g. dwarf shrubs).
3. *Hemicryptophytes:* periodic shoot reduction to a remnant shoot system, so that buds in the 'harsh season' are close to the ground surface (e.g. many grasses and rosette forbs).
4. *Geophytes:* annual reduction of the complete shoot system to storage organs below the soil surface (e.g. many bulb flowers and bracken *Pteridium*).
5. *Therophytes:* plants whose shoot and root system dies after seed production and which complete their whole life cycle within one year (e.g. many annuals in arable fields).
6. *Helophytes:* vegetative buds for surviving the harsh season are below the water surface, but the shoot system is mostly above the water surface (e.g. many bright-flowered monocotyledons such as *Iris pseudacorus*).
7. *Hydrophytes:* the plant shoot remains either entirely under water (e.g. waterweed Elodea) or partly below and partly floating on the water surface (e.g. waterlily *Nymphaea*).

Special cases or extras

Climbers, hemi-epiphytes, and epiphytes may be classified here as phanerophytes or chamaephytes, since their distinct growth forms are classified explicitly above under Growth form.

References on theory and significance

Raunkiaer (1934); Cain (1950); Ellenberg & Müller-Dombois (1967); Whittaker (1975); Box (1981); Ellenberg (1988).

Seed mass

Brief trait description

Seed mass, also called seed size, is the oven-dry mass of an average seed of a species, expressed in mg. Small seeds tend to be dispersed further away from the mother plant (although this relationship is very crude), while stored resources in large seeds tend to help the young seedling to survive and establish in the face of environmental hazards (deep shade, drought, herbivory, etc.). Smaller seeds can be produced in larger numbers with the same reproductive effort. Smaller seeds also tend to be buried deeper in the soil, particularly if their shape is close to spherical, which aids their longevity in seedbanks. Interspecific variation in seed mass also has an important taxonomic component, more closely related taxa being more likely to be similar in seed mass.

What and how to collect?

The same type of individuals as for leaf traits and plant height should be sampled, i.e. healthy, adult plants that have their foliage exposed to full sunlight (or otherwise plants with the strongest light exposure for that species). The seeds should be mature and alive. If the shape of the dispersal unit (seed, fruit, etc.) is measured too (see above), do not remove any parts until measurement (see below). We recommend

continued

Box 2.1 *Continued*

collecting at least five seeds from each of three plants of a species, but more plants per species are preferred. Depending on the accuracy of the balance available, 100 or even 1000 seeds per plant may be needed for species with tiny seeds (e.g. orchids). In some parts of the world, e.g. some tropical rainforest areas, it may be efficient to work in collaboration with local people specialized in tree climbing to help with the collecting (and identification).

Storing and processing

If dispersule shape is also measured, then store cool in sealed plastic bags, whether or not wrapped in moist paper [. . .], and process and measure as soon as possible. Otherwise air-dry storage is also appropriate.

Measuring

After dispersule shape measurements (if applicable), remove any accessories (wings, comas, pappus, elaiosomes, fruit flesh), but make sure not to remove the testa in the process. In other words, first try to define clearly which parts belong to the fruit as a whole and which strictly to the seed. Only leave the fruit intact in cases where the testa and the surrounding fruit structure are virtually inseparable. Dry the seeds (or achenes, single-seeded fruits) at 80°C for at least 48 hours (or until equilibrium mass in very large or hard-skinned seeds) and weigh. Be aware that, once taken from the oven, the samples will take up moisture from the air. If they cannot be weighed immediately after cooling down, put them in the desiccator until weighing, or else back in the oven to dry off again. Note that the average number of seeds from one plant (whether based on five or 1000 seeds) counts as one statistical observation for calculations of mean, standard deviation, and standard error.

Special cases or extras

- Be aware that seed size may vary more within an individual than between individuals of the same species. Make sure to collect 'average-sized' seeds from each individual, and not the exceptionally small or large ones.
- Be aware that a considerable amount of published data are already available in the literature, while some of the large unpublished databases may be accessible under certain conditions. Many of these data can probably be added to the database, but make sure the methodology used is compatible.
- For certain (e.g. allometric) questions, additional measurements of the mass of the dispersule unit or the entire infructescence (reproductive structure) may be of additional interest. Both dry and fresh mass may be useful in such cases.

References on theory and significance

Salisbury (1942); Grime & Jeffrey (1965); MacArthur & Wilson (1967); Silvertown (1981); Mazer (1989); Jurado & Westoby (1992); Thompson et al. (1993); Leishman & Westoby (1994); Allsopp & Stock (1995); Hammond & Brown (1995); Leishman et al. (1995); Saverimuttu & Westoby (1996); Seiwa & Kikuzawa (1996); Swanborough & Westoby (1996); Hulme (1998); Reich et al. (1998); Westoby (1998); Cornelissen (1999); Gitay et al. (1999); Weiher et al. (1999); Thompson et al. (2001); Westoby et al. (2002).

More on methods

Food and Agriculture Organization (1985); Hendry & Grime (1993); Thompson et al. (1993); Hammond & Brown (1995); Thompson et al. (1997); Westoby (1998); Weiher et al. (1999).

section devoted to a continuous trait, seed mass. The protocols described in the manual of Pérez-Harguindeguy et al. (2013) are accessible on the internet via the DiverSus[4] and PrometheusWiki[5] websites. The latter was, developed to share and easily access information regarding measurement protocols in the field of plant ecophysiology, and to provide a forum for the discussion of technical matters of interest to this community (Sack et al., 2010). The widely adopted use of these protocols has led to them being considered as 'standards'[6] for the measurement of a certain number of plant traits. It should be acknowledged however that the number of traits for which these protocols are available remains relatively limited,

[4] DiverSus stands for "bioDIVERsity, ecosystems and SUStainability" (http://www.nucleodiversus.org)

[5] Prometheus is an acronym for 'PROtocols, METHods, Explanations, and Updated Standards' (http://prometheuswiki.publish.csiro.au).

[6] A standard is a published reference whose diffusion and use are widespread, and which is recognized by a large proportion of a given community. Standards differ from norms, which are a reference established by consensus and approved by an official organization (see e.g. http://www.wikipedia.org).

and it is necessary to continue the process of measurement standardization for a larger number of traits.

In contexts where traits are being used to understand the relationships between organisms and environmental factors or to respond to questions concerning levels of organization above that of a given organism, the development of standardized protocols to measure variables describing the environment, community structure, etc. would also be necessary. Some progress has been made in this direction (see for example Bartholomeus et al., 2008; Knevel et al., 2005, for soil characteristics), with a particular emphasis on the standardization and description of environments used in laboratory experiments (Hannemann et al., 2009). Nevertheless, the generalization and use of such protocols remains marginal in most ecological studies. The case of the multiplicity of different methods used to measure ecosystem primary productivity (Scurlock et al., 2002, and see Box 6.1 in Chapter 6) or the difficulties encountered in attempting to formalize protocols for the measurement of defoliation in grazed pastures (Garnier et al., 2007) are indicative of these shortcomings. This lack of standardization significantly constrains our ability to generalize from current studies, and the removal of this limitation should be considered a priority for ecology (see in particular the discussion in Austin, 1980; Bradshaw, 1987a; Garnier et al., 2007).

2.3 Traits and the comparative approach in ecology

The 'trait' concept allows us to understand the expression of different functions in organisms, and thus is widely used in different disciplines concerned with the study of biological diversity, including that of comparative functional ecology, the core discipline covered by this book. Below we first describe the nature and utility of the comparative approach in ecology, before presenting the logic of the use of traits in this context.

2.3.1 Comparison in ecology

Amongst the different approaches used in ecology, the approach called 'comparative' has long been of central importance (Grime, 1965; Parsons, 1968). In a somewhat provocative fashion, Bradshaw (1987b) even suggested that comparative ecology and functional ecology are synonymous. When it is focused at the level of the individual, the central question of comparative ecology pertains to the understanding of the correspondence between organisms and their environment. As Bradshaw (1987a, 1987b) has underlined, this match is never perfect: the distribution of plants among different habitats being simply a result of the fact that some organisms better match a given environment than others. An organism's adaptation is, as a consequence, a purely comparative concept. The two following examples illustrate this assertion.

The first example comes from a pioneering study in plant ecophysiology carried out by Bonnier (1899). As a starting point for this study, Bonnier observed that 'the plants of the Mediterranean region have, in general, characteristics which seem to be in keeping with the climate of this region'. These characteristics in keeping with the climate (which are in fact, traits) correspond to 'adaptations', even if this term was not used by Bonnier. Nevertheless, Bonnier was able to make such a statement because he knew plants which are not characteristic of the climate of Mediterranean regions and *compared* them (implicitly in this case) to plants which do not grow in these climatic conditions. The experiment carried out by Bonnier consisted of a *comparison* between individuals transplanted into botanic gardens located in the Paris region (temperate climate) and in southern France near Toulon (Mediterranean climate). One of the conclusions of this study was that the leaves of plants cultivated in the south of France were 'thicker, tougher, had less apparent veins, [. . .], and were often more persistent'[7], and tended towards the attributes known from Mediterranean species which can be considered as adaptations. The second example is drawn from a bibliographic review of the components of plants' carbon budgets, in which Mooney (1972) reported that the maximal rate of photosynthesis (a leaf *trait*) of herbaceous plants characteristic of shaded environments varied

[7] These characteristics are leaf traits such as leaf thickness or longevity, the ecological and functional significance of which will be presented in Chapters 3 and 4.

between 4 and 16 mg CO_2 dm^{-2} h^{-1}. The *comparison* of these values, which are two to three times lower than those from herbaceous plants growing in open environments, led Mooney to consider the adaptive significance of differences in the rate of photosynthesis between species, and based on this to develop cost-benefit analyses of carbon acquisition to explain plant distribution in different habitats.

Even if the adaptive argument advanced by Bradshaw is not applicable at levels of organization above those of the organism, the comparative approach remains nonetheless relevant to these levels. A first example relates to the distribution of the major types of biomes[8] found on Earth, a subject studied ever since naturalists began to travel across the world in the nineteenth century. Understanding the factors that control this distribution requires by its very essence a comparative approach, as the aim is to determine the factors which lead to the observed differences between biomes. As a first step—more descriptive than explanatory—towards this understanding, Holdridge (1947) proposed a classification based on differences in temperatures and annual average rainfall between sites: for example, tropical forests are found in areas where temperatures and rainfall are both high, while deserts are found in areas where rainfall is very low but over a large range of annual average temperatures (e.g. hot and cold deserts). This classification of major biomes, based on a combination of two climatic variables, is still considered authoritative today. A second example comes from the synthesis of Odum (1969) of community structure and ecosystem functioning during the process of plant succession[9]. From a comparison of early and late successional stages of the types of species present (defined on the basis of their size and the components of their life cycle), the species diversity (richness and evenness), the productivity and various components of nutrient cycles, Odum contrasted in a visionary manner two major types of ecosystems: 1) those that are modified over the short term by human activity for the production of food and fibre and have characteristics comparable to those of early successional stages, and 2) those, less modified by humans, whose functioning regulates long-term planetary biogeochemical cycles (atmospheric CO_2 and O_2 concentrations, water quality) and have characteristics comparable to those of advanced successional stages. It was a comparative approach which allowed Odum to identify these basic concepts, which some fifteen years later became known as 'services' provided by ecosystems to human society (Ehrlich & Mooney, 1983; and see Chapter 7): provisioning services in ecosystems with rapid growth in which nutrient cycles are open vs regulatory services which maintain quality and stability in slower growing ecosystems in which nutrient cycles are more closed.

These few examples illustrate how the value of many ecological observations is enriched when they are placed in a comparative context. They show the power of this approach at different levels of organization in identifying the key factors controlling the distribution and functioning of ecological systems and their components.

2.3.2 Implementation of the comparative approach

The first stage of a comparative approach using the individual as an entry point to study functional diversity consists of synthesizing information relating to one or a number of traits for a large number of species (or populations) in matrices of species (or populations) crossed by traits (Figure 2.4: Keddy, 1992b). The order of magnitude of the number of species can vary from tens of species for questions addressed at the community level, to multiple thousands of species when addressing questions at the scale of the planet.

As emphasized by Keddy (1992b) and Duarte et al. (1995), this approach is literally orthogonal to that used in autecological studies, where the

[8] A biome is defined as a group of terrestrial ecosystems similar in terms of: 1) vegetation structure and physiognomy; 2) the environmental factors responsible for this structure; and 3) certain characteristic animal communities found there (Whittaker, 1975). This author recognizes six types of major biomes: forests, grasslands, woodlands, shrub-lands, semi-arid shrub-lands, and deserts.

[9] Succession is the phenomenon of the colonization of a habitat by living organisms and the changes over time in the faunal and floristic composition of a given site after a disturbance has either partially, or totally, destroyed the previous ecosystem (Lepart & Escarré, 1983).

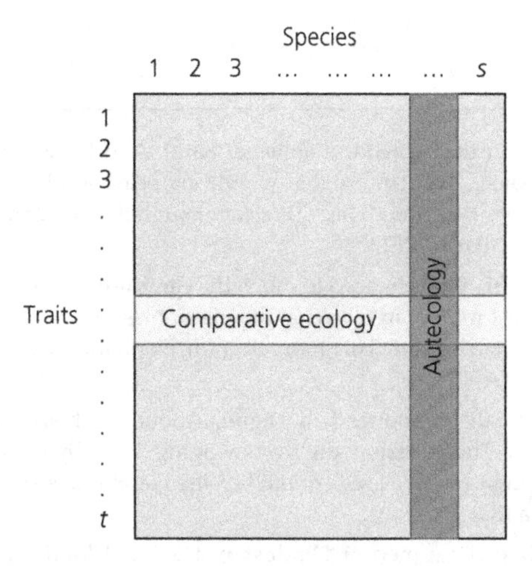

Figure 2.4 Schematic representation of a species X traits matrix for *s* species and *t* traits. Until the middle of the decade 1990–2000, few of this type of matrix existed as ecologists were primarily focused on autecological studies of single species (symbolized by the vertical band on the figure) rather than on comparative studies of traits involving multiple species (symbolized by the horizontal band). The creation of such matrices is an essential first step to exploring the relationships between traits, the distribution of organisms along environmental gradients, the functional structure of communities, and the effects of this structure on ecosystem functioning (see also Figure 2.5). From Keddy (1992b). Reproduced with permission from Wiley.

objective is to understand as precisely as possible the requirements of a single species in respect to the biotic or abiotic environmental factors, and requiring the measurement of a large number of traits for this particular species (Figure 2.4).

Autecology can provide relatively precise predictions for the small number of species that can be studied in this manner. This approach is still used extensively, both in ecological (see for example the approximately 300 monographs on species of the flora of the British Isles published in the *Journal of Ecology*[10] since 1941) and agronomic studies. In this last case, the number of species of interest is relatively small (103 plant species provide 90% of the world's food supply, excluding forage species: Prescott-Allen & Prescott-Allen, 1990), and the objective is often to favour only one or two species in a cultivated field in which agricultural practices can lead to an extreme simplification of environmental conditions. By its very nature, an autecological approach can only be of limited generality. For instance, is it possible to predict the response of *Nardus stricta* (Matgrass), a perennial grass found in heathlands and low productivity grasslands, to nitrogen availability from that of *Lolium perenne* (Perennial Ryegrass), a perennial grass often used as a pasture plant? As demonstrated by the study of Bradshaw et al. (1964), the reply is clearly no: by varying the concentration of nitrogen in a nutrient solution used to water plants of *L. perenne* from 1 to 243 mg l^{-1}, their biomass is multiplied by 12 and we see no sign of response saturation at high concentrations; under the same experimental conditions, the biomass of *N. stricta* only changes by a factor of 2, and is maximal at a concentration of 27 mg l^{-1}, with higher concentrations than this appearing toxic.

In order to go beyond the limits fixed by the particularities of each individual species, a comparative approach should be based on a sufficiently high number of taxa[11], which also has the advantage of increasing the statistical power of any statistical tests used (Duarte et al., 1995; Keddy, 1992b). This objective of generalization is fundamental in order to respond to numerous questions concerning ecological systems in which the number of species to be considered can be relatively high. As it is currently unrealistic to constitute a species X traits matrix such as that in Figure 2.4 for the approximately 315 000 species found on the planet, the comparative approach is based on a simple choice: rather than collect a large amount of precise information for a small number of species, the objective is to collect a smaller amount of less precise information for a large number of species. As was concluded by Duarte et al. (1995) in an article presenting the principles and strengths of the field of comparative plant functional ecology: 'Traditional autecology yields precise predictions for limited scenarios (one, or a few, species in a specific habitat), whereas broad-scale comparative plant ecology yields imprecise predictions, but is

[10] These monographs are compiled together in a collection called the 'Biological Flora of the British Isles'(http://www.britishecologicalsociety.org/journals_publications/journalofecology/biologicalflora.php).

[11] The method by which a large number of traits are measured simultaneously for a large number of species is called 'screening' (Grime & Hunt, 1975; Keddy, 1992b).

Box 2.2 The parable of the blind men and the elephant

This parable originally comes from the Anekāntavāda, one of the most important and fundamental doctrines of the Hindu religion of Jainism. It alludes to the principles of pluralism and of multiple points of view, with a central idea being that as 'truth' or 'reality' are perceived differently by different people, only one point of view cannot define the truth. In the middle of the nineteenth century, the American poet John Godfrey Saxe wrote a poem based on the parable, which is presented below:

It was six men of Hindustan, To learning much inclined, Who went to see the Elephant, (Though all of them were blind), That each by observation, Might satisfy his mind

The First approached the Elephant, And happening to fall, Against his broad and sturdy side, At once began to bawl: 'God bless me! but the Elephant, Is very like a WALL!'

The Second, feeling of the tusk, Cried, 'Ho, what have we here, So very round and smooth and sharp? To me 'tis mighty clear, This wonder of an Elephant, Is very like a SPEAR!'

The Third approached the animal, And happening to take, The squirming trunk within his hands, Thus boldly up and spake: 'I see', quoth he, 'the Elephant, Is very like a SNAKE!'

The Fourth reached out an eager hand, And felt about the knee, 'What most this wondrous beast is like, Is mighty plain', quoth he: ''Tis clear enough the Elephant, Is very like a TREE!'

The Fifth, who chanced to touch the ear, Said: 'Even the blindest man, Can tell what this resembles most; Deny the fact who can, This marvel of an Elephant, Is very like a FAN!'

The Sixth no sooner had begun, About the beast to grope, Than seizing on the swinging tail, That fell within his scope, 'I see', quoth he, 'the Elephant, Is very like a ROPE!'

And so these men of Hindustan, Disputed loud and long, Each in his own opinion, Exceeding stiff and strong, Though each was partly in the right, And all were in the wrong!

Moral of the story: Too often in theological arguments, the parties in conflict, I think, argue with each other in the most complete ignorance of what their opponents are trying to say, and debate endlessly with each other about an elephant that none of them have ever seen!

Reich (1993) appears to have been the first to use this parable in an ecological context, on the subject of relationships between leaf traits.

Figure Box 2.2 Illustration of the parable of the blind men and the elephant.

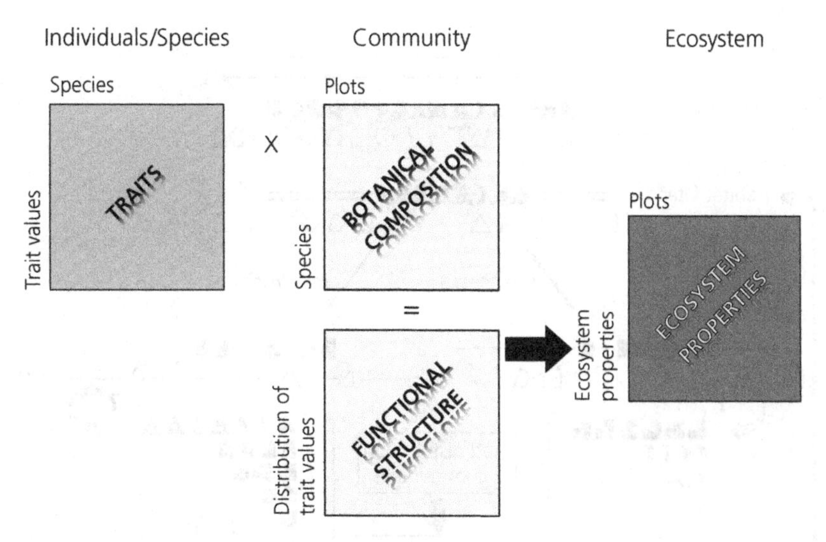

Figure 2.5 Example of the different types of matrices that can be combined to respond to different ecological questions, from the level of the organism (species X traits matrix) to plant communities (matrices of community composition and functional structure by plot: cf. Chapter 5) and ecosystems (matrix of ecosystem properties by plot: cf. Chapter 6). Inspired by Díaz & Cabido (1997).

applicable to most possible scenarios'. This debate between precision and generalization is not new in ecology (Bradshaw, 1987a; Grime, 2001; Grubb, 1998; Harper, 1982), and is certainly far from over. The position taken here is that the comparative approach is necessary to detect general patterns, but that these patterns may not necessarily take into account processes at a high degree of resolution. This tension is well illustrated by the parable of the blind men and an elephant (Box 2.2): seen in the perspective of our current discussion, comparative ecology can be understood as allowing us to see the whole of the elephant, while autecology can be seen as allowing us to precisely describe each of its parts.

If the objective is to understand and predict how the expression of the functional diversity of organisms translates to the level of communities and ecosystems, the establishment of species X traits matrices is only the first stage to be completed. To extend the approach described above to these other levels of organization (Díaz & Cabido, 1997), and in order to describe the functional structure of communities (Chapter 5) and to understand how organisms affect ecosystem properties (Chapter 6), it is necessary to combine the information available for the traits of organisms with those that describe, for example, community floristic composition (Figure 2.5) and the pools and fluxes within and between components of an ecosystem, respectively. The conceptual framework that allows us to articulate the links between these different levels is presented in Section 2.4.

2.4 Conceptual framework: organisms, communities, ecosystems

The conceptual framework around which a large part of this book is organized involves a number of components and stages, shown schematically in Figure 2.6. In this figure, species are represented by two groups of connected traits (same symbol, but different colours), taking into account that some traits can vary independently between species (for example the case between vegetative and regeneration traits: cf. Grime, 2001). Identifying the appropriate traits to use in a given ecological context is the objective of Chapter 3.

The upper part of Figure 2.6 down to the two curved side arrows shows a sketch of the different stages involved in the process of community assembly. It is comparable to the 'twin-filter' model recently proposed by Grime and Pierce (2012), except that it is based on individual traits rather than on suites of traits. The top of Figure 2.6 shows the range of

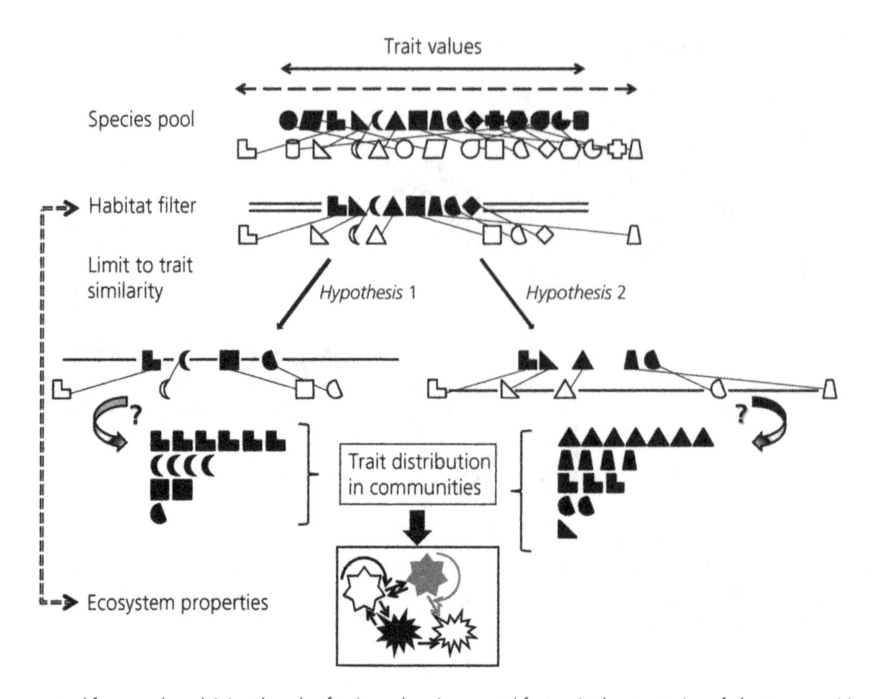

Figure 2.6 A conceptual framework explaining the role of traits and environmental factors in the structuring of plant communities and in ecosystem properties. In this figure, each species is represented by two connected symbols with the same shape but different colours, corresponding to two putative groups of traits (e.g. vegetative and reproductive). The upper part of the figure (above the curved side arrows) represents hypotheses concerning the effect of processes underlying community assembly and determining the distribution of traits within communities (after Woodward & Diament, 1991; Keddy, 1992a; Grime, 2006; Suding et al., 2008; Cornwell & Ackerly, 2009). The lower part of the figure (below the curved side arrows) represents hypotheses concerning the relationships between community structure and ecosystem properties (box at the bottom of the figure): the compartments (for example standing biomass or soil water reserves) are represented by different symbols, and the fluxes (for example net primary productivity or litter decomposition) within or between compartments are represented by different types of arrows (after Chapin et al., 2000; Loreau et al., 2001; Lavorel & Garnier, 2002; Suding et al., 2008). The direct effects of habitat filters on ecosystem properties and any potential feedbacks are symbolized by the double arrows on the left hand side of the figure. See the text of section 2.4 for further details. Modified from Garnier & Navas (2012).

values covered by the two types of traits in the pool of species found in a given geographical region, and from which local community assembly takes place (Belyea & Lancaster, 1999). Community assembly is assumed to take place under the effect of two processes: habitat filtering and limiting similarity. The first one corresponds to the effects of biotic and/or abiotic components of the habitat, which 'filter' species on the basis of their trait values (Cornwell & Ackerly, 2009; Grime, 2006; Keddy, 1992a; Suding et al., 2008; Woodward & Diament, 1991). The existence of these filters leads to a reduction in the range of trait values—'trait convergence'—represented by the selected species (double horizontal line on Figure 2.6) relative to that present in the regional pool: in Figure 2.6, this habitat filtering is assumed to act on one type of traits (e.g. vegetative) represented by the black symbols. The second process, 'limiting similarity', implies that species can more readily coexist if the competition among them is reduced, which is assumed to occur with increased trait spacing among species within a community—'trait divergence'—(single horizontal lines on Figure 2.6). Habitat filtering and limiting similarity may act on the same traits (hypothesis 1 on Figure 2.6), which is the case when the filtering of species is due to, for example, resource limitation or when the competitive interactions between species are strong. Different groups of traits can however respond to each of these processes (hypothesis 2 on Figure 2.6), for example when limiting similarity is induced by disturbance (Grime, 2006; Navas & Violle, 2009;

and see Chapter 5 for a more detailed discussion). The relative strength of each of these two processes on trait distribution depends both on the nature of the traits being filtered and on particular site conditions. Traits associated with the response of organisms to these biotic and abiotic factors are called 'response traits' (Lavorel & Garnier, 2002), and will be discussed in Chapters 4 and 8 for certain selected environmental factors. The rules themselves which regulate the assembly process in communities will be discussed in Chapters 5 and 8.

The lower part of Figure 2.6 (below the two curved side arrows) presents hypotheses concerning the relationships between community structure and ecosystem properties (Chapin et al., 2000; Lavorel & Garnier, 2002; Loreau et al., 2001; Suding et al., 2008). 'Ecosystem properties' refers equally well to both compartments (represented by different symbols in the box at the bottom of Figure 2.6), such as standing biomass or soil water reserves, and fluxes either within or between compartments, such as primary productivity or litter decomposition (arrows in the box at the bottom of Figure 2.6). This framework is based on the hypothesis that ecosystem properties depend on at least two components of the functional structure of communities: the mean value and the divergence of attributes, both of which are descriptors of the distribution of trait values in a community (see Chapters 5 and 6, and Reich (2014), for a comparable approach which does not explicitly take community structure into account). Those particular traits which determine the effects of plants on ecosystem properties, such as biogeochemical cycles or resistance to disturbance, are called 'effect traits' (Lavorel & Garnier, 2002); these will be examined in detail for a number of these properties and related services in Chapters 6, 7, and 8. The extremities of the two curved side arrows point to two examples of the distribution of species abundances filtered on the basis of a considered trait value, leading to different trait value distributions in the communities. The question marks alongside these arrows indicate that currently there exist neither solid hypotheses nor sufficient understanding to establish quantitative relationships between the presence/absence of species in a community and their actual abundance in these communities: we might be able to predict when a species is present, e.g. on the basis of its response traits, but what its local abundance will be remains difficult to predict (Shipley et al., 2006; Webb et al., 2010). Finally, the direct effects of habitat filters on ecosystem properties and potential feedbacks are represented by the dashed double arrow on the left of the figure.

2.5 Conclusions

At the centre of the primary ecological and evolutionary processes occurring in the living world, the individual is the primary entry point for studying functional diversity at different levels of organization. Traits, measurable characteristics of individuals, are tools extensively used in comparative ecological studies aimed at describing and understanding this functional diversity, and to evaluate its impacts on the structure and functioning of ecological systems. In the following chapters, we will illustrate and discuss different aspects of the conceptual framework based on traits presented in Section 2.4. We will begin with a currently very active research area concerned with functional traits, the objective of which is to identify a restricted list of dimensions (Westoby et al., 2002), allowing an understanding of the diversity of organisms from a functional perspective.

2.6 Key points

1. Functional diversity can be studied at different levels of organization of biological systems. Here we will consider the individual as a key level to understanding this functional diversity, as it is at the centre of the primary ecological and evolutionary processes occurring in the living world. Adaptation occurs at this level, and the responses of individuals determine those of populations and communities. These responses, in conjunction with the rates of resource and energy exchanges between organisms and their environment, regulate the functioning of ecosystems.
2. The 'trait' concept allows one to assess the different expressions of functions performed by organisms (e.g. photosynthesis, growth, reproduction),

and as such, it is extensively used in the various fields of study concerned with biological diversity. A trait is defined as 'any morphological, physiological, or phenological heritable feature measurable at the level of the individual, from the cell to the whole organism, without reference to the environment or any other level of organization'. The standardization of the measurement of plant trait values has been the object of considerable effort over the last few decades.

3. Trait-based ecology, the discipline at the core of this book, extensively uses the trait concept at different levels of organization. A fundamental stage of this approach consists of synthesizing information relative to one or a number of traits for a large number of species (or populations) in species (or populations) X traits matrices. If the aim is to understand and predict how the expression of organismic functional diversity translates to the levels of community structure and ecosystem functioning, the information collected at this stage needs to be combined with information relating to these other levels of organization, such as, for example, community botanical composition or variables describing the stocks and fluxes within and between the compartments of ecosystems.
4. A conceptual framework of response and effect applied to the functional approach allows for the linkage of the response of plants to environmental factors to the potential effects of this on ecosystem properties and services. Environmental factors act as filters which select species as a function of their trait values (called 'response traits'), resulting in a community functional structure defined on the basis of the value, the range, and the relative abundance of trait values, which depends on the type and strength of these filters. This functional structure has, in turn, impacts on ecosystem properties and the services delivered by these to human societies through the intermediary of certain traits called 'effect traits'. The hypothesis defended here is that it is the components of the functional structure of communities which have impacts on ecosystem properties and services, and not the number of species present in these communities.

2.7 References

Ackerly, D. D., & Monson, R. K. (2003). Waking the sleeping giant: the evolutionary foundations of plant functions. *International Journal of Plant Sciences, 164 (3 Suppl.)*, S1–S6.

Albert, C. H., Thuiller, W., Yoccoz, N. G., Soudant, A., Boucher, F., Saccone, P., & Lavorel, S. (2010). Intraspecific functional variability: extent, structure, and sources of variation. *Journal of Ecology, 98*, 604–613.

Allsopp, N., & Stock, W. D. (1995). Relationships between seed reserves, seedling growth, and mycorrhizal responses in 14 related shrubs (*Rosidae*) from a *low-nutrient environment. Functional Ecology*, 9(2), 248–254. doi:10.2307/2390571

Arnold, S. J. (1983). Morphology, performance, and fitness. *American Zoologist, 23*, 347–361.

Austin, M. P. (1980). Searching for a model for use in vegetation analysis. *Vegetatio*, *42*(1–3), 11–21. doi:10.1007/bf00048865

Bartholomeus, R. P., Witte, J.-P.M., van Bodegom, P.M., & Aerts, R. (2008). The need of data harmonization to derive robust empirical relationships between soil conditions and vegetation. *Journal of Vegetation Science, 19*, 799–808.

Belyea, L. R., & Lancaster, J. (1999). Assembly rules within a contingent ecology. *Oikos, 86*, 402–416.

Bonnier, G. (1899). Cultures expérimentales sur l'adaptation des plantes au climat méditerranéen. *Comptes-Rendus Hebdomadaires des Séances de l'Académie des Sciences, 129*, 1207–1213.

Box, E. O. (1981). *Macroclimate and Plant Forms: An Introduction to Predictive Modeling in Phytogeography* (Vol. 1). The Hague: Dr. W. Junk Publishers.

Bradshaw, A. D. (1987a). Comparison - Its scope and limits. *New Phytologist, 106* (Suppl.), 3–21.

Bradshaw, A. D. (1987b). Functional ecology: comparative ecology? *Functional Ecology, 1*, 71.

Bradshaw, A. D., Chadwick, M. J., Jowett, D., & Snaydon, R. W. (1964). Experimental investigations into the mineral nutrition of several grass species. IV. Nitrogen level. *Journal of Ecology, 52*, 665–677.

Cain, S. A. (1950). Life-forms and phytoclimate. *The Botanical Review, 16*, 1–32.

Chapin, F. S., III, Zavaleta, E. S., Eviner, V. T., Naylor, R. L., Vitousek, P. M., Reynolds, H. L.,... Díaz, S. (2000). Consequences of changing biodiversity. *Nature, 405*, 234–242.

Chave, J., Coomes, D., Jansen, S., Lewis, S. L., Swenson, N. G., & Zanne, A. E. (2009). Towards a worldwide wood economics spectrum. *Ecology Letters, 12*, 351–366.

Cornelissen, J. H. C. (1999). A triangular relationship between leaf size and seed size among woody species:

allometry, ontogeny, ecology, and taxonomy. *Oecologia* (Berl.), *118*, 248–255.

Cornelissen, J. H. C., Cerabolini, B., Castro-Díez, P., Villar-Salvador, P., Montserrat-Martí, G., Puyravaud, J. P., . . . Aerts, R. (2003a). Functional traits of woody plants: correspondence of species rankings between field adults and laboratory-grown seedlings? *Journal of Vegetation Science*, *14*, 311–322.

Cornelissen, J. H. C., Lavorel, S., Garnier, E., Díaz, S., Buchmann, N., Gurvich, D. E., . . . Poorter, H. (2003b). A handbook of protocols for standardised and easy measurement of plant functional traits worldwide. *Australian Journal of Botany*, *51*, 335–380.

Cornwell, W. K., & Ackerly, D. D. (2009). Community assembly and shifts in plant trait distributions across an environmental gradient in coastal California. *Ecological Monographs*, *79*, 109–126.

Craine, J. M., Engelbrecht, B. M. J., Lusk, C. H., McDowell, N., & Poorter, H. (2012). Resource limitation, tolerance, and the future of ecological plant classification. *Frontiers in Plant Science*, *3*. doi:10.3389/fpls.2012.00246

Crisp, M. D., & Cook, L. G. (2012). Phylogenetic niche conservatism: what are the underlying evolutionary and ecological causes? *New Phytologist*, *196*(3), 681–694. doi:10.1111/j.1469-8137.2012.04298.x

Díaz, S., & Cabido, M. (1997). Plant functional types and ecosystem function in relation to global change. *Journal of Vegetation Science*, *8*, 463–474.

Díaz, S., Hodgson, J. G., Thompson, K., Cabido, M., Cornelissen, J. H. C., Jalili, A., . . . Zak, M. R. (2004). The plant traits that drive ecosystems: evidence from three continents. *Journal of Vegetation Science*, *15*, 295–304.

Duarte, C. M., Sand-Jensen, K., Nielsen, S. L., Enríquez, S., & Agustí, S. (1995). Comparative functional ecology: rationale and potentials. *Trends in Ecology and Evolution*, *10*, 418–421.

Ehrlich, P. R., & Mooney, H. A. (1983). Extinction, substitution, and ecosystem services. *Bioscience*, *33*(4), 248–254.

Ellenberg, H. (1988). *Vegetation Ecology of Central Europe* (4th ed.). Cambridge: Cambridge University Press.

Ellenberg, H., & Müller-Dombois, D. (1967). A key to Raunkiaer plant life forms with revised subdivisions. *Berichte des Geobotanischen Institutes der ETH, Stiftung Rübel*, *37*, 56–73.

Ellenberg, H., Weber, H. E., Düll, R., Wirth, V., Werner, W., & Paulissen, D. (1992). *Zeigerwerte von Pflanzen in Mitteleuropa* (2nd ed., Vol. 18).

Fenner, M., & Thompson, K. (2005). *The Ecology of Seeds*. Cambridge: Cambridge University Press.

Fitter, A. H., & Hay, R. K. M. (2002). *Environmental Physiology of Plants* (3rd ed.). London: Academic Press.

Food and Agriculture Organization (1985). *A guide to forest seed handling* (Vol. 20–2). Rome: FAO.

Freschet, G. T., Cornelissen, J. H. C., van Logtestijn, R. S. P., & Aerts, R. (2010). Evidence of the 'plant economics spectrum' in a subarctic flora. *Journal of Ecology*, *98*, 362–373.

Garnier, E., Cortez, J., Billès, G., Navas, M.-L., Roumet, C., Debussche, M., . . . Toussaint, J.-P. (2004). Plant functional markers capture ecosystem properties during secondary succession. *Ecology*, *85*(9), 2630–2637.

Garnier, E., Laurent, G., Bellmann, A., Debain, S., Berthelier, P., Ducout, B., . . . Navas, M.-L. (2001). Consistency of species ranking based on functional leaf traits. *New Phytologist*, *152*, 69–83.

Garnier, E., Lavorel, S., Ansquer, P., Castro, H., Cruz, P., Dolezal, J., . . . Zarovali, M. (2007). Assessing the effects of land use change on plant traits, communities, and ecosystem functioning in grasslands: a standardized methodology and lessons from an application to 11 European sites. *Annals of Botany*, *99*, 967–985.

Garnier, E., & Navas, M.-L. (2012). A trait-based approach to comparative functional plant ecology: concepts, methods, and applications for agroecology. A review. *Agronomy for Sustainable Development*, *32*, 365–399. doi:10.1007/s13593-011-0036-y

Gitay, H., Noble, I. R., & Connell, J. H. (1999). Deriving functional types for rain-forest trees. [; Proceedings Paper]. *Journal of Vegetation Science*, *10*(5), 641–650. doi:10.2307/3237079

Grime, J. P. (1965). Comparative experiments as a key to the ecology of flowering plants. *Ecology*, *46*(4), 513–515.

Grime, J. P. (2001). *Plant Strategies, Vegetation Processes, and Ecosystem Properties* (2nd ed.). Chichester: John Wiley & Sons.

Grime, J. P. (2006). Trait convergence and trait divergence in herbaceous plant communities: Mechanisms and consequences. *Journal of Vegetation Science*, *17*, 255–260.

Grime, J. P., Hodgson, J. G., & Hunt, R. (1988). *Comparative Plant Ecology. A Functional Approach to Common British Species*. London: Unwin Hyman.

Grime, J. P., & Hunt, R. (1975). Relative growth-rate: its range and adaptive significance in a local flora. *Journal of Ecology*, *63*(2), 393–422. doi: Stable URL: http://www.jstor.org/stable/2258728

Grime, J. P., & Jeffrey, D. W. (1965). Seedling establishment in vertical gradients of sunlight. *Journal of Ecology*, *53*(3), 621–642. doi:10.2307/2257624

Grime, J. P., & Pierce, S. (2012). *The Evolutionary Strategies that Shape Ecosystems*. Oxford: Wiley-Blackwell.

Grime, J. P., Thompson, K., Hunt, R., Hodgson, J. G., Cornelissen, J. H. C., Rorison, I. H., . . . Whitehouse, J. (1997). Integrated screening validates primary axes of specialisation in plants. *Oikos*, *79*, 259–281.

Grubb, P. J. (1998). A reassessment of the strategies of plants which cope with shortages of resources.

Perspectives in Plant Ecology, Evolution and Systematics, 1, 3–31.

Hammond, D. S., & Brown, V. K. (1995). Seed size of woody-plants in relation to disturbance, dispersal, soil type in wet neotropical forests. *Ecology, 76*(8), 2544–2561. doi:10.2307/2265827

Hannemann, J., Poorter, H., Usadel, B., Blasing, O. E., Finck, A., Tardieu, F.,... Gibon, Y. (2009). Xeml Lab: a tool that supports the design of experiments at a graphical interface and generates computer-readable metadata files, which capture information about genotypes, growth conditions, environmental perturbations and sampling strategy. *Plant, Cell and Environment, 32*(9), 1185–1200. doi:10.1111/j.1365-3040.2009.01964.x

Harper, J. L. (1982). After description. In E. I. Newman (ed.), *The Plant Community as a Working Mechanism* (pp. 11–25). Oxford: Blackwell Scientific Publications.

Hendry, G. A. F., & Grime, J. P. (eds) (1993). *Methods in Comparative Plant Ecology*. London: Chapman & Hall.

Hodgson, J. G., Wilson, P. J., Hunt, R., Grime, J. P., & Thompson, K. (1999). Allocating C-S-R plant functional types: a soft approach to a hard problem. *Oikos, 85*, 282–294.

Holdridge, L. R. (1947). Determination of world plant formations from simple climatic data. *Science, 105*, 367–368.

Hulme, P. E. (1998). Post-dispersal seed predation: consequences for plant demography and evolution. *Perspectives in Plant Ecology, Evolution and Systematics, 1*(1), 32–46. doi:10.1078/1433-8319-00050

Irschick, D. J. (2003). Measuring performance in nature: implications for studies of fitness within populations. *Integrative and Comparative Biology, 43*(3), 396–407. doi:10.1093/icb/43.3.396

Irschick, D. J., & Garland, T. (2001). Integrating function and ecology in studies of adaptation: investigations of locomotor capacity as a model system. *Annual Review of Ecology and Systematics, 32*, 367–396. doi:10.1146/annurev.ecolsys.32.081501.114048

Jurado, E., & Westoby, M. (1992). Seedling growth in relation to seed size among species of arid Australia. *Journal of Ecology, 80*, 407–416.

Keddy, P. A. (1992a). Assembly and response rules: two goals for predictive community ecology. *Journal of Vegetation Science, 3*, 157–164.

Keddy, P. A. (1992b). A pragmatic approach to functional ecology. *Functional Ecology, 6*, 621–626.

Kleyer, M., Bekker, R. M., Knevel, I. C., Bakker, J. P., Thompson, K., Sonnenschein, M.,... Peco, B. (2008). The LEDA Traitbase: a database of life-history traits of the Northwest European flora. *Journal of Ecology, 96*(6), 1266–1274. doi:10.1111/j.1365-2745.2008.01430.x

Knevel, I. C., Bekker, R. M., Kunzmann, D., Stadler, M., & Thompson, K. (eds) (2005). *The LEDA traitbase. Collecting and measuring standards of life-history traits of the Northern European flora*. Groningen: University of Groningen.

Lavorel, S., & Garnier, E. (2002). Predicting changes in community composition and ecosystem functioning from plant traits: revisiting the Holy Grail. *Functional Ecology, 16*, 545–556.

Lavorel, S., McIntyre, S., Landsberg, J., & Forbes, T. D. A. (1997). Plant functional classifications: from general groups to specific groups based on response to disturbance. *Trends in Ecology and Evolution, 12*, 474–478.

Leishman, M. R., & Westoby, M. (1994). The role of large seed size in shaded conditions: experimental evidence. *Functional Ecology, 8*(2), 205–214. doi:10.2307/2389903

Leishman, M. R., Westoby, M., & Jurado, E. (1995). Correlates of seed size variation: a comparison among 5 temperate floras. *Journal of Ecology, 83*(3), 517–529. doi:10.2307/2261604

Lepart, J., & Escarré, J. (1983). La succession végétale, mécanismes et modèles: analyse bibliographique. *Bulletin d'Ecologie, 14*(3), 133–178.

Loreau, M., Naeem, S., Inchausti, P., Bengtsson, J., Grime, J. P., Hector, A.,... Wardle, D. A. (2001). Biodiversity and ecosystem functioning: current knowledge and future challenges. *Science, 294*, 804–808.

MacArthur, R. H., & Wilson, E. O. (1967). *The Theory of Island Biogeography*. Princeton: Princeton University Press.

Mazer, S. J. (1989). Ecological, taxonomic, and life history correlates of seed mass among Indiana dune Angiosperms. *Ecological Monographs, 59*, 153–175.

McGill, B. J., Enquist, B. J., Weiher, E., & Westoby, M. (2006). Rebuilding community ecology from functional traits. *Trends in Ecology and Evolution, 21*, 178–185.

Mokany, K., & Ash, J. (2008). Are traits measured on pot grown plants representative of those in natural communities? *Journal of Vegetation Science, 19*, 119–126.

Moles, A. T., Ackerly, D. D., Webb, C. O., Tweddle, J. C., Dickie, J. B., & Westoby, M. (2005). A brief history of seed size. *Science, 307*, 576–580.

Mooney, H. A. (1972). The carbon balance of plants. *Annual Review of Ecology and Systematics, 3*, 315–346.

Navas, M.-L., & Violle, C. (2009). Plant traits related to competition: how do they shape the functional diversity of communities? *Community Ecology, 10*(1), 131–137. doi:10.1556/ComEc.10.2009.1.15

Niinemets, Ü. (2001). Global-scale climatic controls of leaf dry mass per area, density, and thickness in trees and shrubs. *Ecology, 82*, 453–469.

Odum, E. P. (1969). The strategy of ecosystem development. *Science, 164*, 262–270.

Parsons, R. F. (1968). The significance of growth-rate comparisons for plant ecology. *The American Naturalist, 102*, 595–597.

Pérez-Harguindeguy, N., Díaz, S., Garnier, E., Lavorel, S., Poorter, H., Jaureguiberry, P.,... Cornelissen, J. H. C. (2013). New handbook for standardised measurement of plant functional traits worldwide. *Australian Journal of Botany*, *61*, 167–234.

Prescott-Allen, R., & Prescott-Allen, C. (1990). How many plants feed the world? *Conservation Biology*, *4*(4), 365–374. doi:10.1111/j.1523-1739.1990.tb00310.x

Raunkiaer, C. (1934). *The Life Forms of Plants and Statistical Plant Geography* (H. Gilbert-Carter & A. Fausbøll, trans. English ed.). Oxford: Oxford University Press.

Reich, P. B. (1993). Reconciling apparent discrepancies among studies relating life-span, structure and function of leaves in contrasting plant life forms and climates: 'the blind men and the elephant retold'. *Functional Ecology*, *7*, 721–725.

Reich, P. B. (2014). The world-wide 'fast–slow' plant economics spectrum: a traits manifesto. *Journal of Ecology*, *102*(2), 275–301. doi:10.1111/1365-2745.12211

Reich, P. B., Tjoelker, M. G., Walters, M. B., Vanderklein, D. W., & Bushena, C. (1998). Close association of RGR, leaf and root morphology, seed mass and shade tolerance in seedlings of nine boreal tree species grown in high and low light. *Functional Ecology*, *12*(3), 327–338. doi:10.1046/j.1365-2435.1998.00208.x

Roche, P., Díaz Burlinson, N., & Gachet, S. (2004). Congruency analysis of species ranking based on leaf traits: which traits are the more reliable? *Plant Ecology*, *174*, 37–48.

Sack, L., Cornwell, W. K., Santiago, L. S., Barbour, M. M., Choat, B., Evans, J. R.,... Nicotra, A. (2010). A unique web resource for physiology, ecology, and the environmental sciences: PrometheusWiki. *Functional Plant Biology*, *37*(8), 687–693. doi:10.1071/fp10097

Salisbury, E. J. (1942). *The Reproductive Capacity of Plants.* London: Bells.

Saverimuttu, T., & Westoby, M. (1996). Seedling longevity under deep shade in relation to seed size. *Journal of Ecology*, *84*(5), 681–689. doi:10.2307/2261331

Scurlock, J. M. O., Johnson, K., & Olson, R. J. (2002). Estimating net primary productivity from grassland biomass dynamics measurements. *Global Change Biology*, *8*, 736–753.

Seiwa, K., & Kikuzawa, K. (1996). Importance of seed size for the establishment of seedlings of five deciduous broad-leaved tree species. *Vegetatio*, *123*(1), 51–64. doi:10.1007/bf00044887

Shipley, B., Vile, D., & Garnier, E. (2006). From plant traits to plant communities: a statistical mechanistic approach to biodiversity. *Science*, *314*, 812–814.

Silvertown, J. W. (1981). Seed size, life-span, and germination date as coadapted features of plant life history. *The American Naturalist*, *118*, 860–864.

Suding, K. N., Lavorel, S., Chapin, F. S., III, Cornelissen, J. H. C., Díaz, S., Garnier, E.,... Navas, M.-L. (2008). Scaling environmental change through the community-level: a trait-based response-and-effect framework for plants. *Global Change Biology*, *14*, 1125–1140.

Swanborough, P., & Westoby, M. (1996). Seedling relative growth rate and its components in relation to seed size: phylogenetically independent contrasts. *Functional Ecology*, *10*, 176–184.

Thomas, H. (2013). Senescence, ageing, and death of the whole plant. *New Phytologist*, *197*(3), 696–711. doi:10.1111/nph.12047

Thompson, K., Bakker, J. P., & Bekker, R. M. (1997). *The soil seed bank of North Western Europe: methodology, density, and longevity*. Cambridge: Cambridge University Press.

Thompson, K., Band, S. R., & Hodgson, J. G. (1993). Seed size and shape predict persistence in soil *Functional Ecology*, *7*, 236–241.

Thompson, K., Hodgson, J. G., Grime, J. P., & Burke, M. J. W. (2001). Plant traits and temporal scale: evidence from a 5-year invasion experiment using native species. *Journal of Ecology*, *89*(6), 1054–1060. doi:10.1046/j.0022-0477.2001.00627.x

Violle, C., Navas, M.-L., Vile, D., Kazakou, E., Fortunel, C., Hummel, I., & Garnier, E. (2007). Let the concept of trait be functional! *Oikos*, *116*, 882–892.

Webb, C. O., Hoeting, J. A., Ames, G. M., Pyne, M. I., & Poff, N. L. (2010). A structured and dynamic framework to advance traits-based theory and prediction in ecology. *Ecology Letters*, *13*, 267–283.

Weiher, E., van der Werf, A., Thompson, K., Roderick, M., Garnier, E., & Eriksson, O. (1999). Challenging Theophrastus: a common core list of plant traits for functional ecology. *Journal of Vegetation Science*, *10*, 609–620.

Westoby, M. (1998). A leaf-height-seed (LHS) plant ecology strategy scheme. *Plant and Soil*, *199*, 213–227.

Westoby, M., Falster, D. S., Moles, A. T., Vesk, P. A., & Wright, I. J. (2002). Plant ecological strategies: some leading dimensions of variation between species. *Annual Review of Ecology and Systematics*, *33*, 125–159.

Whittaker, R. H. (1975). *Communities and Ecosystems* (2nd ed.). New York, USA: Macmillan Publishing Co., Inc.

Woodward, F. I., & Diament, A. D. (1991). Functional approaches to predicting the ecological effects of global change. *Functional Ecology*, *5*, 202–212.

Wright, I. J., Reich, P. B., Westoby, M., Ackerly, D. D., Baruch, Z., Bongers, F.,... Villar, R. (2004). The worldwide leaf economics spectrum. *Nature*, *428*, 821–827.

CHAPTER 3

The functional characterization of plants

3.1 Introduction

Amongst the multitude of traits measurable on an individual, to be useful for comparative functional ecology a trait should present at least four characteristics (Lavorel et al., 2007). It should: (1) be linked to a plant function; (2) be relatively easy and quick to measure ('functional markers' sensu Garnier et al., 2004); (3) be measurable using standardized protocols applicable to a large range of species and growth conditions (see Chapter 2); and (4) allow for the establishment of hierarchies between species which are conserved between contrasting environments, without the absolute values of these traits necessarily remaining constant (Cornelissen et al., 2003; Garnier et al., 2001a; Kazakou et al., 2014; Mokany & Ash, 2008). The identification and use of traits defined using these criteria allows us to compare plant functioning between plants growing in different environments (see Chapter 2 for further details).

Each trait cannot be considered independently of other traits, however, due to the frequent co-variation existing between pairs or suites of traits (Chapin et al., 1993; Grime et al., 1997; Reich et al., 1997; see below and Chapter 4). For example, many plants characteristic of low-resource environments have intrinsically low rates of growth, photosynthesis, and nutrient absorption, low rates of tissue turnover, and high concentrations of secondary metabolites (Chapin et al., 1993, and references therein). The identification of such suites of traits and of their recurrence among environments, has led to the identification of a number of axes of variation representing different plant strategies. These axes are defined on the basis of a small number of traits whose functional significance is considered to be well established (Grime, 2001; Westoby et al., 2002). Box 3.1 presents the advantages for research of the use of systems describing the ecological strategies of plants based on the identification of a limited number of major axes of variation, as initially formulated by Westoby (1998).

Laughlin (2014) discussed the idea of an 'intrinsic dimensionality of plant traits', which represents the number of independent axes of variation necessary to describe adequately plant functioning, as a fundamental quantity in comparative ecology. There is no definitive answer as to what this number is (see discussion in Laughlin, 2014), but there are at least three dimensions which are generally considered as being fundamental for understanding plant functioning and defining ecological strategies (Westoby, 1998; Westoby et al., 2002): the acquisition and use of resources, the stature of the plant, and the plant capacity for sexual regeneration. These three axes of variation are discussed in Section 3.2. We will then present a certain number of other traits whose functional significance is well established, but which have not yet been integrated into a more global vision of plant functioning. Finally, we briefly examine to what extent the differences in trait values that have been observed depend on the degree of evolutionary relatedness among species.

3.2 Traits representative of the three major axes of ecological variation

In his initial system of ecological strategies, Westoby (1998) proposed that specific leaf area, plant height, and seed mass be used as the principal

Box 3.1 Why adopt a common system for the identification of plant ecological strategies?

Taken from Westoby (1998).

The adoption of an identification system for plant ecological strategies based on recognized and accepted criteria should be one of the priorities for research in plant ecology. Such a system would have numerous advantages [. . .], as illustrated by the following examples:

- Facilitating the meta-analysis of field experiments. Over the last few decades, hundreds of experiments investigating competition between plants, herbivory, and other plant interactions have been carried out. In meta-analyses carried out to date aiming to search for generalizations in the conclusions of these studies, species have been grouped into growth forms, biological types, or according to their life cycle. It would be reasonable to assume that if species were grouped using a system based on strategies, this would explain a larger part of pertinent ecological variation, and the results of these meta-analyses would be significantly improved.
- Placing experimental ecophysiological studies in the context of comparative ecology. Measures of gas exchange or other metabolic fluxes can only be carried out on a few species in each individual study. However, the objective of these studies is often to compare different categories of species, for example species growing in shaded environments and these growing in gaps. Any synthesis based on these different studies will remain difficult until a consensus is reached as to the manner of establishing these species categories.
- Predicting the dynamics of vegetation response to global change. Developing projections of the future dynamics of vegetation at the planetary scale has become a major test for the capacity of ecology to respond to applied questions. The primary global changes to be expected are however well known: zones corresponding to a given temperature will move towards higher latitudes, atmospheric CO_2 concentration will continue to rise, and the intensification of land use will continue. Global models of the dynamics of vegetation cannot be parameterized for the 300 000 existing species; it is therefore necessary to use a simplified representation of vegetation using 'functional types' to obtain realistic predictions.
- Developing a truly predictive science of ecology. Data originating from field experiments, ecophysiological studies, or comparative ecology tend to be grouped in electronic databases which are currently reaching a critical mass sufficient to form generalizations much broader than those only a decade ago. These generalizations will be even more apparent and pertinent when they are based on a system of plant strategies explicitly taking into account their functional characteristics.

traits to describe the three fundamental axes of variation amongst plants. These traits are linked respectively to the acquisition and use of resources, the competitive ability, and the capacity for sexual regeneration. Figure 3.1 shows the frequency distributions of these three traits, and the following sections describe the functional dimensions they capture.

3.2.1 Resource acquisition and use

The persistence of an individual at a given site largely depends on the manner by which it manages its resources, that is to say the manner by which it acquires them, conserves them, and loses them over time. The functioning of green leaves is crucial for this management, in that it sets the rate of plant (and ecosystem) carbon acquisition and controls many other aspects of plant metabolism (e.g. uptake of other resources, respiration, etc).

3.2.1.1 The fundamental trade-off between the acquisition and conservation of resources

The existence of a fundamental trade-off between the rapid acquisition and the efficient conservation of resources has been discussed in the ecological literature for more than forty years (Aerts & Chapin, 2000; Berendse & Aerts, 1987; Chapin, 1980; Grime, 1977; Reich, 2014; Reich et al., 1992; Small, 1972). It is, however, only over the course of the last two decades that the availability of large data sets has allowed for its precise quantification and for the identification of the trait syndromes that can be used to characterize this trade-off for a wide variety of plants (Díaz et al., 2004; Grime et al., 1997; Reich et al., 1997; Wright et al., 2004).

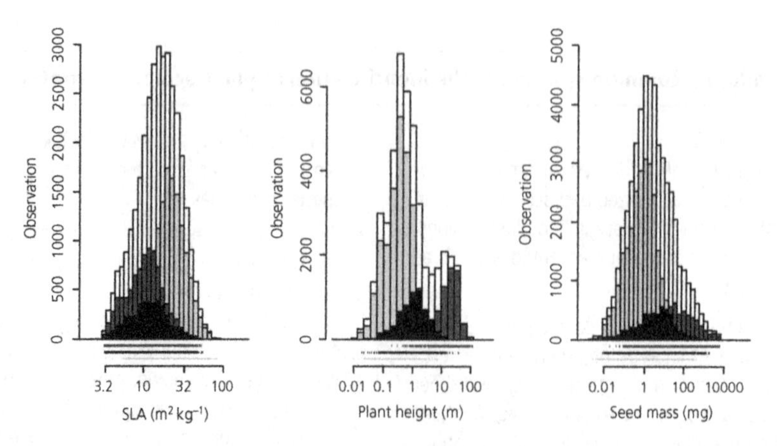

Figure 3.1 Distributions of the values of (a) specific leaf area (SLA), (b) maximum plant height, and (c) seed mass, available in the TRY database. White bars: all data; light grey bars: herbaceous species; dark grey bars: trees; black bars: shrubs. The bars below the x axis represent the range of variation of each trait for each of the above groups (using the same colour code). The number of observations for these different groups are the following : for specific leaf area, total: 33 443; herbaceous plants: 12 989; trees: 8544; shrubs: 4032; for maximum plant height, total: 44 505; herbaceous plants: 24 169; trees: 6357; shrubs: 5845; for seed mass, total: 49 837; herbaceous plants: 25 606; trees: 6458; shrubs: 5081. The sum of the observations for the different groups is different to the total number of observations as some species are not attributed to any of the three groups within TRY. Adapted from Kattge et al. (2011).

A global compilation carried out by 'Glopnet' (Global plant trait network) on individuals belonging to more than 2500 species from natural or semi-natural environments from a wide range of biomes has been carried out for six leaf traits (Wright et al., 2004): the maximum rate of photosynthesis per unit of leaf mass (A_{max}), the rate of dark respiration (R_m), leaf life span (LLS), specific leaf area (SLA: the ratio between leaf area and leaf mass; see Box 3.2), and the leaf concentrations of nitrogen and phosphorus (LNC and LPC, respectively), to which was later added the potassium concentration (Wright et al., 2005).

Figure 3.2 shows three of the fundamental relationships found in this work: negative correlations were found between A_{max} and LLS on the one hand, and between SLA and LLS on the other hand, while a positive correlation was found between SLA and A_{max} (Wright et al., 2004). A key finding is that these relationships were observed to be largely independent of growth form, species functional type, climate, or biome.

From a carbon gain perspective, the negative correlation between A_{max} and LLS expresses a trade-off between the potential rate of carbon return per unit leaf mass and the duration of this return (Westoby et al., 2002). It is consistent with cost—benefits models based on the optimization of leaf carbon gain (Kikuzawa, 1995; Kikuzawa & Ackerly, 1999), which predict that leaf longevity should be short when leaf photosynthetic rate is high, when the photosynthetic rate decreases rapidly through time, or when the construction cost of the leaf is low.

Figure 3.2 shows that high SLA leaves tend to have high photosynthetic capacity per unit leaf mass, which confirms conclusions from previous syntheses (Field & Mooney, 1986; Niinemets, 1999; Reich et al., 1997). Higher SLA corresponds to more light intercepted per unit mass, which generally goes with higher leaf nitrogen concentrations (cf. references in Wright et al., 2004), reflecting higher concentrations of Rubisco and other photosynthetic proteins in the leaf (Evans, 1989; Field & Mooney, 1986). A positive relationship between SLA and LNC was also found in the Glopnet data set, and a multiple regression combining these two traits explained more than 80% of the variance in A_{max}.

Differences in SLA can be generated by differences in leaf thickness, leaf density, or combinations of these (Witkowski & Lamont, 1991, and Box 3.2). The negative correlation between SLA and LLS (Figure 3.2) therefore suggests that leaves with a long life span require high structural strength,

Box 3.2 Specific leaf area and its components

Adapted from Poorter and Garnier, 2007.

The specific leaf area (SLA) and its inverse the leaf mass per unit area (LMA = 1/SLA) are traits that are widely used to describe the morpho-anatomic structure of leaves. SLA is defined as the ratio between the area of a leaf (LA, the area of a single leaf side) and its dry mass ($M_{L(D)}$); the LMA is therefore the ratio between the dry mass of a leaf and its area. In order to understand the different components of these traits, the equation defining SLA can be modified by introducing the water saturated fresh mass of a leaf ($M_{L(F)}$):

$$SLA = \frac{LA}{M_{L(F)}} \times \frac{M_{L(F)}}{M_{L(D)}} \quad (3.2.1)$$

The ratio between the fresh mass and the dry mass of leaves represents the inverse of their leaf dry matter content (LDMC). Additionally, leaf density (ρ_L) is equal to the ratio between the fresh mass of a leaf and its volume (V_L). Equation 3.2.1 can then be written as:

$$SLA = \frac{1}{\rho_L} \times \frac{LA}{V_L} \times \frac{M_{L(F)}}{M_{L(D)}} \quad (3.2.2)$$

For laminar leaves, we can consider that the ratio between the volume and the area of a leaf is equal to its thickness (Roderick et al., 1999), L_{Th}. SLA depends finally on three leaf components: leaf dry matter content, which varies with anatomical structure, leaf thickness and leaf density. As it has been experimentally shown that leaf density is close to 1 for a wide variety of leaves (Vile et al., 2005, and references therein), SLA can finally be expressed as:

$$SLA \approx \frac{1}{L_{Th} \times LDMC} \quad (3.2.3)$$

As LDMC can be considered as an estimate of the density of leaf tissues (the ratio between the dry mass and volume of a leaf, not to be confounded with leaf density ρ_L, which is the ratio between the fresh mass and the volume of the leaf: see above), we can therefore consider that SLA depends on two main components, which are the thickness of the leaf and the density of its tissues (Witkowski & Lamont, 1991).

The SLA can also be expressed as a function of the density of each of the tissues making up the leaf (M_i/V_i, where M_i and V_i are respectively the dry mass and volume of the tissue i): the epidermis, the mesophyll, the vessels and their associated tissues (collenchyma and sclerenchyma), and the volume occupied by each of these per unit of leaf area (Garnier & Laurent, 1994; Poorter & Garnier, 2007):

$$\frac{1}{SLA} = \sum_{i=1}^{5} \frac{V_i}{LA} \times \frac{M_i}{V_i} \quad (3.2.4)$$

The biomass of a leaf can also be subdivided into its different biochemical fractions, which can be roughly regrouped into: lipids, lignin, soluble phenolic components, proteins, structural carbohydrates (cellulose, hemicellulose, and pectin), non-structural carbohydrates (glucose, fructose, sucrose, and starch), organic acids, and mineral nutrients (cf. the review by Poorter & Villar, 1997). The inverse of the SLA is thus the sum of the mass of each of these eight classes of components (M_{Cci}) expressed per unit of leaf area:

$$\frac{1}{SLA} = \sum_{i=1}^{8} \frac{M_{CCi}}{LA} \quad (3.2.5)$$

The three modes of decomposition of the SLA are interrelated: for example, a high density can be the consequence of a high proportion of sclerenchyma, which is reflected in the biochemical composition by a high concentration of cell walls (lignin and structural carbohydrates).

necessary to protect the leaf against physical hazard and deter herbivores (see Section 3.2.1.2 below).

At the whole plant level, high photosynthetic rates tend to be associated with high rates of mineral nutrient acquisition (at least of nitrogen) and of relative growth rate (RGR: Box 3.3, Poorter & Garnier, 2007, and references therein) at least for seedlings (see Poorter et al. 2014, for a correlation network at the whole plant level involving RGR and several traits of the leaf economics spectrum), while a long leaf life span (slow tissue turnover) results in a long residence time of mineral nutrients, a trait which is used to quantify the conservation of resources by a plant (Box 3.4; Aerts & Chapin, 2000; Berendse & Aerts, 1987; Eckstein et al., 1999; Garnier & Aronson, 1998).

The maximum rate of photosynthesis, the leaf life span, and the specific leaf area are consequently three traits that can be used to estimate the position of a species along the continuum between the acquisition and conservation of resources. The analysis of the Glopnet data set also shows positive

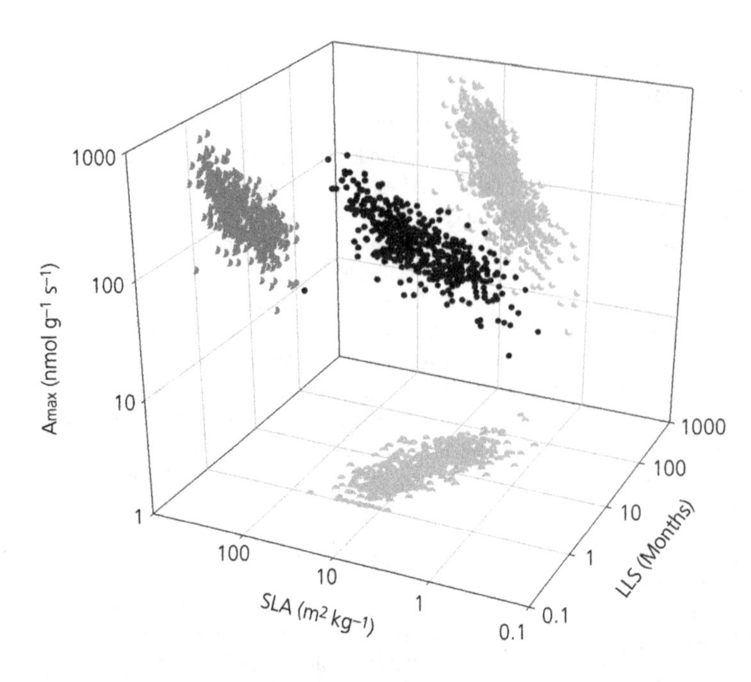

Figure 3.2 Three-way relationships among three fundamental traits of the 'leaf economics spectrum': specific leaf area (SLA), leaf life span (LLS), and the mass-based maximum rate of leaf photosynthesis (A_{max}), established from a database assembled by the Glopnet network. Each point represents a single species. The shadows, which represent the projections from the three-dimensional data cloud onto the different axis planes, show the bivariate relationships between pairs of traits. The three bivariate relationships are significant at $P < 0.001$ ($n = 503$). Data from Wright et al. (2004).

Box 3.3 The relative growth rate and its components

Adapted from Poorter and Garnier, 2007.

The growth rate of a plant can be expressed in different ways, as illustrated below. Assume there are two plants A and B, which have respectively a dry mass of 0.1 g and 1 g at time t_1, and that their mass increases by 0.1 g in 24 hours. Can we consider that the two plants grow at the same rate?

The most straightforward way to express this growth is to calculate the absolute growth rate (AGR), which corresponds to the total biomass increment (M_T) for each of the two plants over the interval of time dt:

$$AGR = \frac{dM_T}{dt} \qquad (3.3.1)$$

The total mass of the plant at time t_2 (M_{Tt2}) can be calculated from its mass at the time t_1 (M_{Tt1}) and the AGR:

$$M_{Tt2} = M_{Tt1} \times AGR\,(t_2 - t_1) \qquad (3.3.2)$$

In our example the two plants A and B have the same AGR. However, plant A has an initial biomass M_{Tt1} much lower than that of plant B. To take into account this initial difference, the biomass increment can be defined relatively to the mass of the plant already present: this is the relative growth rate (RGR):

$$RGR = \frac{1}{M_T}\frac{dM_T}{dt} = \frac{d(\log_e M_T)}{dt} \qquad (3.3.3)$$

RGR thus represents the slope of the relationship between the logarithm of the total biomass of the plant and time. Integrating equation 3.3.3 between t_1 and t_2 leads, after rearrangement, to the following equation:

$$RGR = \frac{\log_e M_{Tt2} - \log_e M_{Tt1}}{t_2 - t_1} \qquad (3.3.4)$$

in which RGR is the average relative growth rate over the considered time interval. From this equation, we can also determine the total biomass of the plant at the time t_2 as a function of its biomass at time t_1:

$$M_{Tt2} = M_{Tt1} x e^{RGR(t_2 - t_1)} \qquad (3.3.5)$$

In the example given above, plant A has a RGR of 0.693 g g^{-1} d^{-1}, while plant B has a RGR of 0.095 g g^{-1} d^{-1}. A unit of biomass of plant A is therefore more efficient than a unit of biomass of plant B to produce new biomass.

When the RGR is determined under 'optimal' controlled experimental conditions (that is, without resource limitation: light, water, mineral nutrients), we call this the maximum relative growth rate (RGR_{max}). RGR_{max} is normally assessed on seedlings, and since the pioneering work of Grime and Hunt (1975) it has been widely used to locate species/populations in relation to each other along a continuum of functioning from slow to fast, a concept related to that of the axis of acquisition /conservation of resources discussed in this text.

The so-called 'growth analysis' technique has been developed in order to determine which components of RGR best explain differences in RGR between plants from different species/populations or growing in different environments. This technique breaks down the components of RGR by first defining and isolating the total plant leaf area (LA_T) in equation 3.3.3:

$$RGR = \frac{1}{M_T}\frac{dM_T}{dt} = \left(\frac{LA_T}{M_T}\right) \times \left(\frac{1}{LA_T} \times \frac{dM_T}{dt}\right) \quad (3.3.6)$$

This reformulation identifies two terms: (1) the ratio between the total plant leaf area and the total plant biomass, called the 'leaf area ratio' (LAR), which represents a morphological component of the whole plant; and (2) the increment in total plant biomass per unit of time, relative to a unit of leaf area, called the 'net assimilation rate' (NAR) or 'unit leaf rate', corresponding to a physiological component which depends on the balance between the rates of photosynthesis and respiration at the whole plant level. The leaf area ratio can itself be broken down into two other terms as:

$$LAR = \frac{LA_T}{M_T} = \left(\frac{LA_T}{M_L}\right) \times \left(\frac{M_F}{M_L}\right) \quad (3.3.7)$$

where M_L represents the total mass of the leaves of a plant. The two terms which appear are: (1) the ratio between the area and the dry mass of leaves, called the 'specific leaf area' (SLA), which represents a morpho-anatomical component of leaves (cf. Box 3.2), expressed here at the whole plant level, and (2) the ratio between leaf biomass and the total plant biomass which is called the 'leaf mass fraction' (LMF), which expresses the proportion of biomass invested in leaves. Finally:

$$RGR = SLA \times LMF \times NAR \quad (3.3.8)$$

which is the classic decomposition of RGR. Other mathematical decompositions can also be used (Garnier, 1991; Hunt, 1982; Lambers & Poorter, 1992), which allow the identification of other variables influencing the RGR.

A first meta-analysis based on experiments carried out on seedlings under optimal growth conditions has shown that between the three terms of Eq. 3.3.8, differences in the specific leaf area had respectively approximately 2.5 times and 5.5 times as much importance as differences in net assimilation rate or in leaf mass fraction to explain interspecific differences in RGR_{max} (Poorter & van der Werf, 1998, cf. Figure 6.3). A second meta-analysis based on experiments where the light level was varied showed that the relative importance of SLA vs NAR to explain differences in RGR_{max} decreased with increasing light (Shipley, 2006).

Although these relationships between SLA and RGR probably do not hold for adult plants, these results have nonetheless led to the idea that SLA could be used as a marker trait for the position of species along the fast growth–slow growth axis, which relates to the acquisition–conservation trade-off described in the main text.

correlations among A_{max}, SLA, LNC, LPC, and the rate of leaf respiration, while these correlations are negative with leaf life span (Wright et al., 2004); leaf potassium concentration is only weakly linked with these different traits (Wright et al., 2005). This combination of relationships has been embedded into the concept of the 'Leaf Economics Spectrum' (LES: Wright et al., 2004). This spectrum runs from plants showing a high rate of photosynthesis and a rapid return on investment in terms of mineral nutrients and leaf dry matter, to others characterized by low photosynthetic rates, with a much slower return on investment. The 'Integrated Screening Programme' carried out under laboratory conditions on a far more limited number of species but for a far larger number of traits has led to similar conclusions (Grime et al., 1997), as has a study carried out by Dìaz et al. (2004) on 640 species from four different floras (Argentina, England, Iran, and Spain).

Reanalysing the Glopnet data in a community rather than a global perspective, Funk and Cornwell (2013) showed that relationships among three common traits of the LES (A_{max}, LNC, and SLA) are weak in communities with low variation in leaf life span (LLS), especially in communities dominated by either herbaceous or deciduous woody species. Although Funk and Conwell acknowledge that more data collected within communities is needed to confirm this pattern, these findings suggest that leaf trait variation within communities, especially those dominated by a limited number of growth forms, should be interpreted with caution.

Amongst the traits of the leaf economics spectrum, specific leaf area is the one most easy, rapid, and efficient to measure, and is therefore the trait most often used to estimate the position of species along the acquisition–conservation continuum

Box 3.4 The conservation of plant nutrients within the plant: the residence time

Modified from Garnier and Aronson (1998).

The mean residence time of a nutrient (MRT) corresponds to the mean time during which a molecule of this nutrient remains within the plant (Berendse & Aerts, 1987). At steady state, this corresponds to the quantity of the nutrient present in the plant divided by the flux; i.e. the quantity that enters or exits (Box 3.4, Figure 1: DeAngelis, 1992; Frissel, 1981).

In such a system, the MRT can be written as:

$$MRT = \frac{M_T[\mathrm{m}]_\mathrm{T}}{eM_T[\mathrm{m}]_\mathrm{Lit}} \quad (3.4.1)$$

where M_T represents the total biomass of the plant, $[\mathrm{m}]_\mathrm{T}$ and $[\mathrm{m}]_\mathrm{Lit}$ are the concentrations of mineral nutrients in the green biomass and litter respectively, and e is the relative rate of biomass lost by the system. The numerator of this equation represents the average quantity of the nutrient present in the plant, and the denominator represents the exit flux of the nutrient via the production of litter (which is equal to the entry flux as we assume steady-state functioning). Under these conditions the MRT can be written as:

$$MRT = \frac{[\mathrm{m}]_\mathrm{T}}{e[\mathrm{m}]_\mathrm{Lit}} \quad (3.4.2)$$

Additionally, the difference between the quantities of a mineral nutrient present in living tissues and in dead material (the litter) corresponds to the process of 'resorption' during tissue senescence. We define the efficiency of this resorption (R_eff) as this difference, relative to the quantity of the nutrient present in the living tissues:

$$R_{eff} = \frac{M_T[\mathrm{m}]_\mathrm{T} - M_L[\mathrm{m}]_\mathrm{Lit}}{M_T[\mathrm{m}]_\mathrm{T}} \quad (3.4.3)$$

To account for mass loss during senescence, we introduce a 'mass loss correction factor' (MLCF: Vergutz et al., 2012) (M_{Lit} = MLCF × M_T), which allows R_{eff} to be expressed using the concentrations of elements rather than their absolute quantity. By introducing R_eff and MLCF into Eq. 3.4.3, and after rearrangement, we obtain:

$$MRT = \left(\frac{MLCF}{e}\right) \times \left(\frac{1}{1 - R_{eff}}\right) \quad (3.4.4)$$

The residence time of a nutrient within a plant thus depends on: (1) the relative rate of biomass loss, which is inversely proportional to organ longevity, (2) the resorption efficiency of nutrients at the time of senescence, and (3) the fraction of biomass lost during senescence. These three traits therefore influence the conservation of nutrients in a plant.

A sensitivity analysis of Eq. 3.4.4 above (which did not take variations in MLCF into account) has shown that, overall, leaf life span appears to have a greater effect on MRT than resorption efficiency, but that improved resorption efficiency is equally beneficial for nutrient conservation for all life spans (e.g. deciduous and evergreen: Eckstein et al., 1999).

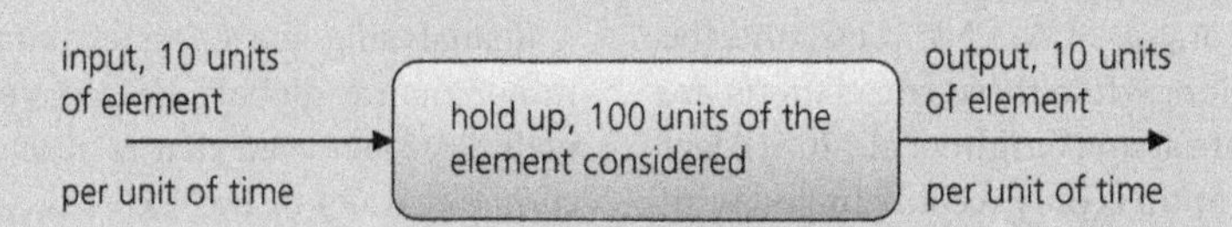

Box 3.4, Figure 1 Schematic representation of a steady state system, in which the fluxes of nutrients entering and exiting the system are equal: a central compartment contains 100 units of an element, into which enter and exit 10 units of the same element per unit of time. In this example, the residence time of an element is equal to 100/10 = 10 per unit of time.

(Weiher et al., 1999; Westoby, 1998; Westoby et al., 2002). Wilson et al., (1999) challenged this conclusion and proposed that leaf dry matter content (LDMC, the ratio between the dry mass and the water-saturated fresh mass of leaves) be used instead of SLA to characterize the position of species along this axis of resource management. LDMC, which is a proxy for leaf density (Box 3.2; Garnier et al., 2001b; Roderick et al., 1999; Vile et al., 2005; Witkowski & Lamont, 1991), is indeed easier to measure as it depends only on the measurements of two masses; this makes it less sensitive to errors and less variable among replicates (Garnier et al., 2001b; Wilson et al., 1999).

It can also be used when comparisons involve species growing in shaded environments, unlike

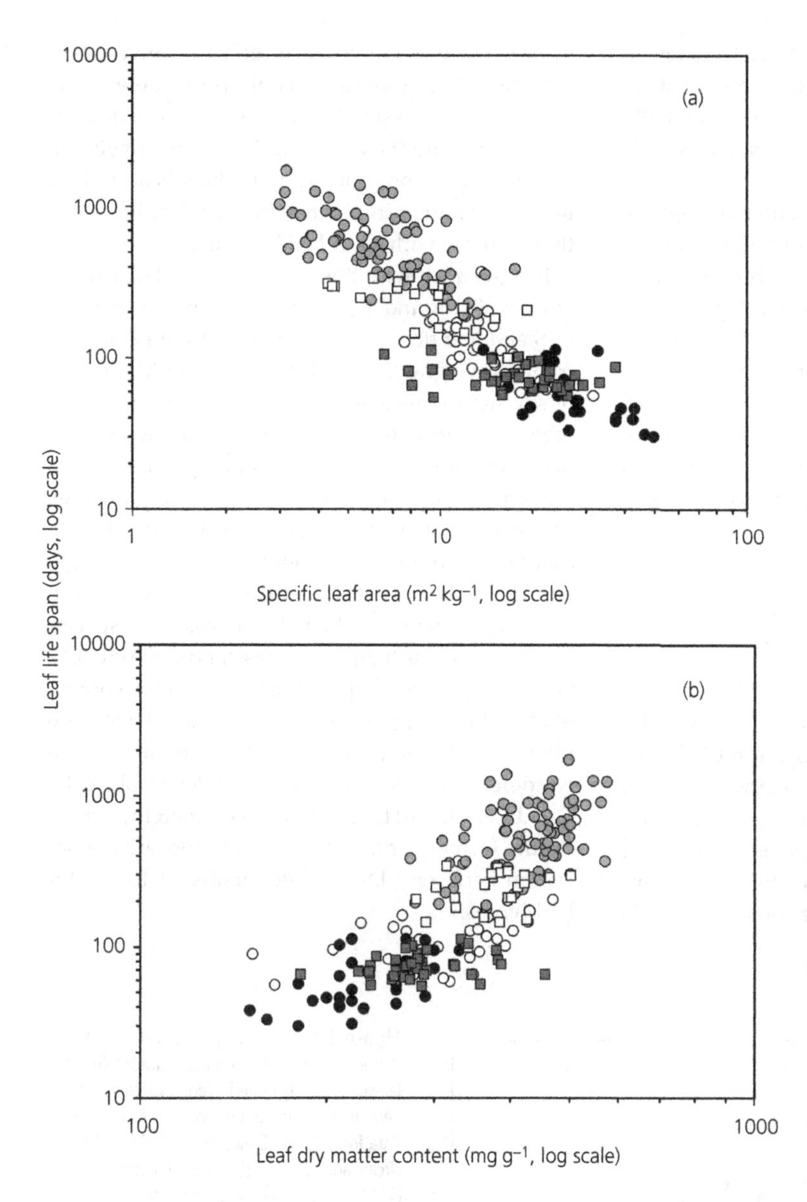

Figure 3.3 Comparisons of relationships established between (a) leaf life span and specific leaf area, and between (b) leaf life span and leaf dry matter content. Correlation coefficients for the overall relationships are: (a): $r = -0.86$ ($P < 0.001$, $n = 203$); (b): $r = 0.79$ ($P < 0.001$, $n = 203$). Dark circles are data from Ryser & Urbas (2000); grey circles: data from Wright (2001); white squares: data from Prior et al. (2003); grey squares: data from Cruz et al. (2010); white circles: data from Navas et al. (2010). Completed on the basis of Garnier & Navas (2012).

SLA which is very sensitive to light intensity (Poorter & de Jong, 1999; Wilson et al., 1999; and see Figure 4.5 in Chapter 4). While all of these different arguments are valid, we currently lack a thorough evaluation of this trait, comparable to that carried out to identify traits involved in the leaf economics spectrum (but see the discussion in Reich, 2014, regarding this point). The limited data available suggest that LDMC is less well correlated with the rate of photosynthesis and the RGR than SLA at the scale of the whole plant (Poorter & Garnier, 2007), and the comparison of five data sets in which SLA, LDMC, and leaf life span are available (Cruz et al., 2010; Navas et al., 2010; Prior et al., 2003; Ryser & Urbas, 2000; Wright, 2001) shows a less strong relationship between LDMC and leaf life span than between SLA and leaf life span (Figure 3.3). Interestingly, in a reanalysis of the Glopnet data set using a path analysis approach, Shipley et al. (2006) showed that the ratio of leaf dry mass to leaf water content, a trait strongly related to LDMC, was potentially underlying the set of trait correlations involved in the leaf economics spectrum.

Specific leaf area and leaf dry matter content are traits that are strongly influenced by the anatomic and structural properties of leaves (Box 3.2). It is thus not surprising that leaf physical properties are also correlated with these two traits. A meta-analysis involving 2819 species coming from 90 sites from all over the planet has shown strong negative correlations between SLA and various measures of leaf resistance to fracture, while these correlations were positive with estimates of leaf tissue density (Onoda et al., 2011).

3.2.1.2 Other functions related to the leaf economics spectrum

Defence against herbivores

An efficient conservation of resources implies—by definition—that mineral nutrients are retained for a long period within a plant's organs (Box 3.4). This requires that leaves with a long potential life-span are adequately protected against herbivores, since the mineral nutrients acquired by these organisms remain potentially available for long periods in order to provide positive return on investment (Chapin et al., 1993; Coley et al., 1985): Coley et al. (1985) have suggested that the quantities of defence metabolites and the concentrations of relatively immobile compounds such as polyphenols and fibres tend to be higher in plants with long leaf life spans than in those with short leaf life spans.

In agreement with this hypothesis, Al Haj Khaled et al. (2006) found a positive relationship between LLS and leaf fibre concentration in 14 grass species from native grasslands of the Pyrenean Mountains of southern France. In this study, a long LLS and a high fibre concentration were associated with a low dry matter digestibility of leaf material, indicating a low amount of energy available for herbivores (cf. Bruinenberg et al., 2002) in these leaves. Dry matter digestibility of leaves or shoots were also found to be negatively related to leaf dry matter content in three experiments conducted on herbaceous species from permanent temperate grasslands (Figure 3.4). Given that leaves with a high dry matter content tend to be long lived (Figure 3.3) and have high physical toughness—which is an extremely effective defence against herbivores (Moles et al., 2013; Onoda et al., 2011, and references therein)—these results further confirm the idea of a negative association between LLS and the quality of leaves for herbivores.

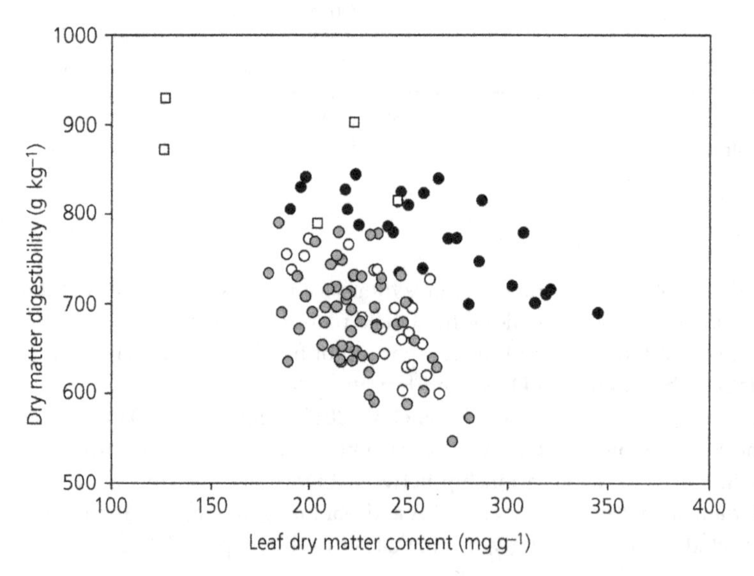

Figure 3.4 Relationships between leaf dry matter content and dry matter digestibility of leaves or shoots in herbaceous species from permanent temperate grasslands. Dark circles: data from Al Haj Khaled et al. (2006); white circles and squares: data from Louault et al. (2005); grey circles: data from Pontes et al. (2007). All species are grasses (circles) except five dicots (squares) in the study by Louault et al. (2005). In the study by Al Haj Khaled et al. (2006), digestibility was measured on leaves, while in the two other studies, digestibility was measured on whole shoots (leaves + stems). Digestibility at a given LDMC is higher in the first study, because digestibility of stems tends to be lower than that of leaves (e.g. Duru, 1997; Bruinenberg et al., 2002). Correlation coefficients: for Al Haj Khaled et al. (2006), $r = -0.65$ ($P < 0.001$, $n = 29$); for the two other studies (pooled): $r = -0.60$ ($P < 0.001$, $n = 85$).

More generally, Herms and Mattson (1992) have postulated the existence of a trade-off between plant growth and investment in defence structures and components (the 'growth–differentiation balance' hypothesis). Agrawal and Fishbein (2006) have proposed the concept of a 'defence traits syndrome', which is based on the idea that strategies of defence against herbivores depend on complex structural traits (e.g. trichrome density, SLA, and leaf resistance to fracture; see Hanley et al., 2007, for a review) and on chemical composition (e.g. carotenoid concentrations, leaf carbon/nitrogen ratios), some of which are also implicated in the leaf economics spectrum (cf. Moles et al., 2013, for a review of existing data). In fact, the resistance of plants to herbivore attack involves three types of mechanisms corresponding to groups of traits that are quite different in nature: escape, defence *sensu stricto*, and tolerance (Figure 3.5: Boege & Marquis, 2005). In addition, these groups of traits depend on the plant's ontogenic stage (see the reviews by Barton & Koricheva, 2010; Boege & Marquis, 2005).

Decomposition of litter

It is now recognized that numerous physiological characteristics of green leaves involved in herbivore defence persist during their senescence, leading to what has been termed 'afterlife' effects (Grime & Anderson, 1986). It also follows that certain properties of living organs are correlated with the properties of dead organs, such as plant litter (Cornelissen et al., 1999; Cornelissen & Thompson, 1997; Díaz et al., 2004; Pérez-Harguindeguy et al., 2000; Wardle et al., 1998). A meta-analysis of 818 species based on 66 experiments investigating decomposition on six continents indeed showed that the effects of interspecific differences in the chemical structure and composition of living leaves on the decomposition rate of dead leaves were as strong as those of differences in climate (Cornwell et al., 2008). This meta-analysis showed that SLA and LNC, two traits involved in the leaf economics spectrum, are positively correlated with the rate of litter decomposition (Figure 3.6). Leaves with high values of SLA and LNC tend to produce litter the chemical and/or

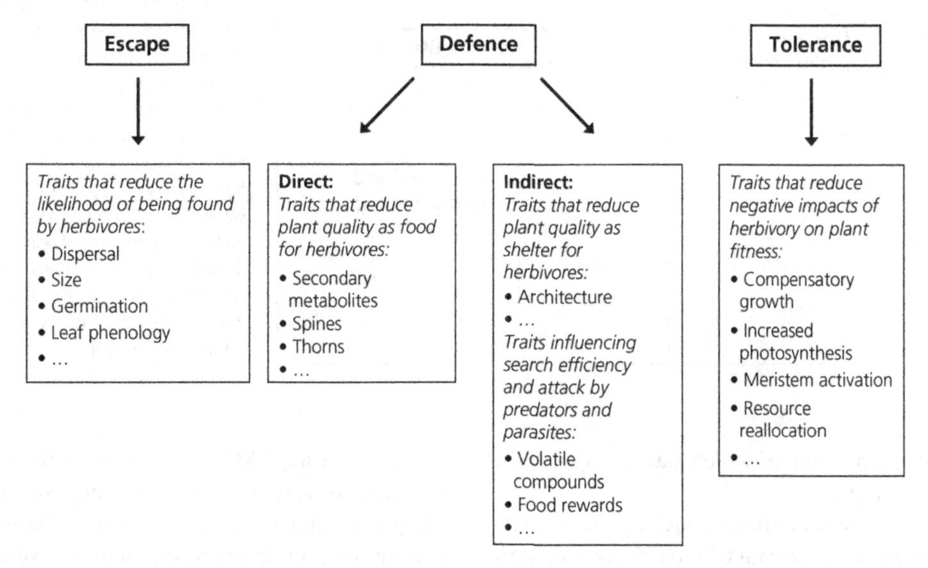

Figure 3.5 Escape, defence, and tolerance are the three mechanisms by which plants reduce the negative impacts of their interactions with herbivores. Herbivory can be minimized by reducing the probability of being found. Its impact can be minimized by avoiding herbivore attacks, either via the direct expression of defence traits which influence the quality of tissues targeted as food by herbivores, or via interactions with a third trophic level (i.e. indirect defences which influence the efficiency of predation by natural enemies of herbivores). Once damage has occurred, the negative impacts of herbivory can be reduced by mechanisms of tolerance which allow a plant to maintain its growth and reproduction despite the damage caused. To the extent that physiological constraints and the selective pressures exerted by herbivores vary over the lifespan of a plant, these three mechanisms and their relative importance can be expected to vary over the course of plant ontogeny. Taken from Boege & Marquis (2005). Reproduced with permission from Elsevier.

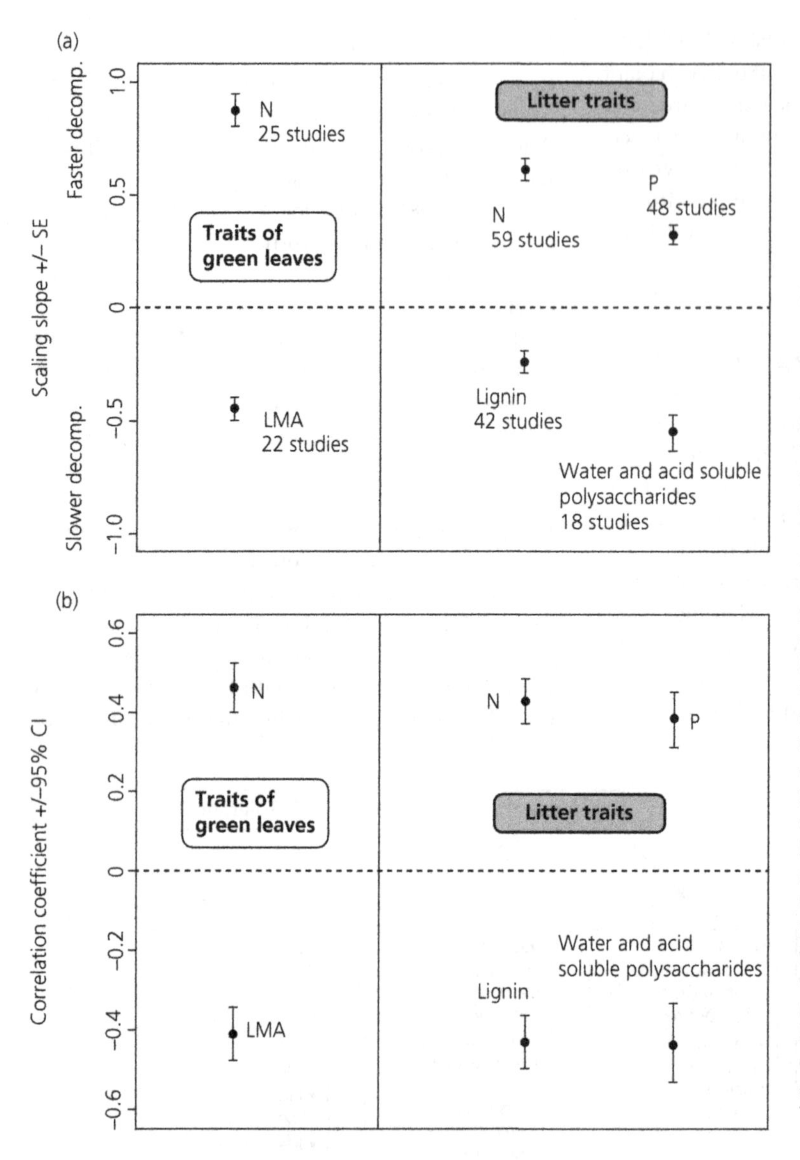

Figure 3.6 Results of a meta-analysis showing relationships between the traits of green leaves, litter, and the rate of litter decomposition, obtained from different studies carried out on species from different ecosystems of the planet. All of the comparisons have been carried out within each study, which allows for the standardization of the climatic conditions and experimental methods. The number of studies in which a given trait has been measured is indicated in the panel. Panel (a) shows the mean slopes (± standard error, SE) of the log-log relationships between each trait and the rate of decomposition: the greater the slope, the greater the rate of decomposition, and reciprocally. Panel (b) shows the mean correlation coefficient (± confidence interval [CI] at 95%) weighted for the size of the sample for these same relationships. N and P are respectively the concentrations of nitrogen and phosphorus per unit of mass; LMA is the leaf mass *per* area (the inverse of the specific leaf area); the fractions of soluble polysaccharides in water and in acid are essentially, but not exclusively, comprised of cellulose and hemicellulose. The associated analyses of covariance show significant relationships at the level $P < 0.01$ for the six traits. Taken from Cornwell et al. (2008).

structural composition of which lead to rapid rates of decomposition.

It appears however that LDMC is always better correlated with the rate of litter decomposition than SLA in the few experiments for which this comparison is possible (Kazakou et al., 2009). This is probably due to the fact that LDMC, which reflects the proportion of dense tissues in the leaf (Garnier & Laurent, 1994, and see Box 3.2) is the SLA component which is best correlated with those initial properties of plant litter which control decomposition (Kazakou, et al., 2009, and references therein). The resistance of leaves to fracture is also strongly correlated to litter decomposition, and plays an important role in interactions between plants and herbivores (references in Onoda et al., 2011).

3.2.1.3 Summary and synthesis

The leaf economic spectrum and its associated traits not only concern the characteristics of living leaves, but are also related to those of dead leaves, further

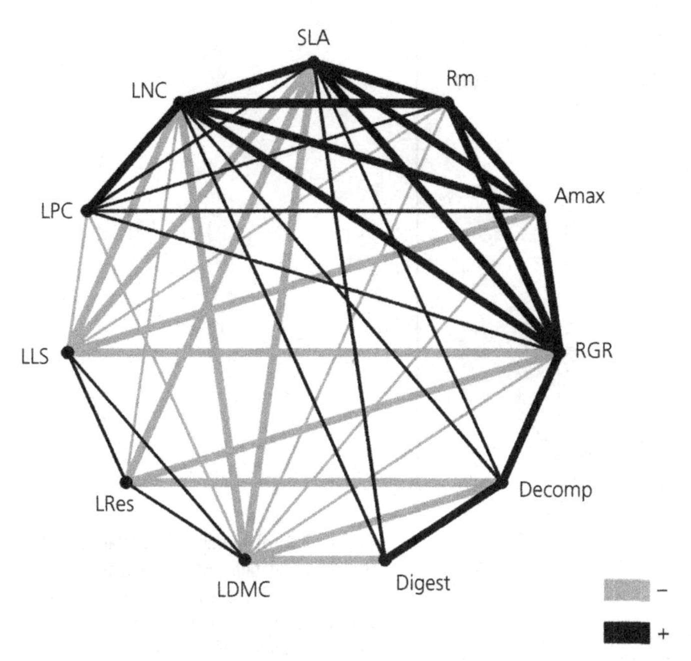

Figure 3.7 Summary diagram of the correlations between traits of the leaf economics spectrum. The black lines link traits that are correlated positively, and the grey lines link traits correlated negatively. The thicknesses of the lines (2 classes per colour) indicate the strength of the correlation. Key: SLA, specific leaf area; LNC, mass-based leaf nitrogen content ; LPC, mass-based leaf phosphorus content; LLS, leaf life span; LRes, physical resistance of leaves; LDMC, leaf dry matter content; Digest, dry matter digestibility; Decomp, rate of litter decomposition; RGR, relative growth rate; A_{max}, mass-based maximum leaf photosynthetic rate; Rm, leaf respiration rate. These correlations are valid when variables are expressed per unit mass.

confirming the leaf economic spectrum's strong degree of generality in explaining plant functioning. An overview of the relationships between the traits of the leaf economics spectrum, as well as the relationships between these and the relative growth rate, are presented in Figure 3.7. This figure emphasizes the large number of traits involved in the axis of resource acquisition and use by the plants.

Resources referred to implicitly in Figure 3.7 are carbon (photosynthesis and respiration) and nutrients (nitrogen and phosphorus). Recent evidence suggests that water economy might also be involved in this general trait syndrome: a high photosynthetic capacity appears to be linked to both a high leaf diffusive conductance for water vapour and a high leaf hydraulic conductance, suggesting coordination between water fluxes in the vapour and liquid phases at the leaf level (Reich, 2014, and references therein). This has led Reich (2014) to postulate that 'being fast for one resource [...]' (e.g. carbon) '[...] requires being fast for the other resources' in the same organ. Furthermore, all traits shown on Figure 3.7 are leaf traits, except RGR. Sections 3.3.2 and 3.3.3 below show that some relationships among equivalent traits also hold for stems and roots.

3.2.2 Plant stature

In the LHS scheme of Westoby (1998), the plant stature axis is represented by plant height (or, more specifically, typical maximum height, at maturity). As an estimator of the general size of a plant, plant height is a quantitative trait which has been used in a large number of studies in comparative ecology (Westoby et al., 2002, and references therein). The height of an individual should be considered comparatively with that of its neighbours: being larger than neighbours confers a competitive advantage via greater access to light, and thus constitutes a fundamental characteristic of the carbon acquisition strategy of an individual (King, 1990; Westoby et al., 2002). The height at which flowers or seeds are produced also has influences on reproductive biology, especially through effects on dispersal (Bazzaz et al., 2000; Greene & Johnson, 1989).

Height has been shown to be the primary driver of light extinction down the canopy of plant species (Figure 3.8) (King, 1990; Violle et al., 2009), explaining most of the variation in growth reduction of competing individuals (Gaudet & Keddy, 1988; Violle et al., 2009). Despite this clear effect on plant performance and its easiness of measurement,

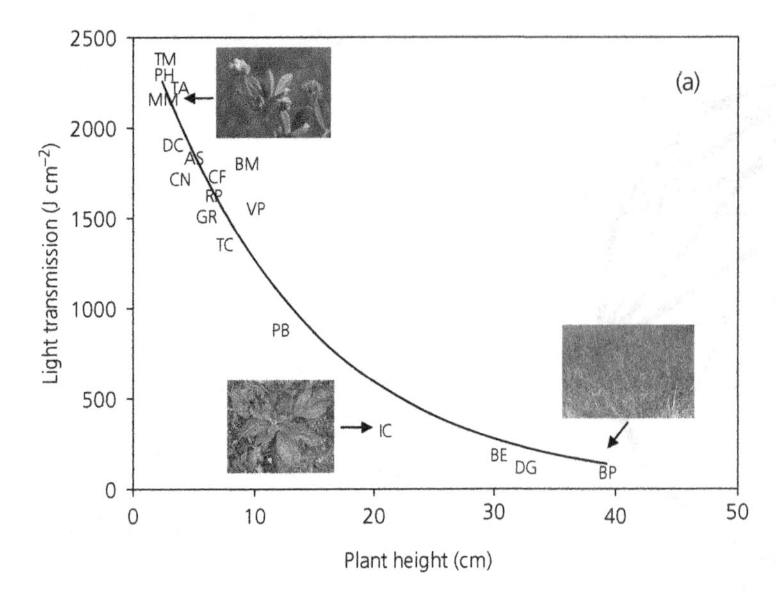

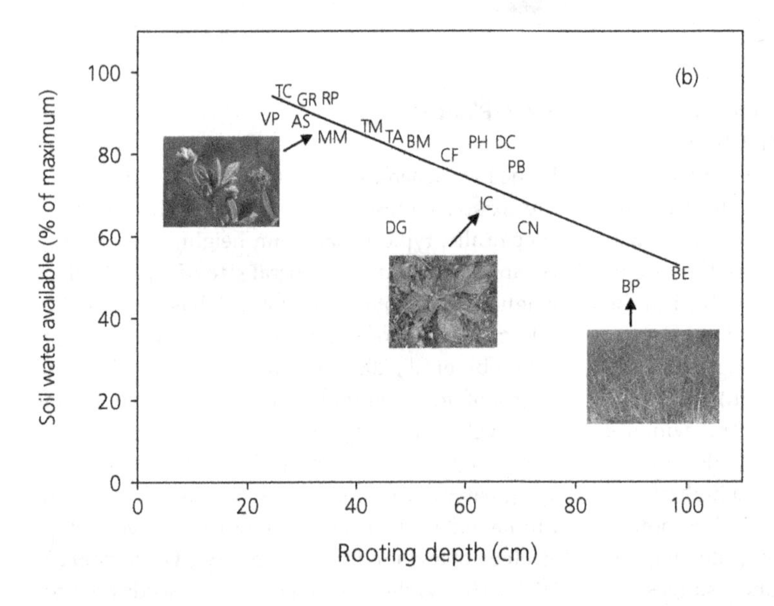

Figure 3.8 (a) Amount of light transmitted through vegetation cover as a function of plant height, and (b) soil water available to plants (expressed as a percentage of maximum water available) as a function of rooting depth, measured for 18 monocultures growing in an experimental garden. The species are represented by their initials. AS: *Arenaria serpyllifolia*; BP (photograph): *Brachypodium phoenicoides*; BE: *Bromus erectus*; BM: *Bromus madritensis*; CN: *Calamentha nepeta*; CF: *Crepis foetida*; DG: *Dactylis glomerata*; DC: *Daucus carota*; GR: *Geranium rotundifolium*; IC (photograph): *Inula conyza*; MM (photograph): *Medicago minima*; PH: *Picris hieracioides*; PB: *Psoralea bituminosa*; RP: *Rubia peregrina*; TC: *Teucrium chamaedrys*; TM: *Tordylium maximum*; TA: *Trifolium angustifolium*; VP: *Veronica persica*. The non-linear correlation coefficient of the relationship in (a) is: $r^2 = 0.93$ ($P < 0.001$, n = 18); the linear correlation coefficient of the relationship in (b) is: $r^2 = 076$ ($P < 0.001$, n = 18). Taken from Garnier & Navas (2012).

the interpretation of plant height can be difficult as it is a highly dynamic variable, which depends greatly on the ontological stage of plant, the disturbance regime that it has experienced, and the nature of neighbouring plants—thus, Westoby's proposal to use the canopy height at maturity (or canopy potential height) as a descriptor of this plant functional dimension.

This concept of potential canopy height is widely used for trees, for which the distinction between vegetative and reproductive heights can be ignored. This is not the case for herbaceous species, for which the upper parts of the reproductive structures can be significantly higher than the vegetative parts. For these species, it has been proposed that vegetative and reproductive heights should be explicitly considered as two different traits (Garnier et al., 2007; McIntyre et al., 1999; Vile et al., 2006). McIntyre et al. (1999) have even proposed that the projection of the inflorescence above the vegetative parts of the plant be considered as a response trait specific to disturbance for grass species, due to its influence on species dispersal and grazing by herbivores. In addition, plant height can sometimes be difficult to define for herbaceous species which have a rapid or continuous transition to their reproductive stage. For example, some annual grasses grow to a given size and then quickly switch to the reproductive stage, rosette species produce a reproductive shoot immediately after a very short vegetative stage, while some species of legumes of intermediate growth form produce leaves and flowers simultaneously. In these different cases, reproductive height at maturity is the only operational estimation of plant height that can be used in the context of large-scale comparisons of herbaceous species differing in their growth form and/or their architecture (discussed in e.g. Pérez-Harguindeguy et al., 2013). How this trait correlates with competitive ability for light remains to be established, however (see the discussion in Violle et al., 2009).

Height is also correlated with other dimensions pertaining to plant stature. For trees, the relationship between height and the basal area of the trunk—the parameters of which depend on the species being considered—is linear above a certain height, before becoming asymptotic (King, 1990; Niklas, 1995b; Wirth et al., 2004). In both the ecological and forestry literatures, a combination of height and basal area in equations of the form *([Basal Area]^2 × Height)* have been frequently used to estimate the above-ground biomass of trees; these estimations are improved when wood density is also taken into account (Chave et al., 2005, and references therein). Similarly, in herbaceous plants, height is positively correlated with stem diameter (Niklas, 1995a) and the average cross-sectional area of the roots (Hummel et al., 2007; Wahl & Ryser, 2000). Such simultaneous increases in the size of a plant along multiple dimensions leads to greater stiffness and better anchorage abilities necessary for enlarged height growth. In fact, plants whose above-ground size is large tend to also have deep root systems (Figure 3.9; Schenk & Jackson, 2002; Violle et al., 2009). Given that soil moisture increases with increasing depth during dry periods (Meinzer et al., 1999), large plants would tend to have access to greater water resources than smaller plants, with less deep rooting systems (Schenk & Jackson, 2002). The consequences of these relationships on the water status of individuals during dry periods will depend in particular on the manner in which the height (and the rooting depth) is linked to the total leaf area of the plant and to the stomatal regulation of transpiration.

Height is also an important component of a coordinated suite of traits pertaining to plant reproduction, including seed mass, the time required

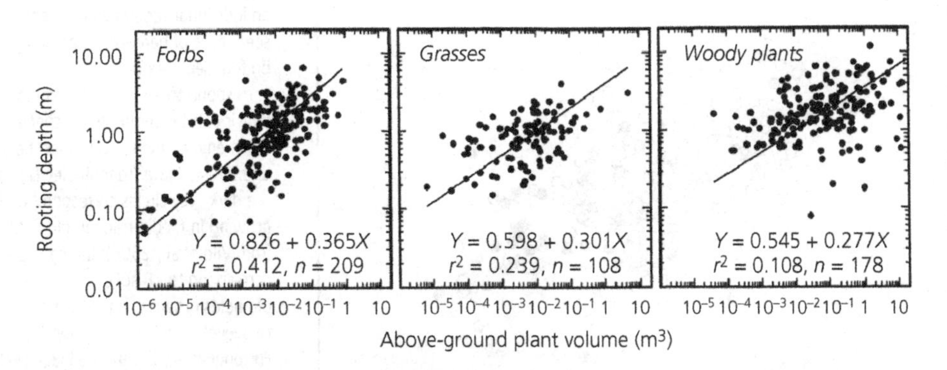

Figure 3.9 Allometric relationships between plant above-ground volume (estimated from height and width of the canopy) and the maximum rooting depth for species of (a) forbs (herbaceous dicotyledons), (b) grasses, and (c) woody plants. The regression lines and equations are determined using 'reduced major axis regressions' carried out on log transformed data, using a general equation of the form: $\log_{10}(D_i) = a + b \log_{10}(V_i)$, where D_i is expressed in m and V_i in m^3. These relationships are based on a bibliographic compilation of more than 1300 data points obtained from individual plants growing in water-limited ecosystems. Taken from Schenk & Jackson (2002). Reproduced with permission from Wiley.

to reach the reproductive stage, longevity, and the number of seeds produced annually (Moles & Leishman, 2008, and references therein; see Figure 3.13). For herbaceous species, a high reproductive height generally confers more efficient pollination and/or seed dispersal (Lortie & Aarssen, 1999; Soons et al., 2004; Verbeek & Boasson, 1995; Waller, 1988), which could explain the important investment in the length of reproductive stems found in many grassland species, rosette plants in particular (Bazzaz et al., 2000). For trees, the height at which seeds are released has a positive effect on the distance that they are dispersed, which explains why this trait is taken into account in most dispersal models (Greene & Johnson, 1989; Kuparinen, 2006).

Finally, a number of questions related to plant height are not yet well understood, such as: what are the factors which limit the maximum height of a tree? Or is there any relationship between maximum height and the time during which stems persist at this maximal height (King, 1990; Westoby et al., 2002)? Nevertheless, plant height used as an estimation of plant stature appears to be a fundamental variable controlling individual plant functioning as well as major interactions between neighbouring plants.

3.2.3 Plant regeneration

This dimension of plant function is represented by seed mass in the LHS scheme proposed by Westoby (1998). Seed mass, which varies by several orders of magnitude among species (Harper et al., 1970; Moles et al., 2005b) is one of the major plant traits. It affects practically every aspect of the ecology of plant regeneration, such as dispersal, establishment, and seedling survival (Fenner & Thompson, 2005; Harper et al., 1970; Weiher et al., 1999; Westoby, 1998).

The trade-off between the mass and number of seeds produced by an individual has been recognized for a long time: for a given carbon investment in reproduction, a plant can either produce a large number of small seeds or a smaller number of large seeds (Figure 3.10; Leishman et al., 2000; Shipley & Dion, 1992; Smith & Fretwell, 1974). The average seed mass is therefore one of the best estimators of the quantity of seeds produced per square metre of canopy cover (Henery & Westoby, 2001; Moles et al., 2004; Westoby, 1998). The two extremes of these combinations correspond to different adaptive strategies, favouring either the production of propagules, and thus their probability of dispersal, or inversely the competitive ability of seedlings

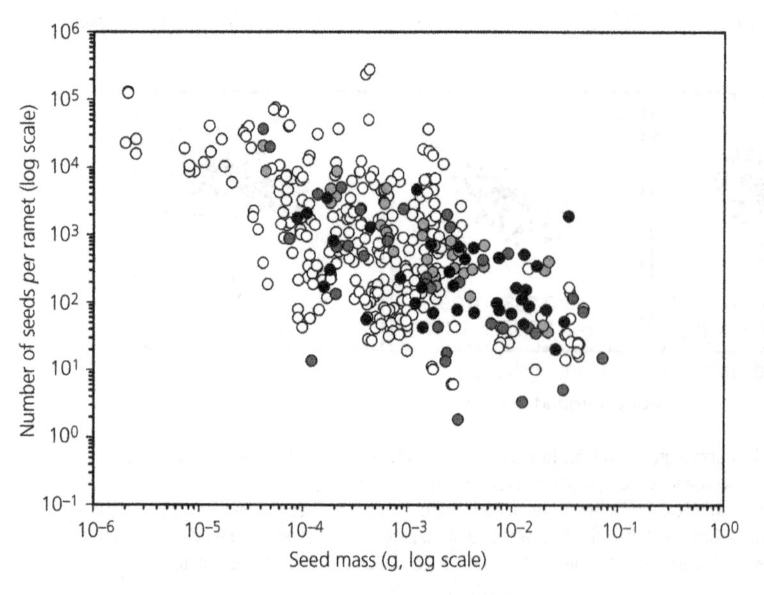

Figure 3.10 Relationship between the number of seeds produced by a ramet (that is, an individual vegetative unit) and the average seed mass of herbaceous angiosperms, based on four separate datasets. The white circles correspond to 285 ramets coming from individuals belonging to 57 species from moist environments, roadsides, fields, ditches, and forests (data from Shipley & Dion, 1992); the dark grey circles correspond to 34 species growing in Mediterranean old fields (data from Vile et al., 2006); the light grey circles correspond to 18 species from Mediterranean old fields grown in an experimental garden at two levels of nitrogen availability (data from Fortunel et al., 2009); the black circles correspond to 37 annual and biannual dicotyledonous crop weeds from the French Mediterranean region (Saatkamp et al., 2009). The global correlation coefficient is $r = -0.61$ ($P < 0{,}001$, $n = 384$). Completed on the basis of Garnier & Navas (2012).

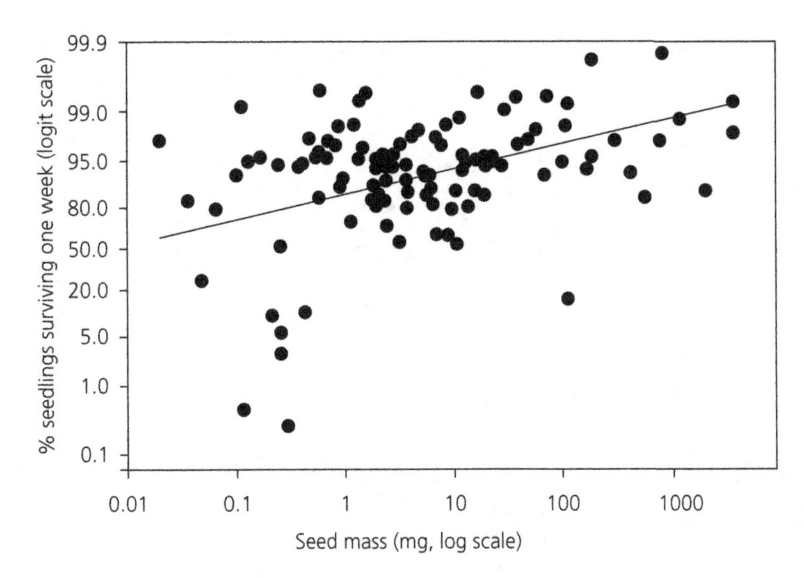

Figure 3.11 Relationship between seed mass and the percentage of seedlings surviving beyond the first week after their emergence. Each point represents the geometric mean of a species. The data is drawn from a worldwide bibliographic compilation (n = 113, $P = 0.003$; slope = 0.85). The regression line has been adjusted using a logistic regression with random effects. Data taken from Moles & Westoby (2004). Reproduced with permission from Wiley.

growing from seeds (Jakobsson & Eriksson, 2000; Moles & Westoby, 2004). However, producing a large number of small seeds does not necessarily result in a larger number of established plants. Actually, whether this is considered within-species or between-species, large seeds tend to give rise to large seedlings (see the review by Leishman et al., 2000), which have a higher probability of survival under unfavourable environmental conditions than seedlings from small seeds (Figure 3.11; Jakobsson & Eriksson, 2000; Leishman et al., 2000, and references therein). Environmental conditions are considered unfavourable for seedlings when they experience competition from existing vegetation and/or other seedlings, or when they germinate under conditions of deep shade, in the soil or under a thick litter layer (reviewed by Moles & Leishman, 2008; Westoby et al., 2002). The advantage of large seeds during establishment relates to the larger amounts of seed reserves, allowing seedlings to sustain respiration during longer periods of carbon deficit than seedlings from small seeds (Westoby et al., 2002).

The relationship between seed mass and the dispersal capacity of species in space and time has also been the subject of much research. Dispersal in space is linked to seed mass: small seeds that are produced in large quantities are often disseminated by wind, a combination favouring their dispersal (Weiher et al., 1999; Westoby et al., 2002). However, the distance of seed dispersal, which depends on other characteristics of seeds and on their dispersal agents (Molinier & Muller, 1938), has not been related to seed mass or to any other trait in a convincing manner (Hughes et al., 1994; Moles et al., 2005a). There does however exist a link between seed mass and the dispersal agent (Moles et al., 2005a): wind dispersed seeds tend to be smaller than those dispersed by animals, while amongst these latter species, those which are dispersed by ants are smaller than those dispersed by birds, with these in turn being smaller than those dispersed by mammals.

From another perspective, dispersal over time can be estimated by the life span of propagules in the seed bank (Thompson et al., 1998). The persistence of a seed in the soil has sometimes been linked to a combination of its mass and shape (Thompson et al., 1993): in the English flora, it has been shown that small seeds with a compact shape (round) persist for longer in the soil seed bank than large elongated seeds (Figure 3.12). This result has since been confirmed for the floras of other European countries, South America, and Iran (references in Fenner & Thompson, 2005), but not for the floras of other continents (Figure 3.12; Leishman & Westoby,

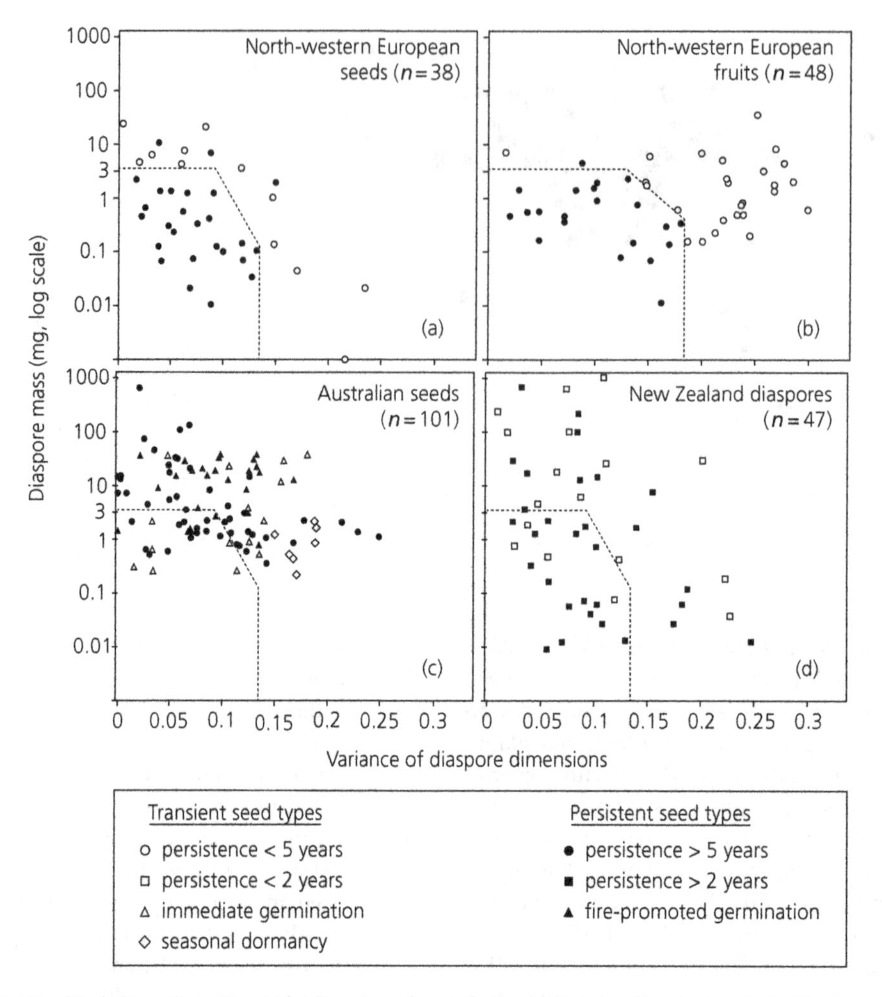

Figure 3.12 Relationships between the variance in the dimensions of seeds, fruits, and diaspores of species from North-West Europe (a & b), Australia (c) and New Zealand (d). The variance is calculated on the basis of the length, the width, and the thickness of seeds (standardized such that the length is equal to 1); a perfect sphere thus has a variance of 0 (the three dimensions are equal), while a needle or a fine disc has a maximum variance value of 0.33. It should be noted that: for the European data, the majority of 'fruits' are complete diaspores; for the Australian data, the 'seed' is defined as the diaspore from which all dispersal structures have been removed (thus including those fruits containing only one seed such as achenes); for the New Zealand data, the measurements have been taken on those parts of the diaspore most likely to be buried in the soil (the dispersal structures have been included if they adhered to the seed). On each figure, the dotted line indicates the region of the graph within which seeds are assumed to be persistent. Taken from Leishman et al. (2000). Reproduced with permission from CAB International.

1998). Fenner and Thompson (2005) argued that this last result may be a consequence of the presence of numerous large, hard-seeded species in the data sets analysed, which persist in the soil due to their physical dormancy. In any case, it is probable that persistence in the soil seed bank depends on traits other than just their simple mass and/or shape. For example, an exponential negative relationship has been found between the thickness of the envelope of certain weed species and the mortality of seeds in the soil (Gardarin et al., 2010), but this relationship remains to be tested for other types of species.

Finally, seed mass is correlated with other traits such as growth form, plant height, minimum time before first reproduction, and the maximum life span of an individual (Moles & Leishman, 2008).

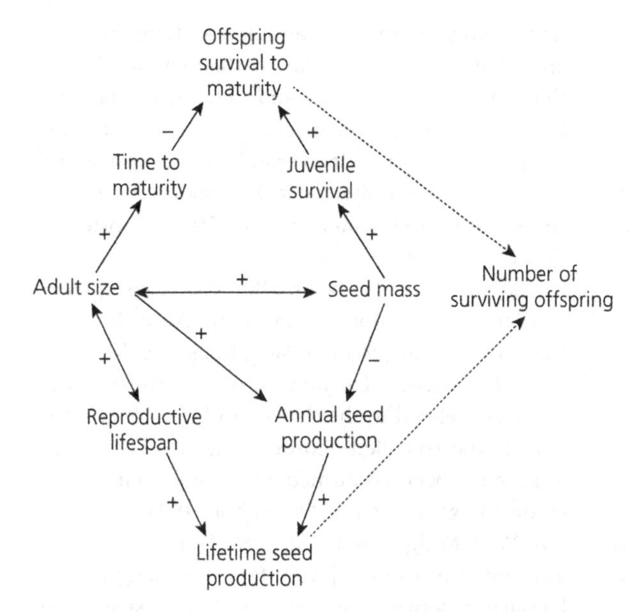

Figure 3.13 Correlations between seed mass and the main traits linked to plant regeneration (+ and – signs indicate positive and negative correlations, respectively). Relationships with the size of the adult plant are also shown on the figure. Arrows with a single point represent causal relationships, while double pointed arrows represent correlations. The dotted lines indicate assumed positive relationships of the total production of seeds, and of the survival of descendants on the total number of viable descendants. Taken from Moles & Leishman (2008). Reproduced with permission from Cambridge University Press.

On average, seed mass is higher in woody than in herbaceous species, in trees than in shrubs, and in perennial herbaceous species as compared to annual species (Moles et al., 2005a; Tautenhahn et al., 2008, and references therein), which corresponds to a trend of large species with long life spans producing larger seeds than small species with short life spans (Rees, 1997; Rees & Venable, 2007; Thompson et al., 1998). The evolutionary implications of this positive relationship between the size of the mother plant and of the seed are currently still strongly debated (Venable & Rees, 2009; Westoby et al., 2009). The arguments remain fairly complex, and here we will simply quote sentences from Venable & Rees (2009): 'So why do large plants have large seeds? [...] The pattern is a weak one. While very small plants never produce seeds as large as those of big seeded trees, plants of any size have seeds that vary approximately 400–650-fold between species. [...], this is a huge interspecific range, independent of adult size'. A synthetic view of the currently understood correlations between traits linked to seed mass is presented in Figure 3.13, showing the central role of this trait in determining different aspects of the plant reproductive phase.

Westoby (1998) initially postulated the independence of the three axes of variation discussed in this section (plant resource use, stature, and regeneration). While the selective pressures acting on the various traits of these axes can largely be considered independent, seed mass and plant height are actually positively correlated (Figure 3.13 and cf. below). This appears to contradict the generally assumed lack of association between traits corresponding to the vegetative and regenerative phases of plants' life cycles (Grime et al., 1987; Shipley et al., 1989; Silvertown et al., 1992). The identification of other relationships between traits of these two phases would require further analyses using databases containing trait information pertaining to the differing phases of plant life cycles for a large number of species. Very few such databases are currently available, however.

3.3 Beyond the three major axes of variation

The small number of traits presented above allows for the simple and rapid understanding of some of the fundamental axes of variation in plant functioning during the vegetative and reproductive phases. In the remainder of this section, we will present those traits for which the functional significance has begun to be established, but which does not necessarily contribute to one of these three axes. We focus on traits which show wide ranges of variations among species, and whose functional and/or ecological significance has been at least partially demonstrated (see introduction of this chapter and Westoby et al., 2002; Westoby & Wright, 2006, for discussion on criteria to select relevant axes). We will discuss the area of individual leaves, then wood density, which appears to be an important trait for tree functioning, before discussing the structural and functional traits of roots. Finally, we will present some elements relative to plant phenology.

3.3.1 Leaf area

Leaf size is a basic physiognomic characteristic of plant foliage. Leaf area, the most common metric for leaf size, is one of the traits most often measured worldwide (cf. Kattge et al., 2011). It spans up to six orders of magnitude among species, from tiny leaves less than 1 mm^2 to gigantic leaves of more than 2.5 m^2 (data from the TRY data base, accessed 7 January 2015: http://www.try-db.org), and this range may remain large among coexisting species within communities (references in Westoby et al., 2002). A number of arguments have been put forward to explain this prominent variation in leaf area.

A small leaf size implies a thin boundary layer of still air above the leaf, which helps to maintain favourable leaf temperatures and high photosynthetic water-use efficiency under the combination of high solar radiation and low water availability or low stomatal conductance (Givnish & Vermeij, 1976; Parkhurst & Loucks, 1972). This suggests that leaf area is involved in a functional strategy associated with water (but also nutrient) availability (Ackerly et al., 2002, and references therein). Smaller leaf sizes in colder, drier, and/or nutrient-poorer habitats have indeed been frequently demonstrated (Niinemets et al., 2007; references in Westoby et al., 2002). Leaf size is therefore one of the traits which has long been used to develop proxies for reconstructing paleoclimate (Peppe et al., 2011, and references therein).

In addition to energy balance, optimization of overall biomass investment in support has been argued to be a selection pressure causing diversification in leaf size, at least in trees: larger leaves allow the plants to gain height more rapidly because fewer woody branches and lower twig biomass are required to support larger than smaller leaves (Niinemets et al., 2007, and references therein). However, large-leafed species tend to invest a larger proportion of their overall biomass in petioles, to insure proper mechanical support and avoid self-shading (Niinemets et al., 2007; Poorter & Rozendaal, 2008).

Leaf size may also be an allometric consequence of plant size, anatomy, and architecture. In trees, Corner (1949) argued that twig thickness, leaf size, and inflorescence size are positively correlated and inversely related to the density of branching in the crown (the so-called 'Corner's rules'). These patterns have been confirmed in several quantitative studies (e.g. Ackerly & Donoghue, 1998; Cornelissen, 1999; Midgley & Bond, 1989; White, 1983). For example, Cornelissen (1999) found a triangular relationship between leaf area and seed size in 58 (semi-)woody species from the British Isles: major variation in leaf size was observed at the small-seeded side of the range, while there was no species combining small leaves with large seeds.

A central tenet of the general allometric model of vascular plants (West et al., 1999) is that the size and functional traits of leaves should remain invariant with plant size. Examining this assumption using the Glopnet data set (see Section 3.2.1.1), Price et al. (2014) actually found a positive—though modest—relationship between leaf area and plant height in woody Angiosperms; this relationship was neither significant in herbaceous Angiosperms nor in woody Gymnosperms.

In woody species, leaf area has also been found to correlate positively with branch-scale leaf mass fraction (the ratio of leaf mass to the [leaf + stem] mass; Pickup et al., 2005) and negatively with wood density (Wright et al., 2007, and references therein). There is still not enough evidence to assess whether these relationships are general, and their functional bases are currently still debated (but see Wright et al., 2007, for a tentative explanation of the leaf area–wood density relationship based on tree hydraulics).

Although attempts to understand interspecific variations in leaf size have been numerous, the adaptive significance of this trait is still not well understood (Niinemets et al., 2007; Westoby et al.,

2002). As recently stated by Price et al. (2014), there is a clear 'need for a better understanding of the drivers of leaf size variation within and across individuals, functional groups, clades, biomes, and habitats'.

3.3.2 Wood density

Wood density, defined as the dry mass per unit volume of fresh wood, is a fundamental property of tree trunks (Chave et al., 2009; Muller-Landau, 2004), which plays a major role in light acquisition strategies. The wood density of species from mature forest stands tend to be relatively high in comparison to those of pioneer species (Muller-Landau, 2004).

As the density of structures composed of lignin without any free spaces is approximately 1.5 g cm^{-3}, wood density is necessarily limited to between 0 and 1.5 g cm^{-3}. It is correlated to a large number of structural and functional characteristics, and represents a central variable in an assumed 'wood economics spectrum', which captures the functional trade-offs and constraints of woody plants (Chave et al., 2009): a high density is associated with a high resistance to mechanical rupture and hydraulic cavitation (the rapid entry into the conductive vessels of the gas phase which expels the liquid phase), but has high costs in terms of tissue construction. The biomechanical basis of these relationships is currently under debate, since the resistance to rupture results from a combination of the density and diameter of trunks: low density trunks with a large diameter may be as resistant as dense trunks with lower diameters for a lower investment in biomass (Anten & Schieving, 2010; Larjavaara & Muller-Landau, 2010, 2012). A low wood density also tends to be associated with high relative growth rates (Figure 3.14a; Chave et al., 2009, and references therein; Wright et al., 2010), while a high density is associated with high resistance to physical damage and to disease, a high survival rate under conditions of low light, and low mortality rates (Figure 3.14b; Chave et al., 2009). The negative relationship between wood density and the surface of individual leaves (Wright et al., 2007) has been discussed in the Section 3.3.1.

At the level of the ecosystem, wood density strongly affects the storage of carbon in standing biomass and in necromass (references in Chave et al., 2009), and tends to be negatively correlated with the rate of trunk decomposition (Freschet et al., 2012; Weedon et al., 2009).

3.3.3 Root traits

So far, only traits measurable on the above-ground parts of plants, which are relatively easy to measure, have been presented. Recent studies have attempted to provide the same types of information for below-ground plant parts, so as to have a better understanding of root functioning and of potential trade-offs between root functions. Research has also been carried out to assess potential relationships with above-ground plant parts, which could

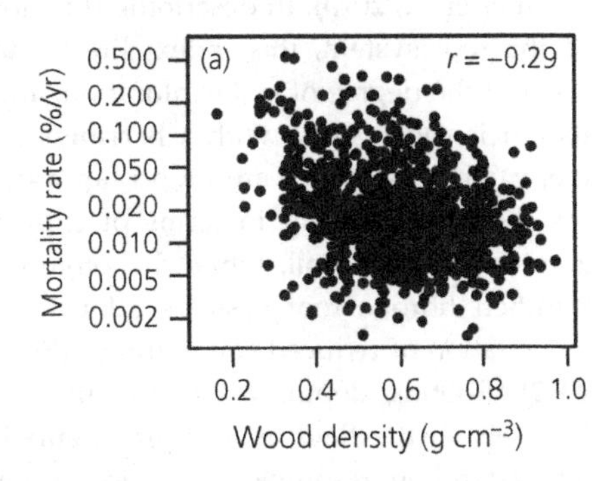

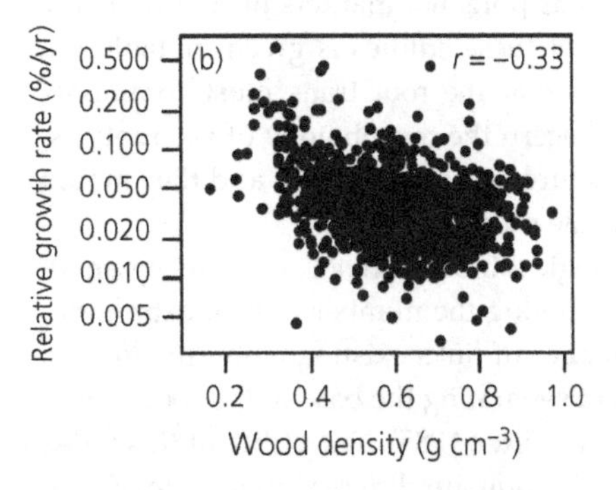

Figure 3.14 Relationships between wood density and (a) relative growth rate (RGR, logarithmic scale), and (b) mortality rate (logarithmic scale), for two tropical forest sites (Barro Colorado Island, Panama, and Pasoh, Malaysia). All of the correlations are highly significant ($P < 0.001$), and the correlation coefficients have values between $r^2 = 0.13$ and 0.19. Taken from Chave et al. (2009). Reproduced with permission from Wiley.

Table 3.1 Principal root traits, functional significance, and relationships with above ground traits. Based on: Fitter & Hay (2002); Roumet et al. (2006); Stokes et al. (2009); Tjoelker et al. (2005); Freschet et al. (2010); Kembel & Cahill (2011); Reich (2014); Freschet et al. (2015).

Trait	Abbreviation (and unit)	Functional significance (and direction of relationship)	Relationships with above-ground traits (and direction of relationship)
Topological index[1]		Exploitation of new zones of soil and the efficiency of this exploration (+) Water transport (-) Anchoring (-)	
Rooting depth	(cm)	Water acquisition (+) Anchoring (+)	Height (+)
Specific root length	SRL (m g^{-1})	Root respiration (+) Rate of water and nutrient acquisition (+) Root elongation rate (+) Rate of root replacement (+) Resistance to tension (+)	Relative growth rate (+) Height (-) Leaf surface area (-) Leaf dry matter content (-) Specific leaf area (+)
Root diameter	d_{root} (mm)	Rate of nutrient acquisition (-) Storage (+) Water transport (+) Force of root penetration into the soil (+)	Height (+) Specific leaf area (-) Leaf surface area (+) Leaf thickness (+) Leaf dry matter content (+)
Root tissue density	RTD (g cm^{-3})	Drought resistance, herbivore resistance (+) Root decomposition rate (-)	Leaf thickness (-)
Root nitrogen concentration	RNC (%)	Root respiration (+) Root growth (+) Root decomposition rate (+)	Leaf nitrogen concentration (+)
Longevity	1/ root turn-over (d)	Root nitrogen concentration (-) Root respiration (-) Root growth (-)	Leaf life span (+)
Mycorrhizal colonization	(% of root length colonized)	Rate of nutrient acquisition (+) Protection against pathogens (+) Resistance to tension (+)	

[1] A high value for the topological index indicates a herringbone root system; a low value indicates a dichotomous root system with a high degree of ramification.

thus be used as potential markers of root function. A summary of these studies is given in Table 3.1, which shows that the root traits most commonly measured concern the morphology of the root system, its structure and composition, and the amount of mycorrhizae present.

The topological index is determined using an algorithm combining the number of root extremities and the number of links existing along the longest root segment separating the base of the root from it furthest point (Fitter, 1985). The utility of this index is its ability to compare different root systems that vary both in their degree of ramification and in size, and has been used to establish a root classification (Bodner et al., 2013). In describing the morphology of the root system, this index directly takes into account the degree of soil exploration, and notably its efficiency: systems with a herringbone morphology, amongst which are found tap root systems, are the most efficient in terms of exploring new zones within the soil, while dichotomous, highly ramified fibrous root systems allow for a better exploitation of reduced soil volumes (Fitter & Hay, 2002). Rooting depth, while difficult to evaluate precisely in the field, is also an essential trait for understanding resource acquisition, especially of water, and the strength of anchoring in the soil of a plant (Jackson et al., 1999). It varies considerably

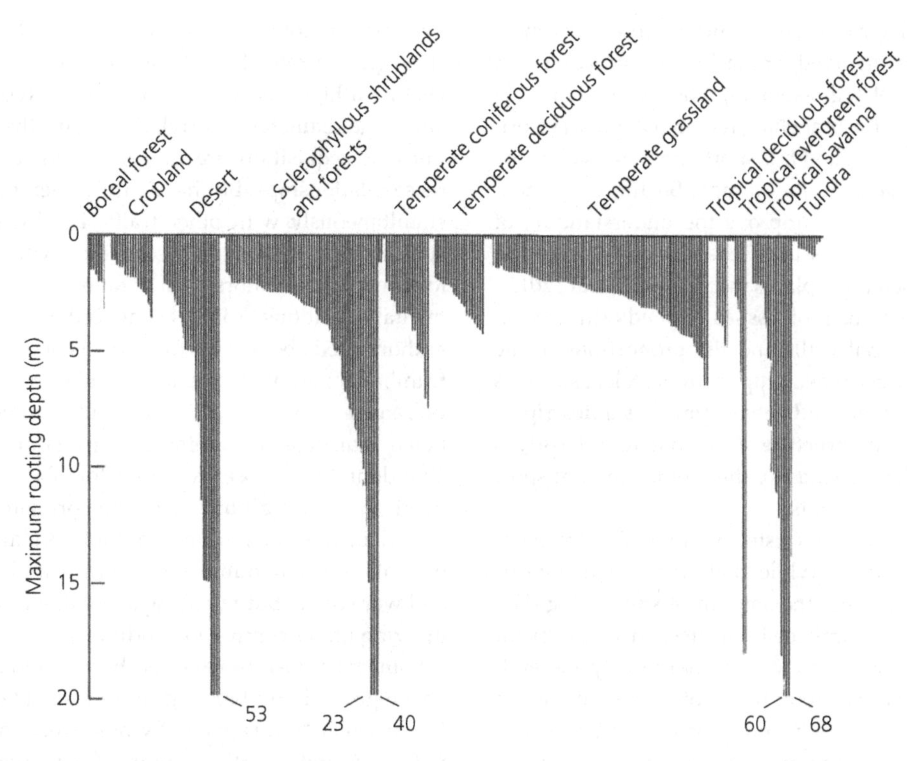

Figure 3.15 Values of maximum rooting depth (in m) observed for different species grouped by biome. Taken from Canadell et al. (1996). Reproduced with permission from Springer.

between species found in the same biome as well as among biomes (Figure 3.15).

The specific root length (SRL), which is the ratio between the length and the mass of the roots, is often considered as the analogue of SLA for the below-ground parts[1]. It is the product of two components which are the root diameter (d_{root}) and the root tissue density (RTD); a low SRL value corresponds to roots that are thick and/or dense. These last two traits are not totally independent as root thickness varies with the amount of tissue forming the root stele. For the same individual, the root diameter has traditionally been used to distinguish three types of roots which have different functions: fine roots (diameter lower than 2 mm), which have a high turnover rate, a high capacity to respond to local resource modifications, and which are responsible for the majority of the acquisition of water and nutrients ; medium-fine roots (from 2 to 10 mm in diameter) which are often lignified in woody plants ; and coarse roots (diameter greater than 10 mm) which have soil anchoring and storage functions, with the oldest roots no longer active for resource acquisition (Stokes et al., 2009). The presence in the same plant of roots of different diameters allows for complementarity in the functions carried out by roots (acquisition, anchorage, storage, and transport) but also for different modalities in carrying out any one given function. For example, the contribution of large coarse roots is important for resource acquisition, in that they allow fine roots located at their extremities to exploit larger volumes of soil over long time scales than would be possible in the absence of these coarse roots.

It has recently been recognized that this approach should be further refined to account for the diversity of forms and functions observed within the fine root class (McCormack et al., 2015). These authors have suggested splitting this single class into two functional groups: absorptive fine roots and

[1] The true analogue of SLA is in fact the specific root area, the ratio between the area and mass of the roots (cf. Chapter 6, Section 6.3.4.)

transport fine roots. The former represent the most distal roots involved primarily in the acquisition and uptake of soil resources, whereas the latter occur higher in the branching hierarchy and serve primarily structural and transport functions with some additional capacity for storage. Such a distinction has been shown to improve the understanding of dynamic root processes, at least in ecosystems dominated by perennial plants (McCormack et al., 2015).

The density of root tissues depends directly on thickness of cell walls and the proportions of the different tissues making up the root stele, as well as the sclerenchyma and aerenchyma. As a descriptor of the anatomic structure of the root, root density is linked to the capacities of the root for the transport of water and nutrients.

The last two traits listed in Table 3.1 are more difficult to study. While their functional importance is recognized, the amount of supporting data is in fact quite limited. Root lifespan is a difficult trait to measure and varies enormously depending on the type of root: fine roots generally have a rapid turnover rate in comparison to coarse roots. For traits relating to mycorrhiza, their presence facilitates the acquisition of nutrients, especially of phosphorus, by increasing the volume of soil in which phosphorus is taken up beyond the resource depletion zone around the roots, as well as the local availability of phosphorus via the exudation of phosphatases and organic acids (Craine, 2009). Mycorrhizas also provide some degree of protection against pathogens (Fitter & Hay, 2002). However, the total amount of mycorrhiza that colonizes roots is not necessarily proportional to its impact in plant functioning. Additionally, the degree of mycorrhizal colonization varies as a function of root age, and can also influence certain other root traits, in particular root length.

As for leaves, a certain number of co-variations have been shown to exist between root traits (Table 3.1; Freschet et al., 2010; Kembel & Cahill, 2011; Roumet et al., 2006; Tjoelker et al., 2005). In general, plants with rapid growth rates have roots with high specific length (SRL), high root nitrogen concentrations, and a short life span. They have high rates of metabolic activity, and in particular high rates of respiration (Figure 3.16; Reich et al., 2008), often associated with high rates of mycorrhizal colonization. In contrast, plants with slow growth rates have long life spans, characterized by a high density of root tissues (Ryser, 1996) and large diameters. Correlations with the rates of nutrient acquisition are in general inferred from these relationships, but have rarely been measured simultaneously with other traits. Finally, root life span is generally positively correlated with a greater tolerance to predators and unfavourable environmental conditions (Roumet et al., 2006). Some of the authors cited above refer to a 'root economics spectrum', comparable to the leaf economics spectrum presented in Section 3.2.1. It should, however, be noted that data concerning root traits are far less abundant than those available for above-ground plant parts, and a number of the correlations presented in this section and in Table 3.1 are based on only a small number of studies and species, and were often, but not always, obtained for plants growing under controlled conditions.

Some relationships have also been found between above-ground and below-ground traits (Table 3.1). In particular SRL is generally positively correlated with SLA and negatively with LDMC; root diameter is positively correlated with LDMC and root longevity varies positively with the RTD. There are also generally strong relationships between the chemical compositions of different organs, in particular their nitrogen concentrations, while leaf and root decomposability are also largely correlated within most ecosystems worldwide (Freschet et al., 2013). Finally, the rooting depth of a plant is positively correlated with its above-ground size (cf. Figure 3.9). These results suggest some kind of trait integration among organs, at least between leaves and roots (Freschet et al., 2010, 2015; Reich, 2014). Whether this integration holds for stems remains an open question and would require further testing on a wide array of species (see the discussion in Reich, 2014).

While the identification of the relationships of traits with key root functions such as nutrient acquisition, plant anchorage, rhizosphere activity, or even the rate of decomposition has been underway for some time (Freschet et al., 2010; Roumet et al., 2006; Stokes et al., 2009), there remains much to be done. In particular, establishing standardized protocols for measuring root traits so as to facilitate

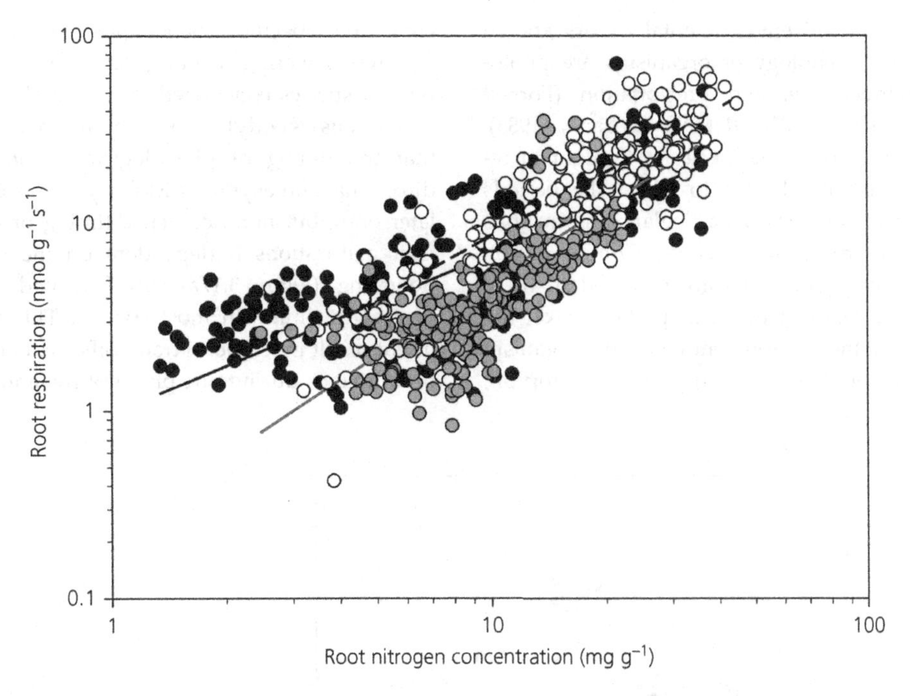

Figure 3.16 Relationship between nitrogen concentration and root respiration rate expressed per unit of mass for different groups of plants. White points and dashed black line, herbaceous plants (n = 103); grey points and line, woody Angiosperms (n = 357); black points and line, Gymnosperms (n = 284). The three regression lines are all significant ($P < 0.001$). Taken from Reich et al. (2008). Reproduced with permission from Wiley.

comparisons between a large range of species and to be able to generalize relationships so far established only for a limited number of species and/or conditions is a challenge yet to be completed. Additionally, due to the difficulties of sampling and the inherent inaccessibility of roots, easily identifiable traits—which could even be above-ground—linked with the different aspects of the functioning of the root system remain to be identified.

3.3.4 Phenology

Phenology, defined as the study of periodic biological events in the animal and plant world (Rathcke & Lacey, 1985; Schwartz, 2003), describes the temporal dimension of organism functioning. The time of the year when a plant produces leaves, flowers, or sets seeds, and the period during which the green foliage is active are examples of events which have long been recorded and used in agriculture, natural history, or in cultural and religious contexts (Lechowicz, 2002; Lieth, 1974, and references therein). The timing of phenological events determines the stage of development reached by an organism at the time when it interacts with biotic and abiotic components of the environment (Forrest & Miller-Rushing, 2010). The seasonal distribution of environmental conditions (e.g. temperature, photoperiod), resources (water, nutrients, pollinators, etc.), and hazards (disturbances, pathogens, frost, etc.) within the period of plant activity generates selection pressures on the timing of the different phases which compose the vegetative and reproductive cycles (Bolmgren & Cowan, 2008, and references therein). Phenology is therefore critical to the adaptation of organisms to their local environment, and as such is an important component of ecological strategies (e.g. Grime, 1977; Lechowicz, 2002). Despite this recognized importance, plant trait research has largely ignored the role of phenology as a trait (Wolkovich & Ettinger, 2014), except in limited instances (e.g. Craine et al., 2012; Weiher et al., 1999).

The best studied environmental factors known to affect the phenology of organisms are photoperiod, temperature, and precipitation (Forrest & Miller-Rushing, 2010; Rathcke & Lacey, 1985). The phenology of plant species is thought to result from natural selection to adjust the periods of activity in relation to particular combinations of these environmental factors (Chuine, 2010). Several complex trade-offs are involved in these adjustments. These depend, in particular, on the seasonality of the environment where the organism thrives (boreal, temperate, dry or wet tropical, etc.), the specific phenological phase studied (leafing, flowering, fruiting, leaf fall, etc.), or the type of species considered (e.g. annual/perennial, herbaceous/woody). A key point to consider is that the timing of phenological events is both directional and asymmetric. Early events can affect later ones, but not vice versa: timing of seed and fruit maturations is dependent on the timing of flowering (Figure 3.17a), itself related to onset of growth timing in most species. This very fact might result in temporal trade-offs, i.e. maximizing performance during one phase of the annual cycle

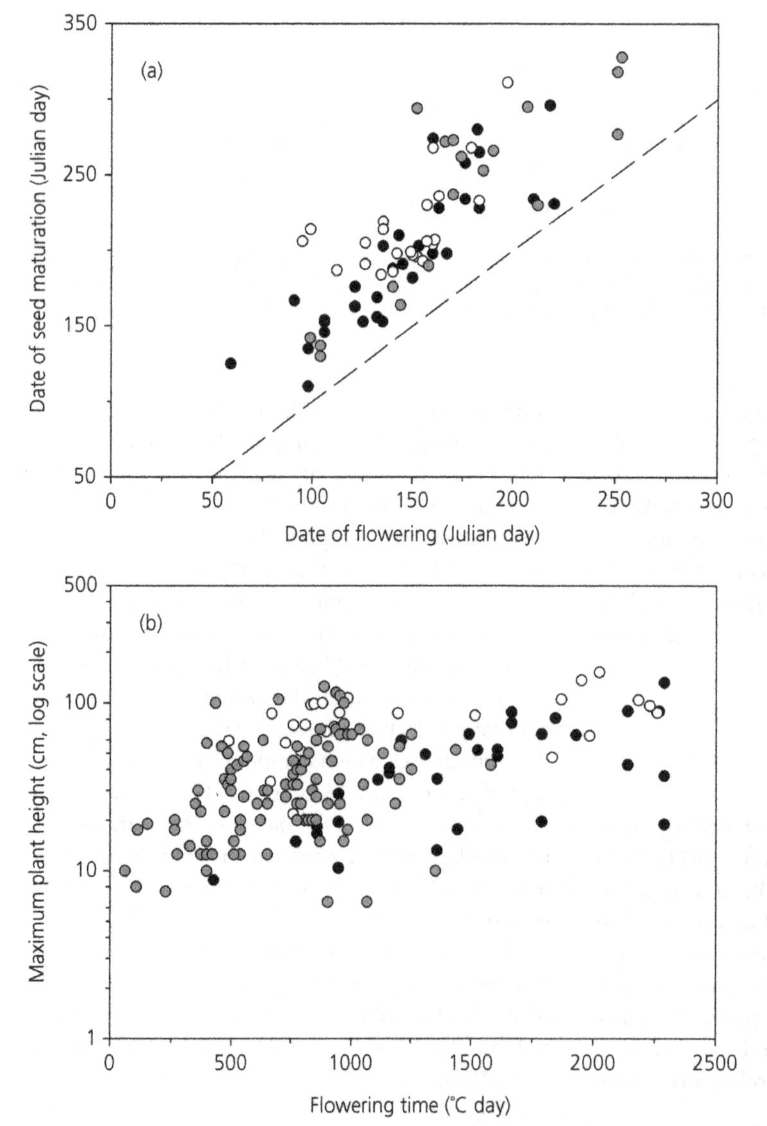

Figure 3.17 Relationships between date of flowering and (a) date of seed maturation, (b) maximum plant height (logarithmic scale). In (a) the date of seed maturation occurs on average 58 days (range: 11–142) after the date of flowering (the vertical distance between each point and the dashed line, which represents the 1:1 relationship). Black points: data for 34 herbaceous species from Mediterranean old-fields of southern France (Vile et al., 2006); grey points: data for 18 species from shortgrass prairie in the USA (Dickinson & Dodd, 1976); white points: data for 31 herbaceous species from mountain grasslands in the French Pyrénées (Ansquer et al., 2009). Correlations within each study are significant; for pooled data: $r = 0.82$, $P < 0.001$, $n = 76$. In (b): flowering time is expressed in degree days above 0°C, to account for the fact that phenology depends on temperature (e.g. Diekmann, 1996); black points: data from Vile et al. (2006); white points: data for 25 herbaceous grassland species growing in north-eastern Connecticut (USA) (Sun & Frelich, 2011); grey points: data for 105 herbaceous species from the north temperate flora (Bolmgren & Cowan, 2008). Correlations within each study are significant; for pooled data: $r = 0.47$, $P < 0.001$, $n = 164$.

may come at the expense of the performance during other phases. In such situations, the optimal timing of an event is the combined result of direct selection on that event and indirect selection acting via effects on other temporally correlated events (Ehrlén, 2015, and references therein). Focusing on one phenological phase in isolation might thus be misleading. For example, the major factor determining flowering time in some woody species in the temperate zone may be constraints on the size of the fruit (e.g. Primack, 1987). As recently advocated by Ehrlén (2015) in the case of the timing of flowering, a better understanding of the selective forces shaping phenological events should consider these events in the context of the plant's annual and whole life cycles.

A detailed account of phenological patterns, and their controls in different species and communities under various types of climate, is beyond the scope of this book, and the reader is referred to e.g. Leith (1974), Rachtke & Lacey (1985), Fenner (1998), Schwartz (2003), and Forrest & Miller-Rushing (2010) for thorough assessments. Here we will illustrate examples of trade-offs involved in selected phases pertaining to vegetative and reproductive phenology, and give an overview of attempts to integrate phenology into trait-based research.

A first example deals with the period during which a plant bears green leaves, which depends on the timing of bud burst, leaf expansion, senescence, and fall of individual leaves, all events crucial to the fitness of plants (e.g. Fenner, 1998). While this sequence of events holds for all types of species, it has been particularly studied in trees. In the boreal and temperate zone, there is a trade-off between maximizing annual carbon assimilation with early leaf production and late leaf senescence and reducing the risk of damage caused by frost to vegetative organs (Chuine, 2010). Global warming appears as a natural experiment to manipulate this trade-off: in some tree species, higher temperatures have been shown to induce both earlier leaf production and delayed leaf senescence, resulting in a longer period of carbon uptake. The impact of earlier production on the net annual CO_2 flux appears to be stronger than that of delayed leaf senescence, owing to less favourable conditions for photosynthesis during the fall (Chuine, 2010, and references therein). It is not likely that all species will respond to increased spring temperature by earlier leaf production however. Körner and Basler (2010) have indeed argued that the mechanism protecting plants from late chilling events differs among species: in some species, generally originating from warm regions, the onset of growth is primarily controlled by temperature (the warmer the temperature, the higher the growth rate), while others from temperate latitudes also require some period of chilling during winter to break bud dormancy. Finally, late successional species tend also to be more sensitive to photoperiod than earlier successional ones, with long photoperiod compensating for insufficient chilling (Caffarra et al., 2011). The shift in terms of the carbon uptake/frost damage trade-off is therefore likely to be very different in these three types of species.

Another major selective force determining the timing of leaf production may be the effect of herbivory. Since the majority of lifetime damage by insects is suffered during the first month of leaf life (Fenner, 1998, and references therein), there is a clear advantage in producing new leaves during a period when herbivores are least abundant. This has been observed in seasonal tropics, where a number of species produce leaves during the dry period. In these situations, the timing of leaf production appears to result from a trade-off between exploiting the most favourable environmental conditions and the avoidance of herbivory (Fenner, 1998).

Another prominent example concerns the onset of flowering, a crucial event in the life cycle of plants assumed to impact several components of their reproductive success. Flowering phenology depends both on abiotic conditions and on interactions with animals such as pollinators and pre-dispersal seed predators (Ehrlén, 2015; Rathcke & Lacey, 1985, and references therein). The constraints on the timing of flowering may differ substantially between species, depending on whether flowering depends on resources acquired during the current growing season or on resources previously stored (e.g. Bolmgren & Cowan, 2008, and references therein). Different sets of constraints are therefore expected between semelparous and iteroparous organisms (e.g. Ehrlén, 2015; Forrest & Miller-Rushing, 2010). For example, in seasonal climates, stored reserves allow a number of woody and herbaceous perennial species to flower before leaf production after winter. In woody species from boreal and temperate zones, there is therefore a trade-off between maximizing fruit set with early flowering and reducing the risk of damage by frost to reproductive organs. Plant species have adapted their phenology to their local environment using primarily temperature and photoperiod to detect such optimal conditions (Chuine, 2010).

In annuals and some herbaceous perennials, flowering phenology is involved in the so-called 'age–size' or 'time–size' trade-off, which relates to the relative benefits of reaching reproductive maturity earlier in the season or growing larger before reproducing (Forrest & Miller-Rushing, 2010). As summarized by Bolmgren & Cowan (2008), 'earlier reproduction implies fewer resources allocated for maternal plant growth and smaller size at time of reproduction, and thus fewer resources available for seed production. On the other hand, earlier flowering will allow for a longer development time of seeds, larger seeds, and a longer period available for germination and juvenile growth. [...] Flowering onset time may thus represent the outcome of the parent vs offspring development time partitioning problem', usually described at the intraspecific level. Positive relationships between flowering time and biomass (Mooney et al., 1986) or plant height (e.g. Bolmgren & Cowan, 2008; Sun & Frelich, 2011; Vile et al., 2006) at time of reproduction (Figure 3.17b) confirm that early flowering species tend to be smaller than late flowering species (but see Du & Qi, 2010 for mixed results). Relationships between flowering time and seed or fruit mass have been found to be negative in some studies (e.g. Bolmgren & Cowan, 2008; Du & Qi, 2010; Primack, 1985; Vile et al., 2006), but not in others (e.g. Craine et al., 2012; Eriksson & Ehrlén, 1991; Kolb et al., 2006). The postulated time-size trade-off has thus not been verified in all situations. As mentioned above, a clear distinction should be made between semelparous and iteroparous species to understand this trade-off, which has not been done systematically in the studies mentioned.

The relationship between flowering time and seed/fruit size was one of the three linkages involving phenology which was predicted by Primack (1987) in his review on relationships among flowers, fruits, and seeds (Table 3.2, prediction 1). The rationale and experimental evidence concerning predictions 2 and 3 (Table 3.2) can be found in Ratchke & Lacey (1985) and Primack (1987).

More recently, Wolkovich and Cleland (2014) have synthesized the results of studies combining data on phenological events and a number of

Table 3.2 Overview table of the predicted linkages involving phenological events postulated by Primack (1987). The right-hand side column indicates whether these predictions concern the reproductive system only or combine the reproductive and vegetative systems of the plant.

Prediction	Statement	Phases/structures involved
1	Species with large fruits will require a greater period of time for fruit maturation than will species with smaller fruits. In the temperate zones, large-fruited species will be forced to flower early in the spring in order to have sufficient time for fruit maturation before the onset of cold weather in the fall. Small-fruited species may flower anytime during the growing season but will tend to flower and fruit at the times when carbohydrate levels in the plant are highest[1]	**Reproductive and vegetative**: fruits, flowering time, time for fruit ripening, carbohydrate storage
2	Fruit development will normally be as rapid as possible to minimize exposure to seed predators and to minimize metabolic cost	**Reproductive**: time for fruit ripening and interactions with seed predators
3	The time of year that is best for seed germination will influence fruiting and flowering times, particularly in species lacking seed dormancy and in tropical habitats with a pronounced dry season	**Reproductive**: timing of germination, flowering and fruiting times

[1] The modelling approach followed by Guilbaud et al. (2015) suggests that in annual plants, the trigger for flowering should occur at a time in the growing season when whole-plant nitrogen uptake rate reaches a maximum (a point called 'peak N'). After this time is reached, the plant invests all further new growth directly into the reproductive structures.

plant traits. The relationships between flowering time, plant height, and seed size involved in the time–size trade-off are by far the most studied (see above). Correlations with traits involved in the leaf economics spectrum have been tested in a smaller number of studies.

Both flowering and seed maturation dates were found to be negatively related to SLA in 34 species from Mediterranean old-fields (Vile et al., 2006). By contrast, this relationship was found to be not significant for the 25 species studied by Sun and Frelich (2011), while these authors found a negative relationship between flowering date and plant height and RGR. In grassland species, flowering date was found to be positively related to LDMC (Ansquer et al., 2009) and to leaf density (Craine et al., 2012). In this latter study, flowering date was also found to be marginally related to leaf thickness and leaf nitrogen concentration. Positive relationships between LDMC and the dates of stem elongation and seed maturation were found by Ansquer et al. (2009), pointing to an overall delayed phenology in high LDMC species. Another interesting result is that deep-rooted species tend to flower late and have a long growing season (references in Wolkovich & Cleland, 2014). Other relationships are also presented by Wolkovich and Cleland, often based on single studies conducted on a limited number of species.

The picture that emerges from these studies, although still a bit fuzzy, is that fast-growing species (traits described in Section 3.2.1) of small stature above- and below-ground (cf. sections 3.2.2 and 3.3.3) tend to flower and set seed earlier than taller, slower-growing species, which also grow longer during the season. The number of studies on which these conclusions are drawn remains limited however, and care should be taken to clearly separate conclusions for herbaceous and woody species. As advocated by Wolkovich and Ettinger (2014), there is a clear need for a more integrated view of phenology as a functional trait that would improve our understanding of 'the causes of, consequences for, and constraints on phenology and phenological shifts, at individual, physiological, and functional levels'. Such an understanding would not only return phenology to its role in basic physiology and development, but also bring it more explicitly into the broader context of ecological strategies.

3.4 Traits and phylogeny

The approach presented so far has been focused on the trait concept with an emphasis placed on the contribution of traits to the different functions of individual plants. This section will briefly present another perspective on traits, which consists of assessing whether evolutionary relationships among species have an influence on the values of their traits.

The evolutionary links between species can be determined by the construction of phylogenetic trees based on the analysis of sequences of nuclear or cytoplasmic genes (Davies et al., 2004): an example of such a phylogenetic tree at a low level of resolution for vascular plants is shown on the left hand side of Figure 3.18. These relationships between species imply that at some stage of their evolutionary history, these species all have a common ancestor that is more or less distant: for example, the common ancestor of the Rosids and the Asterids is less distant than that of all Dicotyledons, which is less distant than that between Dicotyledons and Monocotyledons (Figure 3.18). It has thus been suggested that there will always some degree of similarity between species based on the degree of their phylogenetic proximity (Silvertown et al., 1997, and references therein). This is what is termed the 'phylogenetic conservatism' of traits, defined as 'the tendency of species to conserve ancestral ecological characteristics' (references and a discussion can be found in Mouquet et al., 2012). Trait conservatism might arise at different depths of the phylogeny, that is for species sharing more or less distant ancestors. However certain traits which have major ecological impacts can be very poorly conserved (these are termed 'labile' traits) at the extremities of the phylogenetic tree (Cavender-Bares et al., 2009).

The detection of a phylogenetic signal on traits[2] is only in its infancy. This requires the availability of robust phylogenetic trees similar to the large comparative databases for traits, an area which is currently the focus of much active research. As an example, Figure 3.18 presents the distribution of

[2] A phylogenetic signal on a trait corresponds to a statistical sample in which evolutionarily related species contain trait values closer to each other than would be expected from species drawn at random from the phylogeny (Mouquet et al., 2012).

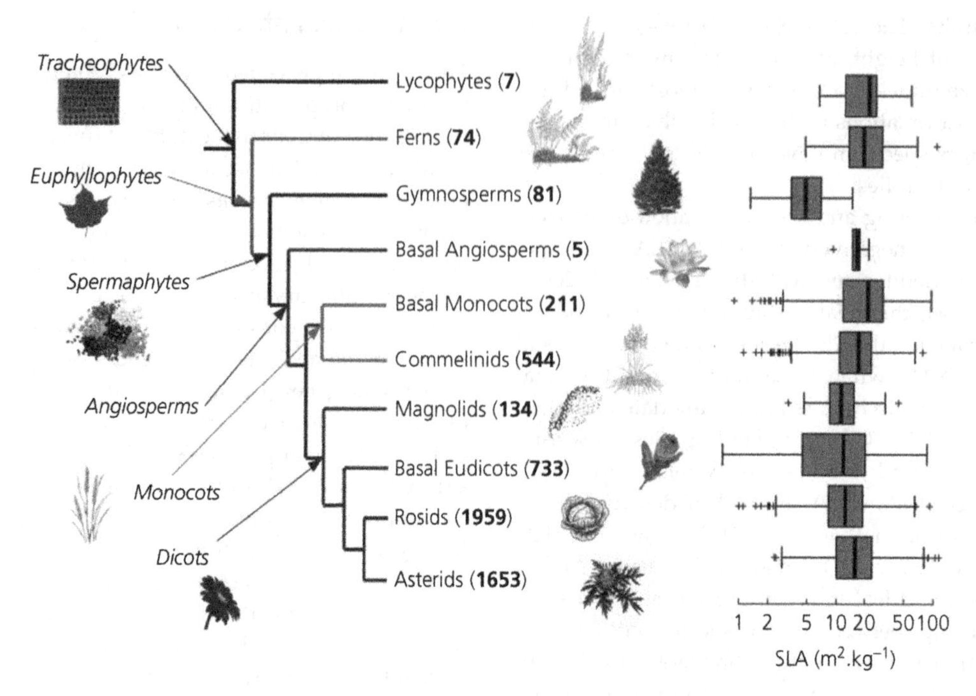

Figure 3.18 Phylogeny of specific leaf area (the ratio between the area and mass of a leaf—see Box 3.2 for a detailed presentation of this trait) for the major clades of vascular plants. The number shown in parentheses corresponds to the number of species considered per clade in the analysis. Based on data from Flores et al. (2014).

specific leaf area between some of the main clades[3] of vascular plants, determined on the basis of trait values for 5401 plant species (Flores et al., 2014). This figure shows for example how the divergence between Gymnosperms and Angiosperms strongly contributes to trait variability among currently existing species. Comparable studies have been carried out for a number of other traits, such as seed mass (Moles et al., 2005b), dates of flowering and leaf emergence (Davies et al., 2013), and leaf chemical composition (Watanabe et al., 2007). This last study, in particular, showed that more than 25% of the total variation in leaf nutrient concentrations can be attributed to the level of the family or even to a higher phylogenetic level for 21 of these elements, suggesting quite a high degree of conservatism for these traits.

In a recent study, Cornwell et al. (2014) took a different perspective to this phylogenetic issue. These authors quantified the relative contributions of various clades to the global distribution of five functional traits: specific leaf area, leaf nitrogen concentration, leaf area, maximum adult height, and seed mass. They showed, for example, that three phylogenetically distant lineages (Proteaceae, Ericaceae and some closely related families, and Acrogymnospermeae) contributed to the slow end of the leaf economics spectrum (low SLA and low nitrogen): in the absence of Proteaceae in particular, the span of the LES would be substantially shorter. Another finding of this study is that the Monocot clade was that with the smaller maximum height (when Acorales are excluded), while it was not exceptional for any other trait. In terms of leaf area, large-leaved clades were primarily found in tropical environments, with Magnoliidae and Rosids showing large leaf sizes compared to other clades.

The generalization of comparable studies on the major traits determining the performance of individuals will increase our ability to dissociate the phylogenetic signal from that of the ecological effect upon the attributes of existing species, as well

[3] A clade, or monophyletic group, is a group of taxa which includes exhaustively all of the descendants of a given ancestor.

as to evaluate their relative variation as a function of the evolutionary history and current conditions of the environment. We will return to these evolutionary relationships between species in Chapter 5, where we will discuss to what extent the distribution of traits within communities depends on their phylogenetic position.

3.5 Conclusions

This chapter has presented evidence for the existence of three fundamental axes of variation in plant function. While the emphasis has been placed on these three well-known major axes and their associated traits, they do not, however, constitute an exhaustive list allowing for the understanding of all the different aspects of plant and vegetation function. A number of other dimensions have been explored in Section 3.3 (see also Westoby & Wright, 2006), corresponding to some of the suggestions put forward by Laughlin (2014). In addition, understanding the responses of vegetation to disturbance (McIntyre et al., 1999; Weiher et al., 1999), to land-use modifications (Garnier et al., 2007), or water availability (Ackerly, 2004), requires taking into account specific traits relevant to the environmental gradient of interest. However the argument put forward by Westoby (1999) and Westoby et al. (2002) is that those traits which describe the major axes should be measured in a systematic manner, as 'absolute' calibration variables allowing for the comparison of species among different studies and study locations (cf. Box 3.1), with further information on more specific traits being collected in relation to targeted questions. The four studies cited above constitute a good illustration of this approach, since specific leaf area, plant height, and seed mass have indeed been measured in all cases, but they were completed by other, different traits relevant to the specific gradient under study.

This chapter has focused mainly on the contribution of traits to the different functions of individuals, but an increasing number of studies have shown that the evolutionary relatedness of species may have an influence on the values of their traits. The importance of this phylogenetic signal remains, however, to be established for the major traits which determine the functioning of individuals.

Using this understanding of the functional significance of traits, the following chapter will address the subject of the ecological significance of some of these in relation to environmental factors.

3.6 Key points

1. Amongst the multitude of traits measurable on an individual, those that are the most useful for comparative functional ecology should have at least four characteristics. They should: (a) be clearly linked to a plant function; (b) be relatively easy and quick to measure; (c) be measurable using standardized protocols applicable to a wide variety of species and environmental conditions; and (d) allow for the establishment of hierarchies among species which are conserved between contrasting environments. The identification and use of traits defined according to these criteria would allow plant functioning to be compared over a wide range of species and environmental conditions.
2. The existence of co-variation between traits which recurs between environments has led to the identification of a number of axes of variation characteristic of the different strategies of plant functioning. A first axis is linked to resource acquisition and use, and is captured in particular by traits linked to the 'leaf economics spectrum', with some of these also linked to the rate of litter decomposition and/or protection from herbivores. A second axis, described by plant size, is linked primarily to their competitive ability and, to a lesser extent, to their dispersal capability. A third axis is described by seed characteristics and is linked in particular with the sexual regeneration of individuals.
3. These three fundamental axes of variation were initially considered as being independent. However, while the selective pressures which act on the traits of these axes can be considered as such, empirical data shows that seed mass and plant height are in fact positively correlated.
4. It is entirely possible that axes of variation which have not yet been clearly identified will be so in the future. This could be the case, for example, for traits linked to a root economics spectrum,

those correlated with wood density, or to plant vegetative and reproductive phenology.

5. The different axes of variation discussed in this chapter and their associated traits do not constitute an exhaustive list allowing us to understand all of the different aspects of plant functioning. However, the systematic determination of traits linked to these fundamental axes, used as 'absolute' calibration variables, should allow for the comparison of species between different studies and sites at the global scale. A more detailed understanding of the mechanisms involved at finer scales will require increasing this limited list with the addition of other traits associated with local, more specific factors.

3.7 References

Ackerly, D. (2004). Functional strategies of chaparral shrubs in relation to seasonal water deficit and disturbance. *Ecological Monographs, 74*, 25–44.

Ackerly, D. D., & Donoghue, M. J. (1998). Leaf size, sapling allometry, and Corner's rules: phylogeny and correlated evolution in maples (*Acer*). *The American Naturalist, 152*(6), 767–791. doi:10.1086/286208

Ackerly, D. D., Knight, C. A., Weiss, S. B., Barton, K., & Starmer, K. P. (2002). Leaf size, specific leaf area, and microhabitat distribution of chaparral woody plants: contrasting patterns in species level and community level analyses. *Oecologia (Berlin), 130*, 449–457.

Aerts, R., & Chapin, F. S., III. (2000). The mineral nutrition of wild plants revisited: a re-evaluation of processes and patterns. *Advances in Ecological Research, 30*, 1–67.

Agrawal, A. A., & Fishbein, M. (2006). Plant defense syndromes. *Ecology, 87*(7), S132–S149.

Al Haj Khaled, R., Duru, M., Decruyenaere, V., Jouany, C., & Cruz, P. (2006). Using leaf traits to rank native grasses according to their nutritive value. *Rangeland Ecology and Management, 59*, 648–654.

Ansquer, P., Al Haj Khaled, R., Cruz, P., Theau, J.-P., Therond, O., & Duru, M. (2009). Characterizing and predicting plant phenology in species-rich grasslands. *Grass and Forage Science, 64*, 57–70.

Anten, N. P. R., & Schieving, F. (2010). The role of wood mass density and mechanical constraints in the economy of tree architecture. *The American Naturalist, 175*, 250–260.

Barton, K. E., & Koricheva, J. (2010). The ontogeny of plant defense and herbivory: characterizing general patterns using meta-analysis. *The American Naturalist, 175*(4), 481–493. doi:10.1086/650722

Bazzaz, F. A., Ackerly, D. D., & Reekie, E. G. (2000). Reproductive allocation in plants. In M. Fenner (ed.), *Seeds. The Ecology of Regeneration in Plant Communities* (2nd ed., pp. 1–29). Wallingford, UK: CAB International.

Berendse, F., & Aerts, R. (1987). Nitrogen-use-efficiency: a biologically meaningful definition? *Functional Ecology, 1*, 293–296.

Bodner, G., Leitner, D., Nakhforoosh, A., Sobotik, M., Moder, K., & Kaul, H.-P. (2013). A statistical approach to root system classification. *Frontiers* in *Plant Science, 4*. doi:10.3389/fpls.2013.00292

Boege, K., & Marquis, R. J. (2005). Facing herbivory as you grow up: the ontogeny of resistance in plants. *Trends in Ecology and Evolution, 20*(8), 441–448. doi:10.1016/j.tree.2005.05.001

Bolmgren, K., & Cowan, P. D. (2008). Time-size tradeoffs: a phylogenetic comparative study of flowering time, plant height, and seed mass in a north-temperate flora. *Oikos, 117*, 424–429.

Bruinenberg, M. H., Valk, H., Korevaar, H., & Struik, P. C. (2002). Factors affecting digestibility of temperate forages from seminatural grasslands: a review. *Grass and Forage Science, 57*(3), 292–301. doi:10.1046/j.1365-2494.2002.00327.x

Caffarra, A., Donnelly, A., Chuine, I., & Jones, M. B. (2011). Modelling the timing of *Betula pubescens* budburst. I. Temperature and photoperiod: a conceptual model. *Climate Research, 46*(2), 147–157. doi:10.3354/cr00980

Canadell, J., Jackson, R. B., Ehleringer, J. R., Mooney, H. A., Sala, O. E., & Schulze, E. D. (1996). Maximum rooting depth of vegetation types at the global scale. *Oecologia, 108*(4), 583–595. doi:10.1007/bf00329030

Cavender-Bares, J., Kozak, K. H., Fine, P. V. A., & Kembel, S. W. (2009). The merging of community ecology and phylogenetic biology. *Ecology Letters, 12*, 693–715.

Chapin, F. S., III, Autumn, K., & Pugnaire, F. (1993). Evolution of suites of traits in response to environmental stress. *The American Naturalist, 142*, S78–92.

Chapin, F. S., III. (1980). The mineral nutrition of wild plants. *Annual Review of Ecology and Systematics, 11*, 233–260.

Chave, J., Andalo, C., Brown, S., Cairns, M. A., Chambers, J. Q., Eamus, D.,.. Yamakura, T. (2005). Tree allometry and improved estimation of carbon stocks and balance in tropical forests. *Oecologia, 145*, 87–99.

Chave, J., Coomes, D., Jansen, S., Lewis, S. L., Swenson, N. G., & Zanne, A. E. (2009). Towards a worldwide wood economics spectrum. *Ecology Letters, 12*, 351–366.

Chuine, I. (2010). Why does phenology drive species distribution? *Philosophical Transactions of the Royal Society B: Biological Sciences, 365*(1555), 3149–3160. doi:10.1098/rstb.2010.0142

Coley, P. D., Bryant, J. P., & Chapin, F. S., III. (1985). Resource availability and plant antiherbivore defense. *Science, 230*, 895–899.

Cornelissen, J. H. C. (1999). A triangular relationship between leaf size and seed size among woody species: allometry, ontogeny, ecology, and taxonomy. *Oecologia (Berl.), 118*, 248–255.

Cornelissen, J. H. C., & Thompson, K. (1997). Functional leaf attributes predict litter decomposition rate in herbaceous plants. *New Phytologist, 135*, 109–114.

Cornelissen, J. H. C., Cerabolini, B., Castro-Díez, P., Villar-Salvador, P., Montserrat-Martí, G., Puyravaud, J. P.,... Aerts, R. (2003). Functional traits of woody plants: correspondence of species rankings between field adults and laboratory-grown seedlings? *Journal of Vegetation Science, 14*, 311–322.

Cornelissen, J. H. C., Pérez-Harguindeguy, N., Díaz, S., Grime, J. P., Marzano, B., Cabido, M., ... Cerabolini, B. (1999). Leaf structure and defence control litter decomposition rate across species and life forms in regional flora on two continents. *New Phytologist, 143*, 191–200.

Corner, E. J. H. (1949). The durian theory or the origin of the modern tree. *Annals of Botany, 13*(4), 367–414.

Cornwell, W. K., Cornelissen, J. H. C., Amatangelo, K., Dorrepaal, E., Eviner, V. T., Godoy, O., ... Westoby, M. (2008). Plant traits are the dominant control on litter decomposition rates within biomes worldwide. *Ecology Letters, 11*, 1065–1071.

Cornwell, W. K., Westoby, M., Falster, D. S., FitzJohn, R. G., O'Meara, B. C., Pennell, M. W., ... Zanne, A. E. (2014). Functional distinctiveness of major plant lineages. *Journal of Ecology, 102*(2), 345–356. doi:10.1111/1365-2745.12208

Craine, J. M. (2009). *Resource Strategies of Wild Plants*. Princeton and Oxford: Princeton University Press.

Craine, J. M., Wolkovich, E. M., Towne, E. G., & Kembel, S. W. (2012). Flowering phenology as a functional trait in a tallgrass prairie. *New Phytologist, 193*(3), 673–682. doi:10.1111/j.1469-8137.2011.03953.x

Cruz, P., Theau, J. P., Lecloux, E., Jouany, C., & Duru, M. (2010). Typologie fonctionnelle de graminées fourragères pérennes: une classification multitraits. *Fourrages*, (201), 11–17.

Davies, T. J., Barraclough, T. G., Chase, M. W., Soltis, P. S., Soltis, D. E., & Savolainen, V. (2004). Darwin's abominable mystery: Insights from a supertree of the angiosperms. *Proc. Natl. Acad. Sci. USA, 101*(7), 1904–1909. doi:10.1073/pnas.0308127100

Davies, T. J., Wolkovich, E. M., Kraft, N. J. B., Salamin, N., Allen, J. M., Ault, T. R., ... Travers, S. E. (2013). Phylogenetic conservatism in plant phenology. *Journal of Ecology, 101*(6), 1520–1530. doi:10.1111/1365-2745.12154

DeAngelis, D. L. (1992). *Dynamics of Nutrient Cycling and Food Webs* (1st ed.). London: Chapman and Hall.

Díaz, S., Hodgson, J. G., Thompson, K., Cabido, M., Cornelissen, J. H. C., Jalili, A., ... Zak, M. R. (2004). The plant traits that drive ecosystems: Evidence from three continents. *Journal of Vegetation Science, 15*, 295–304.

Dickinson, C. E., & Dodd, J. L. (1976). Phenological pattern in the shortgrass prairie. *American Midland Naturalist, 96*, 367–378.

Diekmann, M. (1996). Relationship between flowering phenology of perennial herbs and meteorological data in deciduous forests of Sweden. *Canadian Journal of Botany, 74*(4), 528–537. doi:10.1139/b96-067

Du, G., & Qi, W. (2010). Trade-offs between flowering time, plant height, and seed size within and across 11 communities of a QingHai-Tibetan flora. *Plant Ecology, 209*(2), 321–333. doi:10.1007/s11258-010-9763-4

Duru, M. (1997). Leaf and stem *in vitro* digestibility for grasses and dicotyledons of meadow plant communities in spring. *Journal of the Science of Food and Agriculture, 74*(2), 175–185.

Eckstein, R. L., Karlsson, P. S., & Weih, M. (1999). Leaf life span and nutrient resorption as determinants of plant nutrient conservation in temperate-arctic regions. *New Phytologist, 143*, 177–189.

Ehrlén, J. (2015). Selection on flowering time in a life-cycle context. *Oikos, 124*, 92–101. doi:10.1111/oik.01473

Eriksson, O., & Ehrlén, J. (1991). Phenological variation in fruit characteristics in vertebrate-dispersed plants. *Oecologia, 86*(4), 463–470. doi:10.1007/bf00318311

Evans, J. R. (1989). Photosynthesis and nitrogen relationships in leaves of C_3 plants. *Oecologia (Berlin), 78*, 9–19.

Fenner, M. (1998). The phenology of growth and reproduction in plants. *Perspectives in Plant Ecology, Evolution and Systematics, 1*(1), 78–91. doi:10.1078/1433-8319-00053

Fenner, M., & Thompson, K. (2005). *The Ecology of Seeds*. Cambridge: Cambridge University Press.

Field, C. B., & Mooney, H. A. (1986). The photosynthesis-nitrogen relationship in wild plants. In T. J. Givnish (ed.), *On the Economy of Plant Form and Function* (pp. 25–55). Cambridge: Cambridge University Press.

Fitter, A. H. (1985). Functional significance of root morphology and root system architecture. In A. H. Fitter, D. Atkinson, D. J. Read & M. B. Usher (eds), *Ecological Interactions in Soil* (pp. 87–106.). Oxford, UK: Blackwell Scientific.

Fitter, A. H., & Hay, R. K. M. (2002). *Environmental Physiology of Plants* (3rd ed.). London: Academic Press.

Flores, O., Garnier, E., Wright, I. J., Reich, P. B., Pierce, S., Díaz, S., ... Weiher, E. (2014). An evolutionary perspective on leaf economics: phylogenetics of leaf mass per area in vascular plants. *Ecology and Evolution, 4*(14), 2799–2811. doi:10.1002/ece3.1087

Forrest, J., & Miller-Rushing, A. J. (2010). Toward a synthetic understanding of the role of phenology in ecology and evolution. *Philosophical Transactions of the*

Royal Society B: Biological Sciences, 365(1555), 3101–3112. doi:10.1098/rstb.2010.0145

Fortunel, C., Violle, C., Roumet, C., Buatois, B., Navas, M.-L., & Garnier, E. (2009). Allocation strategies and seed traits are hardly affected by nitrogen supply in 18 species differing in successional status. *Perspectives in Plant Ecology, Evolution and Systematics, 11*(4), 267–283. doi:10.1016/j.ppees.2009.04.003

Freschet, G. T., Aerts, R., & Cornelissen, J. H. C. (2012). A plant economics spectrum of litter decomposability. *Functional Ecology, 26*(1), 56–65. doi:10.1111/j.1365-2435.2011.01913.x

Freschet, G. T., Cornelissen, J. H. C., van Logtestijn, R. S. P., & Aerts, R. (2010). Evidence of the 'plant economics spectrum' in a subarctic flora. *Journal of Ecology, 98*, 362–373.

Freschet, G. T., Cornwell, W. K., Wardle, D. A., Elumeeva, T. G., Liu, W., Jackson, B. G., ... Cornelissen, J. H. C. (2013). Linking litter decomposition of above- and below-ground organs to plant–soil feedbacks worldwide. *Journal of Ecology, 101*(4), 943–952. doi:10.1111/1365-2745.12092

Freschet, G. T., Swart, E. M., & Cornelissen, J. H. C. (2015). Integrated plant phenotypic responses to contrasting above- and below-ground resources: key roles of specific leaf area and root mass fraction. *New Phytologist, 206*(4), 1247–1260. doi:10.1111/nph.13352

Frissel, M. J. (1981). The definition of residence times in ecological models. In F. E. Clark & T. Rosswall (eds), *Terrestrial nitrogen cycles. Processes, ecosystem strategies and management impacts* (pp. 117–122). Stockholm: Swedish Natural Science Research Council.

Funk, J. L., & Cornwell, W. K. (2013). Leaf traits within communities: context may affect the mapping of traits to function. *Ecology, 94*(9), 1893–1897. doi:10.1890/12-1602.1

Gardarin, A., Dürr, C., Mannino, M. R., Busset, H., & Colbach, N. (2010). Seed mortality in the soil is related to seed coat thickness. *Seed Science Research, 20*(4), 243–256. doi:10.1017/s0960258510000255

Garnier, E. (1991). Resource capture, biomass allocation and growth in herbaceous plants. *Trends in Ecology and Evolution, 6*, 126–131.

Garnier, E., & Aronson, J. (1998). Nitrogen use efficiency from leaf to stand level: clarifying the concept. In H. Lambers, H. Poorter & M. M. I. van Vuuren (eds), *Inherent Variation in Plant Growth. Physiological Mechanisms and Ecological Consequences* (pp. 515–538). Leiden: Backhuys Publishers.

Garnier, E., & Laurent, G. (1994). Leaf anatomy, specific mass and water content in congeneric annual and perennial grass species. *New Phytologist, 128*, 725–736.

Garnier, E., & Navas, M.-L. (2012). A trait-based approach to comparative functional plant ecology: concepts, methods, and applications for agroecology. A review. *Agronomy for Sustainable Development, 32*, 365–399. doi:10.1007/s13593-011-0036-y

Garnier, E., Cortez, J., Billès, G., Navas, M.-L., Roumet, C., Debussche, M., ... Toussaint, J.-P. (2004). Plant functional markers capture ecosystem properties during secondary succession. *Ecology, 85*(9), 2630–2637.

Garnier, E., Laurent, G., Bellmann, A., Debain, S., Berthelier, P., Ducout, B., ... Navas, M.-L. (2001a). Consistency of species ranking based on functional leaf traits. *New Phytologist, 152*, 69–83.

Garnier, E., Lavorel, S., Ansquer, P., Castro, H., Cruz, P., Dolezal, J., ... Zarovali, M. (2007). Assessing the effects of land use change on plant traits, communities, and ecosystem functioning in grasslands: a standardized methodology and lessons from an application to 11 European sites. *Annals of Botany, 99*, 967–985.

Garnier, E., Shipley, B., Roumet, C., & Laurent, G. (2001b). A standardized protocol for the determination of specific leaf area and leaf dry matter content. *Functional Ecology, 15*, 688–695.

Gaudet, C. L., & Keddy, P. A. (1988). A comparative approach to predicting competitive ability from plant traits. *Nature (London), 334*, 242–243.

Givnish, T. J., & Vermeij, G. J. (1976). Sizes and shapes of liane leaves. *The American Naturalist, 110*(975), 743–778. doi:10.1086/283101

Greene, D. F., & Johnson, E. A. (1989). A model of wind dispersal of winged or plumed seeds. *Ecology, 70*, 339–347.

Grime, J. P. (1977). Evidence for the existence of three primary strategies in plants and its relevance to ecological and evolutionary theory. *The American Naturalist, 111*, 1169–1194.

Grime, J. P. (2001). *Plant Strategies, Vegetation Processes, and Ecosystem Properties* (2nd ed.). Chichester: John Wiley & Sons.

Grime, J. P., & Anderson, J. M. (1986). Environmental controls over organism activity. In K. Van Cleve, F. S. Chapin III, P. W. Flanagan, L. A. Viereck & C. T. Dyrness (eds), *Forest ecosystems in Alaskan taïga: a synthesis of structure and function* (Vol. 57, pp. 89–95). Berlin: Springer-Verlag.

Grime, J. P., & Hunt, R. (1975). Relative growth-rate: its range and adaptive significance in a local flora. *Journal of Ecology, 63*(2), 393–422. doi: Stable URL: http://www.jstor.org/stable/2258728

Grime, J. P., Hunt, R., & Krzanowski, W. J. (1987). Evolutionary physiological ecology of plants. In P. Calow (ed.), *Evolutionary Physiological Ecology* (pp. 105–125). Cambridge: Cambridge University Press.

Grime, J. P., Thompson, K., Hunt, R., Hodgson, J. G., Cornelissen, J. H. C., Rorison, I. H., ... Whitehouse, J. (1997). Integrated screening validates primary axes of specialisation in plants. *Oikos, 79*, 259–281.

Guilbaud, C. S. E., Dalchau, N., Purves, D. W. & Turnbull, L. A. (2015). Is 'peak N' key to understanding the timing of flowering in annual plants? *New Phytologist, 205*, 918–927.

Hanley, M. E., Lamont, B. B., Fairbanks, M. M., & Rafferty, C. M. (2007). Plant structural traits and their role in anti-herbivore defence. *Perspectives in Plant Ecology, Evolution and Systematics, 8*(4), 157–178. doi:10.1016/j.ppees.2007.01.001

Harper, J. L., Lovell, P. H., & Moore, K. G. (1970). The shapes and sizes of seeds. *Annual Review of Ecology and Systematics, 1*, 327–356.

Henery, M. L., & Westoby, M. (2001). Seed mass and seed nutrient content as predictors of seed output variation between species. *Oikos, 92*(3), 479–490. doi:10.1034/j.1600-0706.2001.920309.x

Herms, D. A., & Mattson, W. J. (1992). The dilemma of plants: to grow or defend. *Quarterly Review of Biology, 67*, 283–335.

Hughes, L., Dunlop, M., French, K., Leishman, M. R., Rice, B., Rodgerson, L., & Westoby, M. (1994). Predicting dispersal spectra: a minimal set of hypotheses based on plant attributes. *Journal of Ecology, 82*, 933–950.

Hummel, I., Vile, D., Violle, C., Devaux, J., Ricci, B., Blanchard, A., ... Roumet, C. (2007). Relating root structure and anatomy to whole plant functioning: the case of fourteen herbaceous Mediterranean species. *New Phytologist, 173*, 313–321.

Hunt, R. (1982). *Plant Growth Curves. The Functional Approach to Plant Growth Analysis*. London: Edward Arnold.

Jackson, R. B., Moore, L. A., Hoffmann, W. A., Pockman, W. T., & Linder, C. R. (1999). Ecosystem rooting depth determined with caves and DNA. *Proc. Natl. Acad. Sci. USA, 96*, 11387–11392.

Jakobsson, A., & Eriksson, O. (2000). A comparative study of seed number, seed size, seedling size, and recruitment in grassland plants. *Oikos, 88*, 494–502.

Kattge, J., Díaz, S., Lavorel, S., Prentice, I. C., Leadley, P., Bönisch, G., ... Wirth, C. (2011). TRY – a global database of plant traits. *Global Change Biology 17*, 2905–2935. doi:10.1111/j.1365-2486.2011.02451.x

Kazakou, E., Violle, C., Roumet, C., Navas, M.-L., Vile, D., Kattge, J., & Garnier, E. (2014). Are trait-based species rankings consistent across data sets and spatial scales? *Journal of Vegetation Science, 25*(1), 235–247. doi:10.1111/jvs.12066

Kazakou, E., Violle, C., Roumet, C., Pintor, C., Gimenez, O., & Garnier, E. (2009). Litter quality and decomposability of species from a Mediterranean succession depend on leaf traits but not on nitrogen supply. *Annals of Botany, 104*, 1151–1161.

Kembel, S. W., & Cahill, J. F. (2011). Independent evolution of leaf and root traits within and among temperate grassland plant communities. *Plos One, 6*(6), e19992. doi:10.1371/journal.pone.0019992

Kikuzawa, K. (1995). The basis for variation in leaf longevity of plants. *Vegetatio, 121*, 89–100.

Kikuzawa, K., & Ackerly, D. (1999). Significance of leaf longevity in plants. *Plant Species Biology, 14*, 39–45.

King, D. A. (1990). The adaptive significance of tree height *The American Naturalist, 135*, 809–828.

Kolb, A., Barsch, F., & Diekmann, M. (2006). Determinants of local abundance and range size in forest vascular plants. *Global Ecology and Biogeography, 15*(3), 237–247. doi:10.1111/j.1466-8238.2005.00210.x

Körner, C., & Basler, D. (2010). Phenology under global warming. *Science, 327*(5972), 1461–1462. doi:10.1126/science.1186473

Kuparinen, A. (2006). Mechanistic models for wind dispersal. *Trends in Plant Sciences, 11*, 296–301.

Lambers, H., & Poorter, H. (1992). Inherent variation in growth rate between higher plants: a search for physiological causes and ecological consequences. *Advances in Ecological Research, 23*, 187–261.

Larjavaara, M., & Muller-Landau, H. C. (2010). Rethinking the value of high wood density. *Functional Ecology, 24*(4), 701–705. doi:10.1111/j.1365-2435.2010.01698.x

Larjavaara, M., & Muller-Landau, H. C. (2012). Still rethinking the value of high wood density. *American Journal of Botany, 99*(1), 165–168. doi:10.3732/ajb.1100324

Laughlin, D. C. (2014). The intrinsic dimensionality of plant traits and its relevance to community assembly. *Journal of Ecology, 102*(1), 186–193. doi:10.1111/1365-2745.12187

Lavorel, S., Díaz, S., Cornelissen, J. H. C., Garnier, E., Harrison, S. P., McIntyre, S., ... Urcelay, C. (2007). Plant functional types: are we getting any closer to the Holy Grail? In J. Canadell, D. Pataki, & L. Pitelka (eds), *Terrestrial Ecosystems in a Changing World* (pp. 149–164). Berlin: Springer-Verlag.

Lechowicz, M. J. (2002). Phenology. In H. A. Mooney & J. G. Canadell (eds), *The Earth System: Biological and Ecological Dimensions of Global Environmental Change* (Vol. 2, pp. 461–465). Chichester, UK: John Wiley & Sons.

Leishman, M. R., & Westoby, M. (1998). Seed size and shape are not related to persistence in soil in Australia in the same way as in Britain. *Functional Ecology, 12*, 480–485.

Leishman, M. R., Wright, I. J., Moles, A. T., & Westoby, M. (2000). The evolutionary ecology of seed size. In M. Fenner (ed.), *Seeds: The Ecology of Regeneration in Plant Communities* (pp. 31–57). Wallingford, UK: CAB International.

Lieth, H. (ed.) (1974). *Phenology and Seasonality Modeling*. New York: Springer-Verlag.

Lortie, C. J., & Aarssen, L. W. (1999). The advantage of being tall: higher flowers receive more pollen in *Verbascum thapsus* L. (Scrophulariaceae). *Ecoscience, 6*, 68–71.

Louault, F., Pillar, V. D., Aufrère, J., Garnier, E., & Soussana, J.-F. (2005). Plant traits and functional types in response to reduced disturbance in a semi-natural grassland. *Journal of Vegetation Science, 16*, 151–160.

McCormack, M. L., Dickie, I. A., Eissenstat, D. M., Fahey, T. J., Fernandez, C. W., Guo, D., ... Zadworny, M. (2015). Redefining fine roots improves understanding of below-ground contributions to terrestrial biosphere processes. *New Phytologist*. doi:10.1111/nph.13363

McIntyre, S., Lavorel, S., Landsberg, J., & Forbes, T. D. A. (1999). Disturbance response in vegetation—towards a global perspective on functional traits. *Journal of Vegetation Science, 10*, 621–630.

Meinzer, F. C., Andrade, J. L., Goldstein, G., Holbrook, N. M., Cavelier, J., & Wright, S. J. (1999). Partitioning of soil water among canopy trees in a seasonally dry tropical forest. *Oecologia, 121*, 293–301.

Midgley, J., & Bond, W. (1989). Leaf size and inflorescence size may be allometrically related traits. *Oecologia, 78*(3), 427–429. doi:10.1007/bf00379120

Mokany, K., & Ash, J. (2008). Are traits measured on pot grown plants representative of those in natural communities? *Journal of Vegetation Science, 19*, 119–126.

Moles, A. T., & Leishman, M. R. (2008). The seedling as part of a plant's life history strategy. In M. A. Leck, V. T. Parker, & R. L. Simpson (eds), *Seedling Ecology and Evolution* (pp. 217–238). Cambridge: Cambridge University Press.

Moles, A. T., & Westoby, M. (2004). Seedling survival and seed size: a synthesis of the literature. *Journal of Ecology, 92*, 372–383.

Moles, A. T., Ackerly, D. D., Webb, C. O., Tweddle, J. C., Dickie, J. B., Pitman, A. J., & Westoby, M. (2005a). Factors that shape seed mass evolution. *Proc. Natl. Acad. Sci. USA, 102*, 10540–10544.

Moles, A. T., Ackerly, D. D., Webb, C. O., Tweddle, J. C., Dickie, J. B., & Westoby, M. (2005b). A brief history of seed size. *Science, 307*, 576–580.

Moles, A. T., Falster, D. S., Leishman, M. R., & Westoby, M. (2004). Small-seeded species produce more seeds per square metre of canopy per year, but not per individual per lifetime. *Journal of Ecology, 92*(3), 384–396.

Moles, A. T., Peco, B., Wallis, I. R., Foley, W. J., Poore, A. G. B., Seabloom, E. W., ... Hui, F. K. C. (2013). Correlations between physical and chemical defences in plants: tradeoffs, syndromes, or just many different ways to skin a herbivorous cat? *New Phytologist, 198*(1), 252–263. doi:10.1111/nph.12116

Molinier, R., & Muller, P. (1938). *La dissémination des espèces végétales* (Vol. 50). Editor Impr. A. Lesot.

Mooney, H. A., Hobbs, R. J., Gorham, J., & Williams, K. (1986). Biomass accumulation and resource utilization in co-occurring grassland annuals. *Oecologia, 70*(4), 555–558. doi:10.1007/bf00379903

Mouquet, N., Devictor, V., Meynard, C. N., Munoz, F., Bersier, L. F., Chave, J., ... Thuiller, W. (2012). Ecophylogenetics: advances and perspectives. *Biological Reviews, 87*(4), 769–785. doi:10.1111/j.1469-185X.2012.00224.x

Muller-Landau, H. C. (2004). Interspecific and intersite variation in wood specific gravity of tropical trees. *Biotropica, 36*, 20–32.

Navas, M.-L., Roumet, C., Bellmann, A., Laurent, G., & Garnier, E. (2010). Suites of plant traits in species from different stages of a Mediterranean secondary succession. *Plant Biology, 12*, 183–196.

Niinemets, Ü. (1999). Components of leaf dry mass per area—thickness and density—alter leaf photosynthetic capacity in reverse directions in woody plants. *New Phytologist, 144*, 35–47.

Niinemets, Ü., Portsmuth, A., Tena, D., Tobias, M., Matesanz, S., & Valladares, F. (2007). Do we underestimate the importance of leaf size in plant economics? Disproportional scaling of support costs within the spectrum of leaf physiognomy. *Annals of Botany, 100*(2), 283–303. doi:10.1093/aob/mcm107

Niklas, K. J. (1995a). Plant height and the properties of some herbaceous stems. *Annals of Botany, 75*, 133–142

Niklas, K. J. (1995b). Size-dependent allometry of tree height, diameter, and trunk-taper. *Annals of Botany, 75*, 217–227.

Onoda, Y., Westoby, M., Adler, P. B., Choong, A. M. F., Clissold, F. J., Cornelissen, J. H. C., ... Yamashita, N. (2011). Global patterns of leaf mechanical properties. *Ecology Letters, 14*(3), 301–312. doi:10.1111/j.1461-0248.2010.01582.x

Parkhurst, D. F., & Loucks, O. L. (1972). Optimal leaf size in relation to environment. *Journal of Ecology, 60*, 505–537.

Peppe, D. J., Royer, D. L., Cariglino, B., Oliver, S. Y., Newman, S., Leight, E., ... Wright, I. J. (2011). Sensitivity of leaf size and shape to climate: global patterns and paleoclimatic applications. *New Phytologist, 190*(3), 724–739. doi:10.1111/j.1469-8137.2010.03615.x

Pérez-Harguindeguy, N., Díaz, S., Cornelissen, J. H. C., Vendramini, F., Cabido, M., & Castellanos, A. (2000). Chemistry and toughness predict leaf litter decomposition rates over a wide spectrum of functional types and taxa in central Argentina. *Plant and Soil, 218*, 21–30.

Pérez-Harguindeguy, N., Díaz, S., Garnier, E., Lavorel, S., Poorter, H., Jaureguiberry, P., ... Cornelissen, J. H. C. (2013). New handbook for standardised measurement of plant functional traits worldwide. *Australian Journal Botany*, *61*, 167–234.

Pickup, M., Westoby, M., & Basden, A. (2005). Dry mass costs of deploying leaf area in relation to leaf size. *Functional Ecology*, *19*, 88–97.

Pontes, L. D. S., Soussana, J.-F., Louault, F., Andueza, D., & Carrère, P. (2007). Leaf traits affect the above-ground productivity and quality of pasture grasses. *Functional Ecology*, *21*, 844–853.

Poorter, H., & de Jong, R. (1999). A comparison of specific leaf area, chemical composition and leaf construction costs of field plants from 15 habitats differing in productivity. *New Phytologist*, *143*, 163–176.

Poorter, H., & Garnier, E. (2007). Ecological significance of inherent variation in relative growth rate and its components. In F. I. Pugnaire & F. Valladares (eds), *Functional Plant Ecology* (2nd ed.) (pp. 67–100). Boca Raton: CRC Press.

Poorter, H., Lambers, H., & Evans, J. R. (2014). Trait correlation networks: a whole-plant perspective on the recently criticized leaf economic spectrum. New Phytologist, *201*, 378–382. doi:10.1111/nph.12547.

Poorter, H., & van der Werf, A. (1998). Is inherent variation in RGR determined by LAR at low irradiance and by NAR at high irradiance? A review of herbaceous species. In H. Lambers, H. Poorter, & M. M. I. Van Vuuren (eds), *Inherent Variation in Plant Growth. Physiological Mechanisms and Ecological Consequences* (pp. 309–336). Leiden: Backhuys Publishers.

Poorter, H., & Villar, R. (1997). The fate of acquired carbon in plants: chemical composition and construction costs. In F. A. Bazzaz (ed.), *Plant Resource Allocation* (pp. 39–72). San Diego: Academic Press.

Poorter, L., & Rozendaal, D. M. A. (2008). Leaf size and leaf display of thirty-eight tropical tree species. *Oecologia*, *158*(1), 35–46. doi:10.1007/s00442-008-1131-x

Price, C. A., Wright, I. J., Ackerly, D. D., Niinemets, Ü., Reich, P. B., & Veneklaas, E. J. (2014). Are leaf functional traits 'invariant' with plant size and what is 'invariance' anyway? *Functional Ecology*, *28*(6), 1330–1343. doi:10.1111/1365-2435.12298

Primack, R. B. (1985). Patterns of flowering phenology in communities, populations, individuals, and single flowers. In J. White (ed.), *Population Structure of Vegetation* (pp. 571–593). Dordrecht: Junk.

Primack, R. B. (1987). Relationships among flowers, fruits, and seeds. *Annual Review of Ecology and Systematics*, *18*(1), 409–430. doi:10.1146/annurev.es.18.110187.002205

Prior, L. D., Eamus, D., & Bowman, D. M. J. S. (2003). Leaf attributes in the seasonally dry tropics: a comparison of four habitats in northern Australia. *Functional Ecology*, *17*, 504–515.

Rathcke, B., & Lacey, E. P. (1985). Phenological patterns of terrestrial plants. *Annual Review of Ecology and Systematics*, *16*, 179–214.

Rees, M. (1997). Evolutionary ecology of seed dormancy and seed size. In J. Silvertown, M. Franco, & J. L. Harper (eds), *Plant life histories. Ecology, phylogeny, and evolution* (pp. 121–142). Cambridge, UK: The Royal Society.

Rees, M., & Venable, D. L. (2007). Why do big plants make big seeds? *Journal of Ecology*, *95*, 926–936.

Reich, P. B. (2014). The world-wide 'fast–slow' plant economics spectrum: a traits manifesto. *Journal of Ecology*, *102*(2), 275–301. doi:10.1111/1365-2745.12211

Reich, P. B., Tjoelker, M. G., Pregitzer, K. S., Wright, I. J., Oleksyn, J., & Machado, J.-L. (2008). Scaling of respiration to nitrogen in leaves, stems and roots of higher land plants. *Ecology Letters*, *11*(8), 793–801. doi:10.1111/j.1461-0248.2008.01185.x

Reich, P. B., Walters, M. B., & Ellsworth, D. S. (1992). Leaf life-span in relation to leaf, plant, and stand characteristics among diverse ecosystems. *Ecological Monographs*, *62*, 365–392.

Reich, P. B., Walters, M. B., & Ellsworth, D. S. (1997). From tropics to tundra: global convergence in plant functioning. *Proc. Natl. Acad. Sci. USA*, *94*, 13730–13734.

Roderick, M. L., Berry, S. L., Noble, I. R., & Farquhar, G. D. (1999). A theoretical approach to linking the composition and morphology with the function of leaves. *Functional Ecology*, *13*, 683–695.

Roumet, C., Urcelay, C., & Díaz, S. (2006). Suites of root traits differ between annual and perennial species growing in the field. *New Phytologist*, *170*, 357–358.

Ryser, P. (1996). The importance of tissue density for growth and life span of leaves and roots: a comparison of five ecologically contrasting grasses. *Functional Ecology*, *10*, 717–723.

Ryser, P., & Urbas, P. (2000). Ecological significance of leaf life span among Central European grass species. *Oikos*, *91*, 41–50.

Saatkamp, A., Affre, L., Dutoit, T., & Poschlod, P. (2009). The seed bank longevity index revisited: limited reliability evident from a burial experiment and database analyses. *Annals of Botany*, *104*, 715–724. doi:10.1093/aob/mcp148

Schenk, H. J., & Jackson, R. B. (2002). Rooting depths, lateral root spreads and below-ground/above-ground allometries of plants in water-limited ecosystems. *Journal of Ecology*, *90*, 480–494.

Schwartz, M. D. (ed.) (2003). *Phenology: An Integrative Environmental Science*. Dordrecht: Kluwer Academic Publishers.

Shipley, B. (2006). Net assimilation rate, specific leaf area and leaf mass ratio: which is most closely correlated with relative growth rate? A meta-analysis. *Functional Ecology*, *20*(4), 565–574. doi:10.1111/j.1365-2435.2006.01135.x

Shipley, B., & Dion, J. (1992). The allometry of seed production in herbaceous angiosperms. *The American Naturalist*, *139*, 467–483.

Shipley, B., Keddy, P. A., Moore, D. R. J., & Lemky, K. (1989). Regeneration and establishment strategies of emergent macrophytes. *Journal of Ecology, 77*, 1093–1110.

Shipley, B., Lechowicz, M. J., Wright, I. J., & Reich, P. B. (2006). Fundamental trade-offs generating the worldwide leaf economics spectrum. *Ecology, 87*, 535–541.

Silvertown, J., Franco, M., & Harper, J. L. (eds). (1997). *Plant life histories. Ecology, phylogeny and evolution*. Cambridge: Cambridge University Press.

Silvertown, J., Franco, M., & McConway, K. (1992). A demographic interpretation of Grime's triangle. *Functional Ecology, 6*, 130–136.

Small, E. (1972). Photosynthetic rates in relation to nitrogen recycling as an adaptation to nutrient deficiency in peat bog plants. *Canadian Journal of Botany, 50*, 2227–2233.

Smith, C. C., & Fretwell, S. D. (1974). The optimal balance between size and number of offspring. *The American Naturalist, 108*, 499–506.

Soons, M. B., Heil, G. W., Nathan, R., & Katul, G. G. (2004). Determinants of long-distance seed dispersal by wind in grasslands. *Ecology, 85*, 3056–3068.

Stokes, A., Atger, C., Bengough, A., Fourcaud, T., & Sidle, R. (2009). Desirable plant root traits for protecting natural and engineered slopes against landslides. *Plant and Soil, 324*(1–2), 1–30. doi:10.1007/s11104-009-0159-y

Sun, S., & Frelich, L. E. (2011). Flowering phenology and height growth pattern are associated with maximum plant height, relative growth rate and stem tissue mass density in herbaceous grassland species. *Journal of Ecology, 99*(4), 991–1000. doi:10.1111/j.1365-2745.2011.01830.x

Tautenhahn, S., Heilmeier, H., Götzenberger, L., Klotz, S., Wirth, C., & Kühn, I. (2008). On the biogeography of seed mass in Germany—distribution patterns and environmental correlates. *Ecography, 31*, 457–468.

Thompson, K., Bakker, J. P., Bekker, R. M., & Hodgson, J. G. (1998). Ecological correlates of seed persistence in soil in the NW European flora. *Journal of Ecology, 86*, 163–169.

Thompson, K., Band, S. R., & Hodgson, J. G. (1993). Seed size and shape predict persistence in soil. *Functional Ecology, 7*, 236–241.

Tjoelker, M. G., Craine, J. M., Wedin, D., Reich, P. B., & Tilman, D. (2005). Linking leaf and root trait syndromes among 39 grassland and savannah species. *New Phytologist, 167*(2), 493–508.

Venable, D. L., & Rees, M. (2009). The scaling of seed size. *Journal of Ecology, 97*, 27–31.

Verbeek, N. A. M., & Boasson, R. (1995). Flowering height and postfloral elongation of flower stalks in 13 species of angiosperms. *Canadian Journal of Botany, 73*, 723–727.

Vergutz, L., Manzoni, S., Porporato, A., Novais, R. F., & Jackson, R. B. (2012). Global resorption efficiencies and concentrations of carbon and nutrients in leaves of terrestrial plants. *Ecological Monographs, 82*(2), 205–220. doi:10.1890/11-0416.1

Vile, D., Garnier, E., Shipley, B., Laurent, G., Navas, M.-L., Roumet, C., . . . Wright, I. J. (2005). Specific leaf area and dry matter content estimate thickness in laminar leaves. *Annals of Botany, 96*, 1129–1136.

Vile, D., Shipley, B., & Garnier, E. (2006). A structural equation model to integrate changes in functional strategies during old-field succession. *Ecology, 87*, 504–517.

Violle, C., Garnier, E., Lecœur, J., Roumet, C., Podeur, C., Blanchard, A., & Navas, M.-L. (2009). Competition, traits, and resource depletion in plant communities. *Oecologia, 160*, 747–755.

Wahl, S., & Ryser, P. (2000). Root tissue structure is linked to ecological strategies of grasses. *New Phytologist, 148*, 459–471.

Waller, D. M. (1988). Plant morphology and reproduction. In J. Lovett Doust & L. Lovett Doust (eds), *Plant reproductive ecology: patterns and strategies* (pp. 203–227). New York: Oxford University Press.

Wardle, D. A., Barker, G. M., Bonner, K. I., & Nicholson, K. S. (1998). Can comparative approaches based on plant ecophysiological traits predict the nature of biotic interactions and individual plant species effects in ecosystems? *Journal of Ecology, 86*, 405–420.

Watanabe, T., Broadley, M. R., Jansen, S., White, P. J., Takada, J., Satake, K., . . . Osaki, M. (2007). Evolutionary control of leaf element composition in plants. *New Phytologist, 174*, 516–523.

Weedon, J. T., Cornwell, W. K., Cornelissen, J. H. C., Zanne, A. E., Wirth, C., & Coomes, D. A. (2009). Global meta-analysis of wood decomposition rates: a role for trait variation among tree species? *Ecology Letters, 12*, 45–56.

Weiher, E., van der Werf, A., Thompson, K., Roderick, M., Garnier, E., & Eriksson, O. (1999). Challenging Theophrastus: a common core list of plant traits for functional ecology. *Journal of Vegetation Science, 10*, 609–620.

West, G. B., Brown, J. H., & Enquist, B. J. (1999). A general model for the structure and allometry of plant vascular systems. *Nature, 400*, 664–667.

Westoby, M. (1998). A leaf-height-seed (LHS) plant ecology strategy scheme. *Plant and Soil, 199*, 213–227.

Westoby, M. (1999). Generalization in functional plant ecology: the species sampling problem, plant ecology strategy schemes, and phylogeny. In F. I. Pugnaire & F. Valladares (eds), *Handbook of Functional Plant Ecology* (pp. 847–872). New York: Marcel Dekker, Inc.

Westoby, M., & Wright, I. J. (2006). Land-plant ecology on the basis of functional traits. *Trends in Ecology and Evolution, 21*, 261–268.

Westoby, M., Falster, D. S., Moles, A. T., Vesk, P. A., & Wright, I. J. (2002). Plant ecological strategies:

some leading dimensions of variation between species. *Annual Review of Ecology and Systematics*, *33*, 125–159.

Westoby, M., Moles, A. T., & Falster, D. S. (2009). Evolutionary coordination between offspring size at independence and adult size. *Journal of Ecology*, *97*(1), 23–26. doi:10.1111/j.1365-2745.2008.01396.x

White, P. S. (1983). Corner's rules in eastern deciduous trees: allometry and its implications for the adaptive architecture of trees. *Bulletin of the Torrey Botanical Club*, *110*(2), 203–212. doi:10.2307/2996342

Wilson, P. J., Thompson, K., & Hodgson, J. G. (1999). Specific leaf area and leaf dry matter content as alternative predictors of plant strategies. *New Phytologist*, *143*, 155–162.

Wirth, C., Schumacher, J., & Schulze, E.-D. (2004). Generic biomass functions for Norway spruce in Central Europe—a meta-analysis approach toward prediction and uncertainty estimation. *Tree Physiology*, *24*, 121–139.

Witkowski, E. T. F., & Lamont, B. B. (1991). Leaf specific mass confounds leaf density and thickness. *Oecologia (Berlin)*, *88*, 486–493.

Wolkovich, E. M., & Cleland, E. E. (2014). Phenological niches and the future of invaded ecosystems with climate change. *AoB PLANTS*, *6*. doi:10.1093/aobpla/plu013

Wolkovich, E. M., & Ettinger, A. K. (2014). Back to the future for plant phenology research. *New Phytologist*, *203*(4), 1021–1024. doi:10.1111/nph.12957

Wright, I. J. (2001). *Leaf economics of perennial species from sites contrasted on rainfall and soil nutrients*. PhD, Macquarie University, Sydney, Australia.

Wright, I. J., Ackerly, D. D., Bongers, F., Harms, K. E., Ibarra-Manriquez, G., Martinez-Ramos, M., ... Wright, S. J. (2007). Relationships among ecologically important dimensions of plant trait variation in seven Neotropical forests. *Annals of Botany*, *99*(5), 1003–1015.

Wright, I. J., Reich, P. B., Cornelissen, J. H. C., Falster, D. S., Garnier, E., Hikosaka, K., ... Westoby, M. (2005). Assessing the generality of global leaf trait relationships. *New Phytologist*, *166*, 485–496.

Wright, I. J., Reich, P. B., Westoby, M., Ackerly, D. D., Baruch, Z., Bongers, F., ... Villar, R. (2004). The worldwide leaf economics spectrum. *Nature*, *428*, 821–827.

Wright, S. J., Kitajima, K., Kraft, N. J. B., Reich, P. B., Wright, I. J., Bunker, D. E., ... Zanne, A. E. (2010). Functional traits and the growth-mortality trade-off in tropical trees. *Ecology*, *91*(12), 3664–3674. doi:10.1890/09-2335.1

CHAPTER 4

Gradients, response traits, and ecological strategies

4.1 Introduction

The previous chapter has shown that species can differ considerably in the values of their traits. These differences can be partially explained by differences in the evolutionary ancestry of species (e.g. Gymnosperms have values of specific leaf area lower than those of Angiosperms: Figure 3.18), but a central hypothesis of comparative ecology is that these differences are the basis of the success of different species in habitats with contrasting environmental conditions in which they can potentially establish, persist, and reproduce. In other terms, the trait values of individuals influence their fitness (cf. Figure 2.2). This perspective, which Westoby and Wright (2006) have termed as a 'world view according to Schimper', stems from the work of Warming and Schimper at the end of the nineteenth century (see Table 1.2 in Chapter 1). In this view of the world, variations in plant functioning—which can be captured by variations in trait values—and differences in species distributions depend on variations in biotic and abiotic factors such as rainfall, temperature, soil type, grazing regime, etc., or 'gradients' of environmental factors, and have local adaptive value.

In this chapter, we will examine the relationships between traits and environment at three levels of organization: 1) the intraspecific level, where the variability of traits in response to environmental factors is analysed within a single species; 2) the species level, where the trait values of species found at different positions along environmental gradients are compared; and 3) the community level, where we show that taking into account the local abundance of species is necessary to improve the detection and understanding of the relationships between traits and the environment. A final section will deal with ecological strategies which encompass an integrative vision of the response of living organisms to environmental factors, rather than the responses of individual traits. We begin first by defining what is meant by the expression 'environmental gradient'.

4.2 Environmental gradients

The concept of gradients was introduced in ecology by Whittaker (1951, 1967). While central to comparative ecology, Shipley (2010, p. 46) notes that there does not seem to exist a precise definition of the term 'environmental gradient' in the ecological literature. Here, we will define an environmental gradient as a gradual change in a given biotic or abiotic environmental factor through space or time. In many ecological studies using gradients, these are actually not made explicit. This fuzziness in the definition of the nature of gradients is also alluded to in the remarks of Austin (1980), who identified the lack of precision and confusion in the identification and description of these gradients as one of the major weaknesses in the development of models in plant ecology.

Based on a review of the literature dealing with the responses of plants to environmental factors, Austin (1980) identified three types of gradients (see also Austin, 2013; Austin & Smith, 1989), to which a fourth one needs to be added. These are:

- *Indirect gradients*: these gradients are those in which the environmental variable used to explain the observations does not have in itself a direct physiological impact on plant functioning. This is notably the case for gradients of geographic location and orientation such as altitude, latitude, distance from shore, and aspect, while successional stage also falls within this category. Multiple environmental variables vary across such spatial or temporal gradients, some of which have a direct impact on plant functioning, while others do not. It is thus essential to determine as precisely as possible which of the environmental variables involved have a direct effect on plant response. In the case of an altitudinal gradient, important variables could include air temperature, soil water content, and the oxygen partial pressure; in the case of a secondary succession, important variables could include soil organic matter content and the amount of light arriving at the soil surface. Austin (1980) emphasized that the extrapolation of results from studies on these types of gradients is not straightforward, as in general they depend strongly on the local conditions of the study zone.

- *Resource gradients*: in such gradients, the factor which varies is a resource consumed by a plant, and which is required for plant growth and reproduction. A resource is an environmental variable which is consumed or used by living organisms such that their quantity and/or quality is altered by the activity of these organisms (e.g. Tilman, 1982). For autotrophic organisms, such resources are quite limited in number and an exhaustive list can be established: light, water, carbon dioxide, oxygen, and the essential mineral elements, of which there are 14 (Epstein & Bloom, 2005). The response of plants to resource gradients is generally taken to be of a limiting response type with a zone in which functioning begins to be hampered beyond a threshold considered to be toxic (Figure 4.1a).
- *Direct gradients*: such gradients are those that have a direct impact on plant growth and functioning, without being consumed. The most frequently studied examples of such gradients are air temperature and soil pH. Responses of plants to these types of factors show optima for functioning which can be offset between differing species (Figure 4.1b).

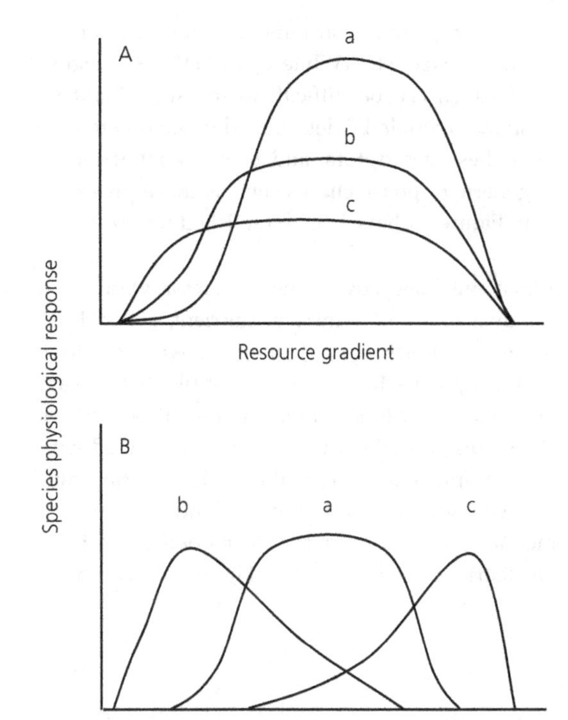

Figure 4.1 Theoretical examples of plant responses to gradients of environmental factors for three species (a, b, and c). (A) resource gradient, (B) direct gradient. Taken from Austin & Smith (1989). Reproduced with permission from Springer.

- *Disturbance gradients:* on disturbance gradients, the factor which varies leads to a destruction of plant biomass by varying magnitudes (Grime, 2001; White & Pickett, 1985). According to White and Pickett (1985), a disturbance is a discontinuous process in time which results in a modification of the structure of an ecosystem, community, or population leading to changes in the availability of resources or substrates, or in the physical environment. This type of gradient was not explicitly taken into account in the classification proposed by Austin and Smith (1989), but is further discussed by Austin (2013). This biomass destruction can be a result of numerous factors such as, for example, grazing, fire, trampling, agricultural practices, pathogenic organisms, or even extreme climatic conditions. The characterization of a disturbance regime potentially requires the quantification of numerous descriptors relative to its frequency, distribution, predictability magnitude, size, etc. (White & Pickett, 1985), some of which can be difficult to measure. Plant responses should be determined in relation to each of these descriptors, and to our knowledge no generic response curves such as those presented in Figure 4.1 have been established for these.

When only one environmental factor varies (e.g. the availability of a mineral nutrient, soil pH, intensity of attack by a pathogen), we refer to it as a 'simple gradient'. When multiple factors vary jointly, we refer to a 'complex gradient' (Whittaker, 1978): this is notably the case for indirect gradients. An example of a conceptual model of relationships between resources and direct and indirect environmental gradients used for the modelling of habitat distributions is shown in Figure 4.2 (Guisan & Zimmermann, 2000). It stresses the many different connections that can exist between variables that have direct and indirect effects on plant growth, functioning, and distribution.

Variation in the performance of organisms along different environmental gradients allows for the identification of their 'realized' and 'fundamental' niches, defined as the n dimensional hypervolumes (relating to n gradients) within which an organism can survive and reproduce, with and without species interactions, respectively (Hutchinson, 1957). For plants, it is the 'realized niche' of the organism which is most often quantified. This niche depends on the plant's performance when it interacts with other organisms present, and is often derived from the distribution of species. The fundamental niche has rarely been quantified for plants, because of the difficulty of assessing the performance of a species

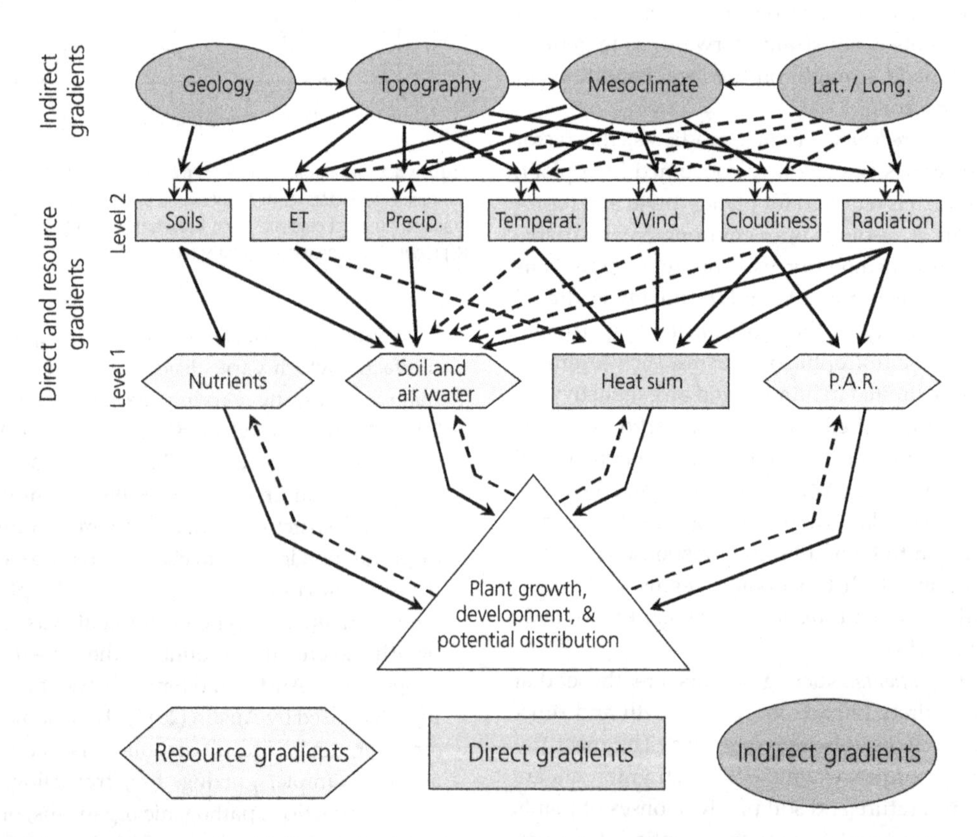

Figure 4.2 Example of a conceptual model of relationships between resources and direct and indirect environmental gradients used in habitat distribution modelling. Abbreviations: Lat.: latitude; Long.: longitude; Precip.: precipitation; Temperat.: temperature; P.A.R.: photosynthetically active radiation. Taken from Guisan & Zimmermann (2000). Reproduced with permission from Elsevier.

over continuous gradients under controlled conditions (Violle & Jiang, 2009, and references therein).

In following sections, we will examine plant responses to differing types of gradients, using traits as indicator response variables of their functioning. If we consider that functional traits, or at least some of them, can be used as proxies for measuring the performance of an individual (Figure 2.2, in Chapter 2, and discussed in Ackerly & Monson, 2003; Craine et al., 2012), such relationships also provide information on the realized niche of the organisms being considered (McGill et al., 2006; Violle & Jiang, 2009).

In the next two sections, we discuss successively the three components of changes in trait values along environmental gradients: those that result from intraspecific variation, those that result from species' intrinsic differences, and those that result from changes in species abundance in communities.

4.3 Environmental factors and trait intraspecific variability

Each species displays some variability in response to environmental factors. The intraspecific variability of traits is the overall variability of traits and of suites of traits expressed by the individuals within a species (Box 4.1). This variability corresponds to the capacity of a species to respond to variations in environmental factors via two complementary mechanisms: (1) genetic variability, which is the phenotypic variability between individual genotypes; and (2) phenotypic plasticity, which represents the potential of each genotype to produce different phenotypes under differing environmental conditions (Albert et al., 2011, and references therein). It is the second of these mechanisms that we will discuss in this section.

Box 4.1 Intraspecific trait variability: definition, mechanisms, and types

(Albert et al., 2011; Lambers et al., 1998; Schilchting, 2002; Violle et al., 2012.)

Intraspecific trait variability corresponds to the range of trait attributes and trade-offs which are expressed by a given species. It results from two processes, genetic differentiation and phenotypic plasticity. The first corresponds to the phenotypic variability expressed by different genotypes of the same species (genotypic polymorphism). Often presented as adaptive and often discussed in the evolutionary literature, it obeys evolutionary processes (genetic drift, mutation, selection, migration, and represents the base material for the evolution of populations and species). Phenotypic plasticity represents the ability of each genotype to express different phenotypes in response to variations in environmental conditions, whether these are predictable or unpredictable.

Intraspecific variability is a result of two types of processes and of their interaction. Thus intraspecific variations measured over time after a change in environmental conditions are first of a phenotypic nature, and then of a genotypic nature. The phenotypic response is expressed more rapidly for physiological traits than for morphological or phenological traits. If the change persists, acclimation, which allows for a return to activity at the initial level, remains a phenotypical change. This can be accompanied by an adaptive modification of the genotype, if this is associated with a modification of fitness.

The intraspecific variability of an individual can correspond to differences in expression during its development (ontogeny) or to plastic responses to spatial or temporal variation in environmental conditions. These differences can be predictable, such as those experienced during the course of a day (circadian rhythm) or between seasons (phenological variations), or otherwise unpredictable, for example after an attack by a parasite or predator.

The intraspecific variability of the individuals of a population is defined as its genotypic polymorphism and the range of phenotypic plasticity shown by the different genotypes (Box 4.1 Figure 1).

Intraspecific variability among populations reflects differences in genotypic composition but also in the degree of plasticity. Populations with a given degree of intraspecific variability can include a small number of highly plastic genotypes, a large number of genotypes with low levels of plasticity, or an intermediate number of genotypes with an intermediate level of plasticity (Box 4.1 Figure 1). Even if their potential responses to environmental factors are equivalent, these populations show distributions of trait values that differ for a given environment and show variable responses to environmental variation over short and medium time scales.

continued

Box 4.1 *Continued*

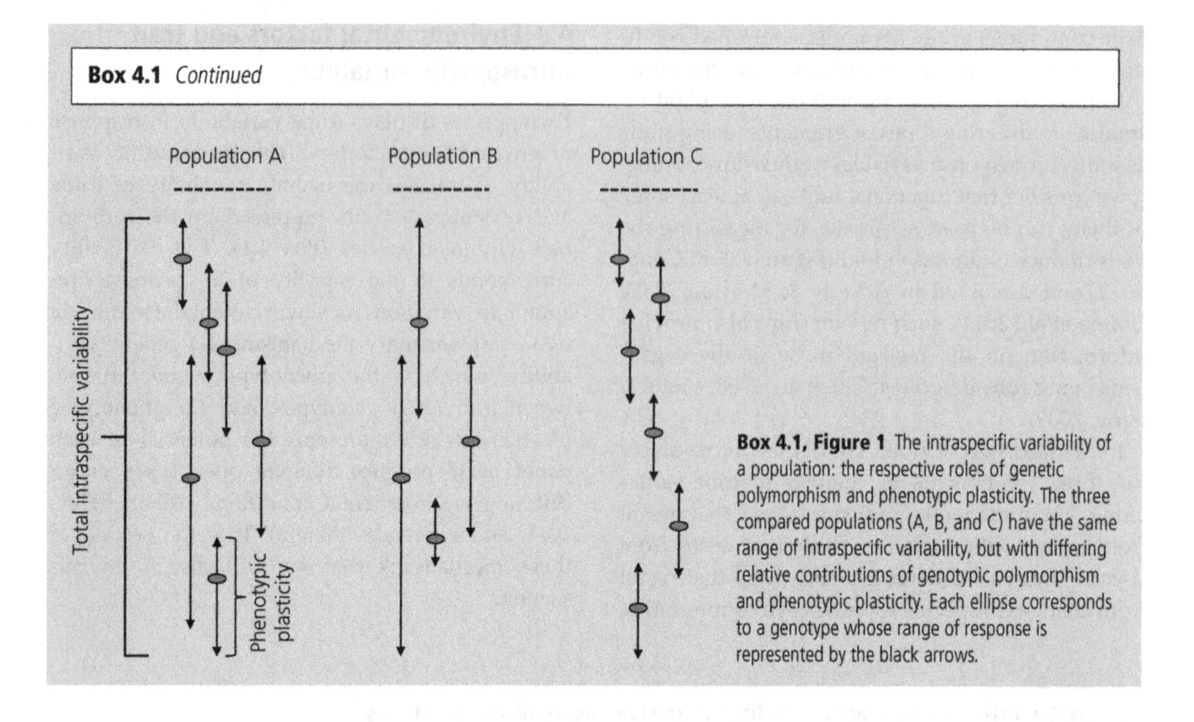

Box 4.1, Figure 1 The intraspecific variability of a population: the respective roles of genetic polymorphism and phenotypic plasticity. The three compared populations (A, B, and C) have the same range of intraspecific variability, but with differing relative contributions of genotypic polymorphism and phenotypic plasticity. Each ellipse corresponds to a genotype whose range of response is represented by the black arrows.

At this stage, two symmetric considerations need to be taken into account. First, there is no reason for trait responses to a particular environmental factor to be the same for all traits (which is also true at the interspecific level: cf. Table 4.1): not all traits are equally plastic. For example, in a study involving four grass species, root diameter was found to vary very little in response to nitrogen availability, in contrast to specific root length (the ratio between root length and root mass: Robinson & Rorison, 1988). Second, there is no reason either that different environmental factors should have the same effect on any given trait; for example, in *Rubia peregrina*, a decrease in the availability of mineral nutrients leads to a decrease in internode length, while in contrast a decrease in the availability of light leads to an increase in this same length (Navas & Garnier, 2002). Consequently, the study of intraspecific trait responses must be carried out for each particular trait × environmental factor combination. This requires a considerable amount of knowledge. If data describing the response of certain traits to environmental factors certainly exist in the literature, no global review is currently available. This observation has recently led to the development of meta-phenomics (Poorter et al., 2010), the objective of which is precisely to establish in a systematic manner the intraspecific response curves of different traits to environmental factors ('dose-response curves') by developing a methodology derived from meta-analyses[1] (Gurevitch et al., 2001).

In the context of this section, we will illustrate the phenotypic component of intraspecific variability (Box 4.1) for three traits relevant to the fundamental axes of variation identified in the previous chapter—specific leaf area, plant height, and seed mass—concentrating in particular on their response to two environmental factors: nutrient availability and light availability. We then compare the intraspecific variability of 12 traits measured in different types of environmental conditions.

The first explicit meta-phenomic study focused on leaf mass per area (LMA, the inverse of specific leaf area, cf. Box 3.2 in Chapter 3; Poorter

[1] A meta-analysis can be defined as the statistical synthesis of results obtained in different studies (Gurevitch et al., 2001).

et al., 2009). This study, which compiled 6100 data records obtained from more than one hundred experiments, showed that LMA varied (i) strongly in response to light availability, temperature, and leaf submersion, (ii) moderately to atmospheric CO_2 concentration, availability of water, and mineral nutrients, and (iii) marginally to six other studied factors (Poorter et al., 2009). Figure 4.3 presents two of these dose-response curves for values of specific leaf area: it shows a moderate increase in the SLA in response to the availability of mineral nutrients, and a major decrease in response to light availability, the factor to which the trait is the most sensitive.

In the absence of comparable reviews for other traits (however see Poorter et al., 2012, for a study of biomass fractions in plants), we present here the intraspecific variability in plant height and seed mass obtained from a common garden experiment where the effects of nitrogen availability on the traits of 18 species drawn from contrasting stages of a secondary succession in the Mediterranean region were studied (Fortunel et al., 2009; Kazakou et al., 2007). The data show that an increase in the availability of nitrogen leads to increases in height for certain species, while for others, the opposite occurs (Figure 4.4a); the global effect on all of the species together is non-significant, but the interaction between species × nitrogen availability is significant (see Boomsma et al., 2009; Hejcman et al., 2012, for other studies). Seed mass shows very little plasticity in response to nitrogen availability for the 18 studied species (Figure 4b, and see Eckstein & Otte, 2004; Hejcman et al., 2012, for other studies).

These 18 species grown in monocultures generated a gradient of light availability that was used to assess the effects of this factor on the traits of individuals of the target species *Bromus madritensis* (an annual grass) grown within each of the monocultures (Violle et al., 2007, 2009). An increase in the amount of light received under the canopies led to a strong decrease in the height of the target plants, regardless of the level of nitrogen fertilization (Figure 4.5a, and see Poorter, 1999; Rice & Bazzaz, 1989, for other studies), while the seed mass of these same target plants varied very little

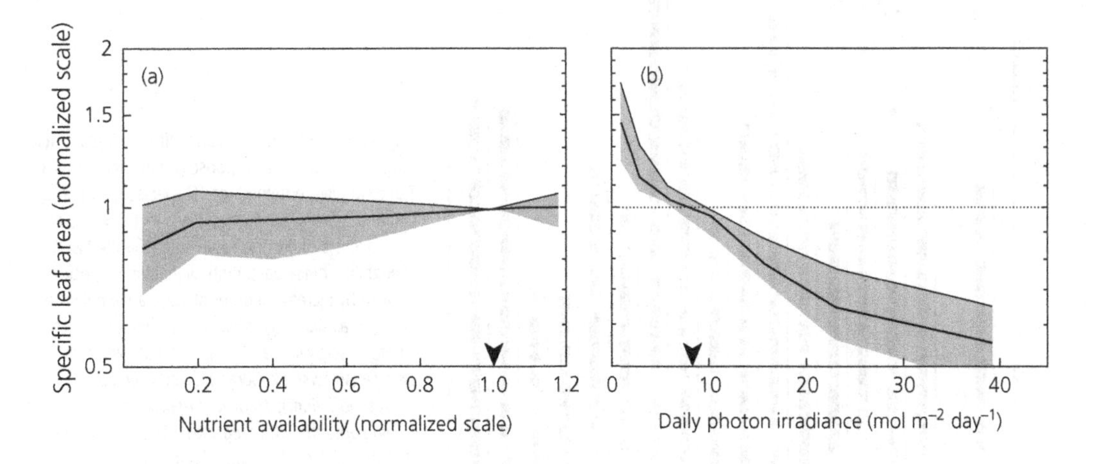

Figure 4.3 Examples of 'dose-response' curves of specific leaf area in response to (a) the availability of mineral nutrients (on a normalized scale), (b) incoming light level. The synthetic curves have each been established on the basis of 720 (panel a) and 1050 (panel b) observations drawn from more than one hundred experiments. For each of these graphs, the solid black line indicates the median response curve and the grey zone represents the inter-quartile interval between the 25th and 75th percentiles. The black arrow indicates the reference value used to normalize the data, which was required to enable comparisons between the different studies. The trait value is normalized to 1 at this reference value. For the response to the availability of mineral nutrients, this value is that at which the maximal biomass of individuals in reached in each experiment; for the response to light this value is fixed at 8 mol m^{-2} day^{-1}, as this value is comprised within the range of irradiances for most of the screened experiments. For more details, see Poorter et al. (2010). Data for the curves: http://www.metaphenomics.org/ MP41a_index.html. Taken from Poorter et al. (2010). Reproduced with permission from Oxford University Press.

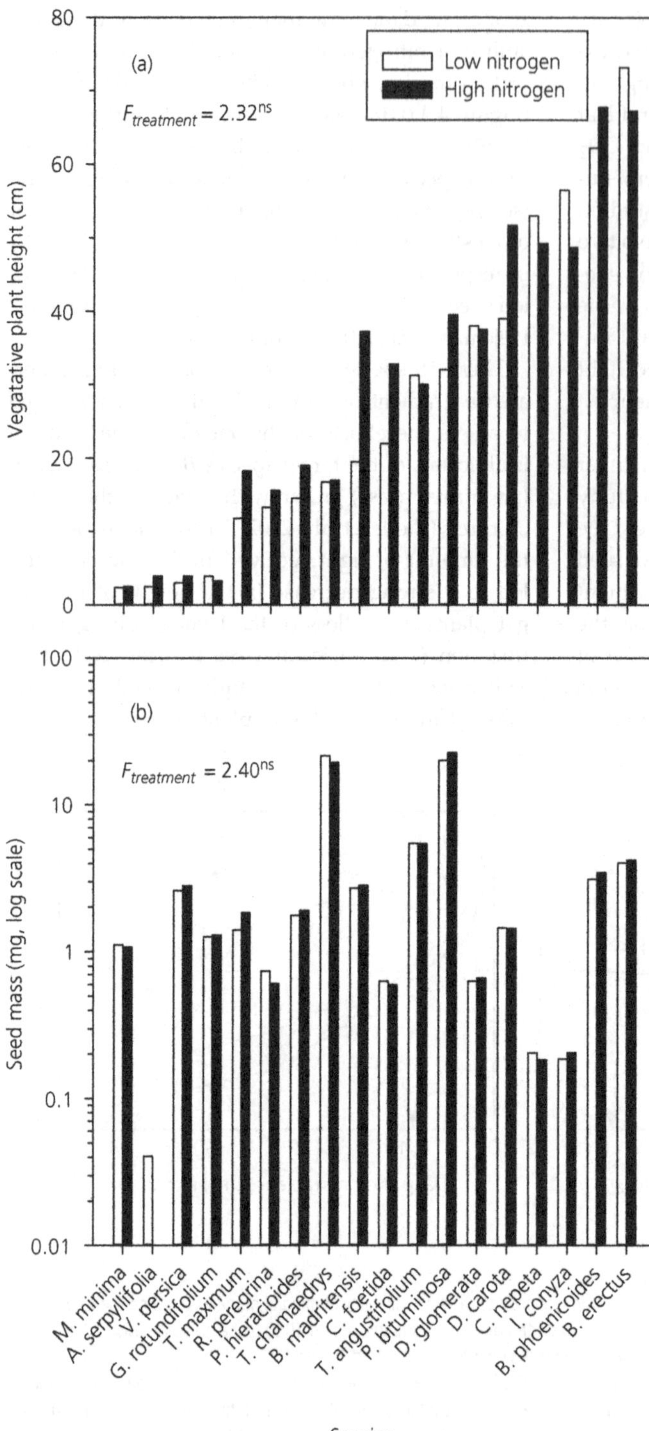

Figure 4.4 Intraspecific variability in (a) plant height and (b) seed mass in response to nitrogen fertilization. The data were obtained for 18 species grown in a common garden as monocultures in 1.2 × 1.2 m plots at two levels of nitrogen availability (white bars: low availability; black bars: high availability). Species (ranked by increasing order of vegetative height at low nitrogen availability): *Medicago minima* (*M. minima*); *Arenaria serpyllifolia* (*A. serpyllifolia*); *Veronica persica* (*V. persica*); *Geranium rotundifolium* (*G. rotundifolium*); *Tordylium maximum* (*T. maximum*); *Rubia peregrina* (*R. peregrina*); *Picris hieracioides* (*P. hieracioides*); *Teucrium chamaedrys* (*T. chamaedrys*); *Bromus madritensis* (*B. madritensis*); *Crepis foetida* (*C. foetida*); *Trifolium angustifolium* (*T. angustifolium*); *Psoralea bituminosa* (*P. bituminosa*); *Dactylis glomerata* (*D. glomerata*); *Daucus carota* (*D. carota*); *Calamintha nepeta* (*C. nepeta*); *Inula conyza* (*I. conyza*); *Brachypodium phoenicoides* (*B. phoenicoides*); *Bromus erectus* (*B. erectus*). Data taken from Fortunel et al. (2009).

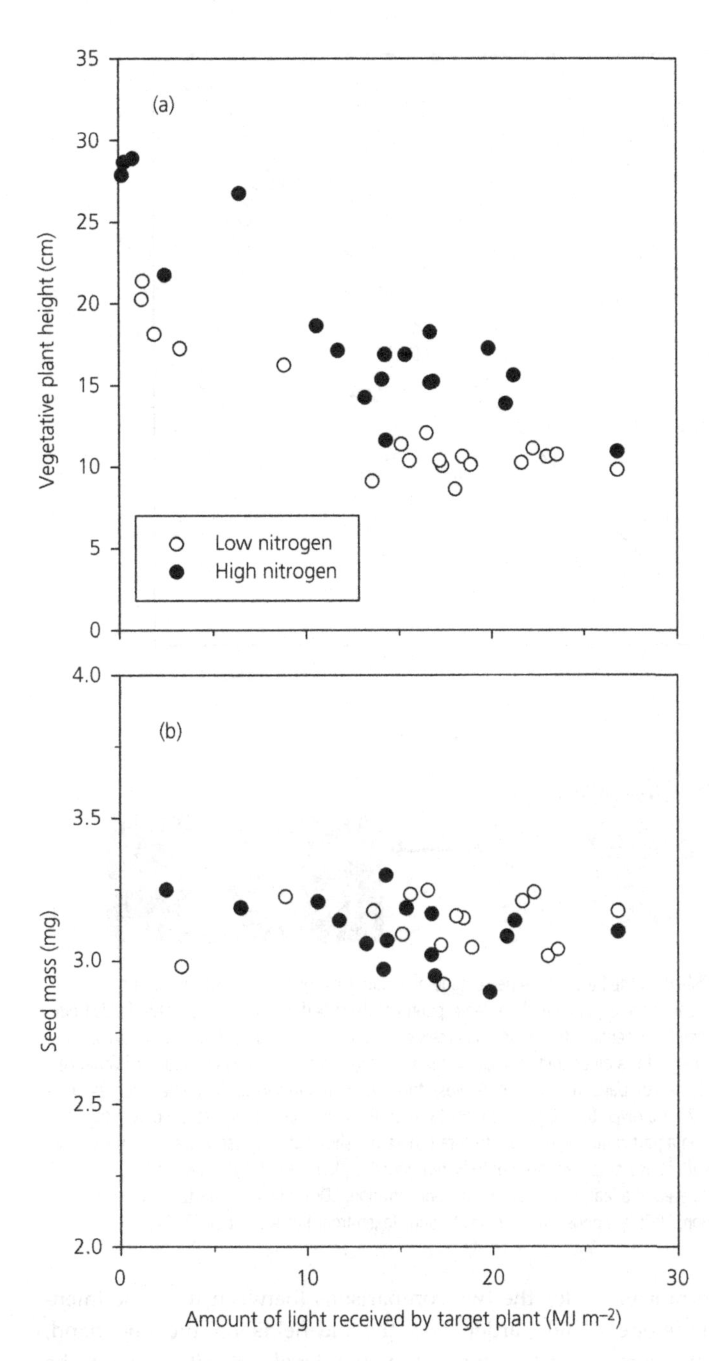

Figure 4.5 Intraspecific variability of (a) plant height and (b) seed mass, in response to light availability in *Bromus madritensis* (an annual grass) grown in an experimental garden at two levels of nitrogen fertilization, in each of the monocultures of 18 species described in Figure 4.4. White points: low availability; dark points: high availability. Data taken from Violle et al. (2009).

(Figure 4.5b, and see Galloway, 2001; Sultan, 2001, for other studies).

Certain traits, finally, appear to be more variable than others in response to a given environmental factor. For example, specific leaf area and plant height are very sensitive to variations in light availability, while this is not the case for seed mass. To what extent can we generalize this conclusion? To answer this question, the intraspecific variability of 12 traits was assessed for 18 species from

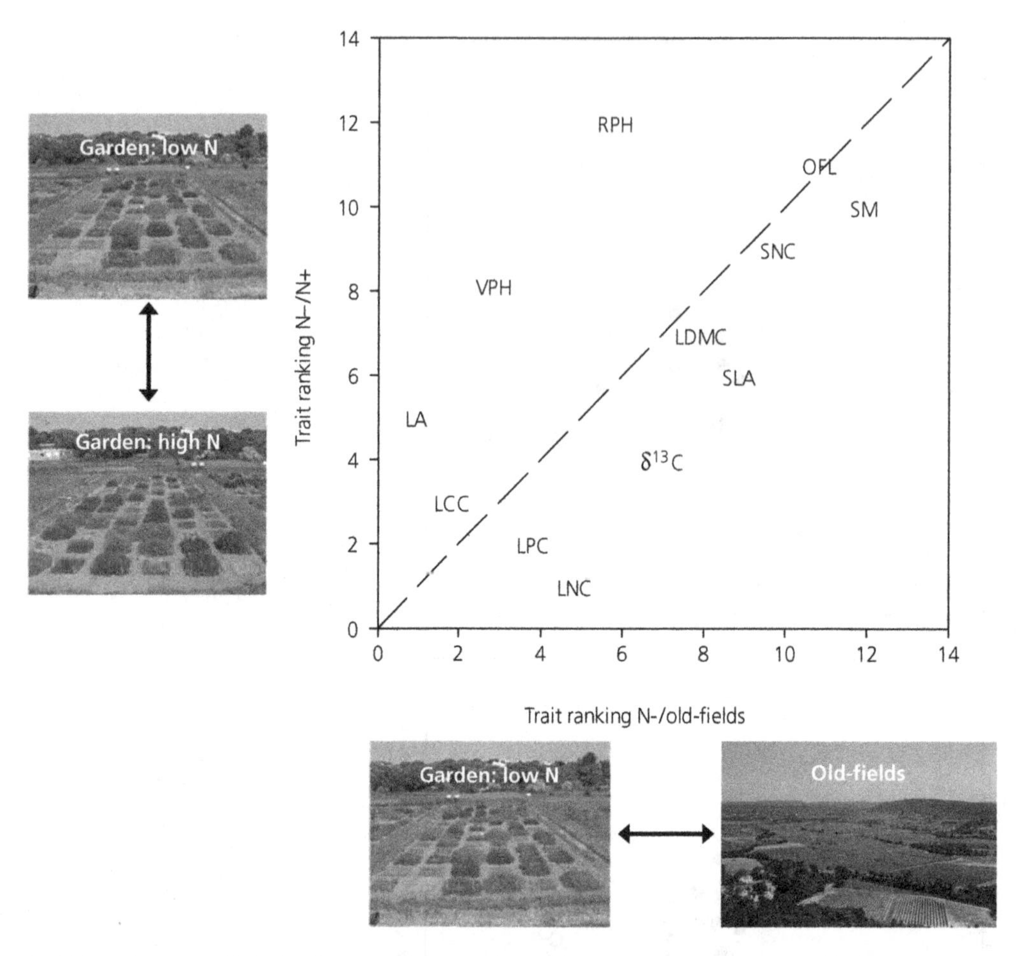

Figure 4.6 Relationship between the trait rankings established on the basis of the percentage of variance assigned to the species in the comparisons conducted between 1) the low nitrogen treatment of the experimental garden experiment described in Figure 4.4 and the fields from which these species originate (x-axis), and 2) the two nitrogen treatments of the common garden experiment described in Figure 4.4 (y-axis). Twelve traits were involved in this study: a classification rank of 1 for a given trait (e.g. LA on the x-axis) indicates that the percentage of variance explained by the species is the lowest for this trait in a particular comparison (e.g. N-/old-fields); this trait therefore shows the highest degree of intraspecific variability. By contrast, a classification rank of 12 for a given trait (e.g. seed mass [SM] on the x-axis) indicates that the percentage of variance explained by the species is the highest for this trait in a particular comparison; this trait therefore shows the lowest degree of intraspecific variability. Trait abbreviations: OFL: flowering date; $\delta^{13}C$: leaf ^{13}C isotopic composition; RPH: reproductive plant height; VPH: vegetative plant height; SM: individual seed mass; SLA: specific leaf area; LA: area of a leaf; LCC: leaf carbon concentration; LDMC: leaf dry matter content; LNC: leaf nitrogen concentration; SNC: seed nitrogen concentration; LPC: leaf phosphorus concentration. Taken from Kazakou et al. (2014).

abandoned vineyards located in the Mediterranean region of southern France in two ways: on the one hand, trait values from field-growing plants were compared to those from the garden experiment described in the previous paragraph (low nitrogen level), and on the other hand, trait values were compared for plants grown under low and high nitrogen availabilities (Kazakou et al., 2014). Traits were then ranked in order of increasing variability for the two comparisons (between the experimental garden and the old-fields on the one hand, and between the two levels of nitrogen on the other hand: Figure 4.6). This study allowed us to identify:

- a group of highly variable traits: the area of a leaf, leaf carbon, phosphorus and nitrogen concentrations, as well as their carbon isotope composition;

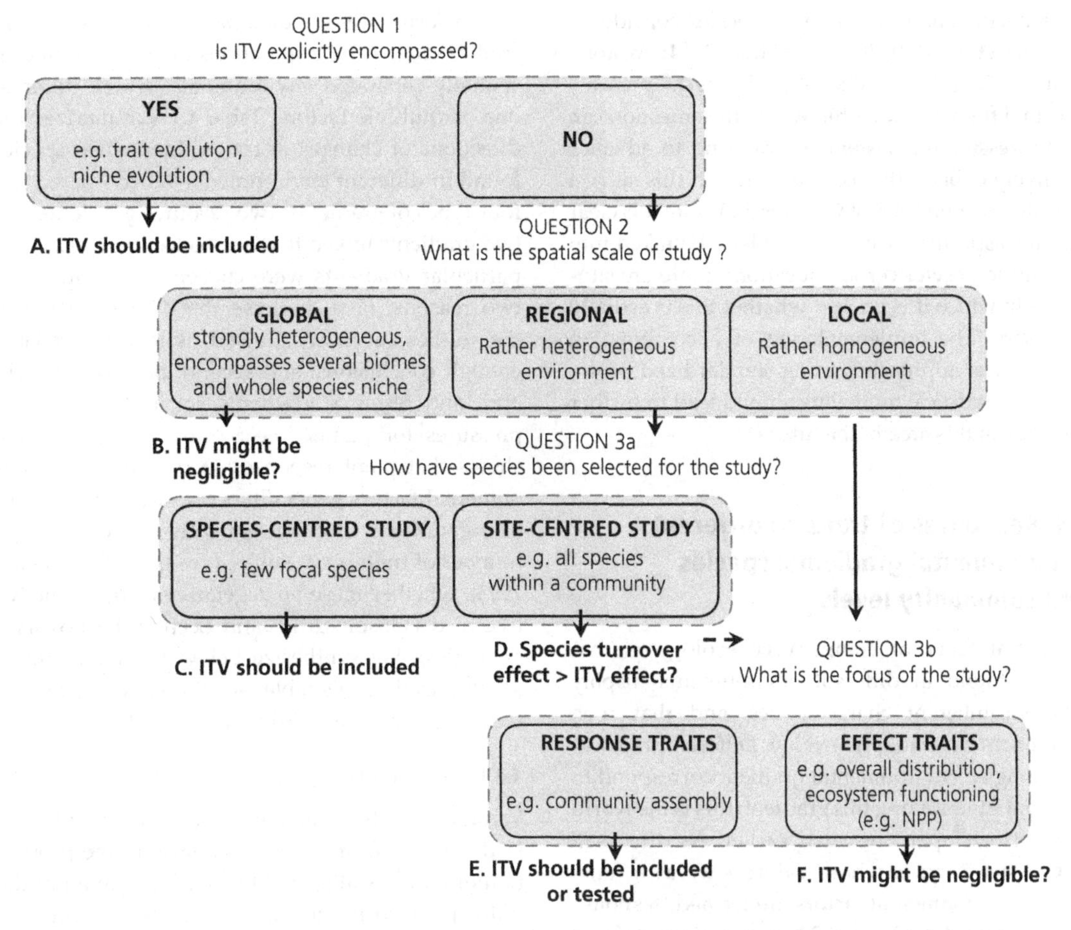

Figure 4.7 A formal framework to be used in deciding in which cases intraspecific trait variability (ITV) needs to be taken into account in ecological studies. This framework is organized around four sequential questions (Qi, symbolized by different levels of grey) which allow for the case-by-case determination of whether intraspecific variability can be ignored in any given study. NPP: net primary productivity of an ecosystem (see Box 6.1 of Chapter 6). Taken from Albert et al. (2011). Reproduced with permission from Elsevier.

- a group of traits of intermediate levels of variability: vegetative height, leaf dry matter content, and specific leaf area; and
- a group of traits showing low variability: reproductive height, seed nitrogen concentration, seed mass, and finally flowering date.

Interestingly, this classification of traits remains largely similar for the two types of comparison (N- in the garden / fields and N+/N- in the experimental garden), generating the positive relationship in Figure 4.6. While it supports the results of other more limited studies (see the discussion in Kazakou et al., 2014), it is still necessary to extend such studies to other ecosystems and other types of environmental factors. For instance, one would expect vegetative height to be much more variable in landscapes with strong but heterogeneous grazing regimes. In this context, further development of the meta-phenomics approach as discussed above would provide essential additional elements.

As this section shows, the response of plants to environmental factors involves a component of intraspecific variability. How important it is to take into account this variability depends greatly on the question being asked (Albert et al., 2011; Messier et al., 2010), and on the relative part of the intraspecific component of variation as compared to the interspecific one in the pool of species considered (Garnier et al., 2001; Kazakou et al., 2014; Westoby et al., 2002, and see Section 4.4). Figure 4.7 presents a formal framework within which this question can be addressed in a systematic way, and to advance the lively debate still occurring around this subject (see also Section 5.4.5 of Chapter 5). While it is clear that intraspecific variability needs to be taken into account to answer certain questions, there are situations in which it is unclear whether this is actually necessary. The implementation of a combination of empirical approaches using standardized protocols and statistical modelling should lead to further progress in this area in the future.

4.4 Responses of traits to different environmental gradients: species and community levels

A central tenet of comparative ecology is that species found in different environments display different inherent characteristics, and that such differences have adaptive value. Different trait values along an environmental gradient correspond to different fitness along this gradient: this is the world view of Schimper presented earlier. As discussed in Chapter 2, traits whose values vary as a function of environmental factors are termed 'response traits' (Lavorel & Garnier, 2002). These are defined at the species level, and will be presented first. Individuals of species usually do not grow isolated in natural environments however. We will thus introduce the idea that taking into account the structure of plant communities, defined as the individuals of all species that potentially interact within a single patch or local area of habitat (e.g. Leibold et al., 2004), is important for a better assessment of the response of vegetation to the environment, beyond that of a single species.

4.4.1 Comparing species

The recurrence of relationships observed between concomitant changes in environmental factors and trait values has allowed for the identification of the principal traits responding in a consistent manner to differing types of environmental gradients. The majority of these data has been obtained from studies comparing populations of species found in spatially separated environments, which differ in one or multiple factors. Table 4.1 summarizes the directions of changes in trait values when species found in different environments are compared, for four types of gradients: two resource gradients and two gradients linked to disturbance regimes. These particular gradients were chosen as examples for two reasons. First, because the data available in the studies surveyed allowed us to reach reliable enough conclusions; and second, because disturbance and resource gradients act as key selective pressures for plants (see Section 4.5). This table shows that plant responses to variation in environmental factors generally occur at multiple levels of organization and consequently concern a large number of traits, or a 'suite of traits' (Chapin et al., 1993), whether these be vegetative or reproductive (see also Gillison, 2013, and Section 4.5 on plant strategies). We synthesize below the main trends summarized in this table (see the table caption for the main references used in these syntheses).

4.4.1.1 Nutrients

Variation in the availability of mineral nutrients leads to modifications in plant size, in the proportion of biomass allocated to roots, and in all of the traits involved in the leaf economics spectrum described in Chapter 3. In terms of adaptations, plants growing in nutrient-poor environments have traits increasing their ability to access soil nutrients (increases in root biomass per unit of plant biomass; root symbioses such as ecto- or ericoid mycorrhiza or N-fixing root nodules) and increasing the efficiency of use and retention of acquired resources (higher photosynthetic nitrogen use efficiency, the ratio of leaf photosynthetic rate to leaf nitrogen concentration, and longer leaf life span: cf. Box 3.4 in Chapter 3). However, traits linked to reproduction appear to be little affected.

4.4.1.2 Light

From a physiological point of view, shade tolerance is defined as the minimal light level below which a plant can no longer maintain a positive carbon balance for survival. The 'carbon gain' hypothesis links this tolerance to a combination of

Table 4.1 Responses of a selection of traits reflecting generalized adaptations to four types of environmental gradients. This table shows the directions of changes in trait values when species found in different environments (e.g. low vs high nutrient availability) are compared. **↗** and **↘** indicate respectively an increase or a decrease in the value of a trait along the gradient (the thickness of the arrows gives an indication of the degree of confidence in this variation). ↔ indicates an absence of variation. ↘ or ↗ indicates that the direction of variation differs amongst studies. Empty cells indicate a lack of information. Key: herb: herbaceous plant; shrb: shrub; tree: tree; A: annual; HP: herbaceous perennial; WP: woody perennial; W,B: wind, bird ; G,M: gravity, mammal. These syntheses are based on: Chapin (1980), Chapin et al. (1993), Aerts & Chapin (2000), Grime (2001), and Craine (2009), for nutrients; Björkman (1981), Givnish (1988), Kitajima (1994), Walters & Reich (1996), and Valladares & Niinemets (2008), for light; Huston & Smith (1987), Bazzaz (1996), Prach et al. (1997), Grime (2001), and Navas et al. (2010), for succession; Briske & Richards (1995), Duru et al. (1998), McIntyre et al. (1999), Hodgson et al. (2005) and Díaz et al. (2007), for grazing.

Trait		**Environmental gradient**			
Type	Name	Nutrients low → high	Light shade → sun	Secondary succession early → late	Grazing weak → intense
Whole plant	Growth form			herb → shrb → tree	erect → prostrate
	Life cycle			A → HP → WP	HP → A
	Height at maturity	**↗**	↘ or ↗	**↗**	**↘**
	Longevity	↗	**↗**	**↗**	**↘**
	Clonality				↗
	Potential relative growth rate	**↗**	↘ or ↗	**↘**	
	Root mass fraction	**↘**	**↗**	**↗**	
Leaves	Maximum photosynthetic rate per unit mass	**↗**	**↗**	**↘**	
	Light compensation point		**↗**	**↘**	
	Respiration rate	**↗**	**↗**	**↘**	
	Stomatal conductance	**↗**	**↗**	**↘**	
	Specific leaf area	**↗**	↘ or ↗	**↘**	**↗**
	Leaf area	**↗**	**↘**	↗	**↘**
	Tissue density or leaf dry matter content	**↘**	↘ or ↗	**↗**	**↘**
	Leaf nitrogen or phosphorus concentration	**↗**	↘ or ↗	↘ or ↗	
	Leaf longevity	**↘**	**↘**	**↗**	↘
Stems and roots	Stem or root density		**↘**	**↗**	
	Nutrient acquisition rate	**↗**		↘	
	Specific root length	↘ or ↗			
	Root tissue density	**↘**			
	Fine root diameter	↘ or ↗			
	Rooting depth			**↗**	**↘**
	Root longevity	↘ or ↗			
Reproduction and regeneration	Seed dispersal mechanism			**W,B → G,M**	
	Seed mass	↔		**↗**	**↘**
	Seed number	**↗**	**↗**	**↗**	
	Flowering date	↔		↗	**↘**

the maximization of light capture and of its use for photosynthesis, and the minimization of carbon losses through maintenance respiration. An extension of this hypothesis postulates the existence of a trade-off between growth and survival, which is evidenced by a negative correlation between the growth rate in sunlight and survival under conditions of deep shade. Numerous other traits are involved in the response of plants to light (Table 4.1, and see Valladares & Niinemets, 2008, for an extensive review). In particular, while numerous plants can tolerate low light levels, only a few can reproduce under such conditions. A true definition of shade tolerance should therefore take into account the whole life cycle of the plants, from seedling survival to the reproductive stage of the adult plant. A second hypothesis called 'stress tolerance' has been proposed to explain the tolerance of species to extremely shaded environments: this is based on the maximization of resistance to biotic and abiotic constraints found in the understory, which involves traits which can differ from those related to the hypothesis of the maximization of carbon gain (Kitajima, 1994). These two hypotheses have been sometimes interpreted as mutually exclusive, but it appears that taking into account aspects relative to the life cycle and ontogeny of individuals allows for the resolution of some of the contradictions that have been identified (see Valladares & Niinemets, 2008, for more details). Finally, trait plasticity (see Box 4.1 and Section 4.3) is generally lower in shade-tolerant plants, but this effect is not necessarily found at all levels of organization (leaf or whole plant): for example, leaf physiological traits are more plastic in light-adapted plants, but internode length is more plastic in plants adapted to intermediate levels of light.

4.4.1.3 Secondary succession

Secondary vegetation succession is the recovery process of vegetation following total or partial destruction of a pre-existing plant community (e.g. Bazzaz, 1996; Lepart & Escarré, 1983). During the course of this recovery, annual species are progressively replaced by perennial species, initially herbaceous and then later woody, leading to increases in the size of plants and consequent increases in competition for light. Species with rapid rates of growth and resource acquisition are replaced by species with slower growth rates, leading to consequent changes in the entire suite of traits linked to the leaf economics spectrum, described in Chapter 3. Finally, species characteristic of late successional stages tend to flower later and produce more seeds (but with a lower ratio of seed biomass / total plant biomass) of a larger size than species from earlier successional stages. Overall, these modifications correspond to the progressive replacement of early successional species whose rapid growth allows the production of large numbers of seeds favouring their dispersal in disturbed or open habitats, by competitive species with slower growth rates which conserve the mineral nutrients that they acquire, and whose larger seeds lead to an increase in survival of seedlings when the vegetation canopy closes.

4.4.1.4 Grazing by large herbivores

Large herbivores consume and/or destroy biomass directly via defoliation or indirectly via trampling: they also influence the availability of mineral elements in space and time via the production of faeces. Tolerance and escape are the two types of responses plants can adopt in the face of grazing, which differ in the types of traits involved. A high grazing pressure favours annuals at the expense of perennials, small plants as opposed to large plants, a prostrate growth form as opposed to an upright growth form, rosette or stoloniferous species as opposed to tussocks, and plant phenology in general with in particular the date of leaf senescence being influenced by the date of defoliation. Grazing also tends to favour species with a high specific leaf area and low leaf dry matter content, which flower early and produce small seeds. It is important to note that the response of certain traits to modifications in grazing pressure is not linear, which makes generalizations sometimes difficult. The type of herbivore also plays a role, as different animals have different dietary specificities as a function of their size and their anatomical particularities: plants of low digestibility and with tough leaves are selectively grazed by cows and in particular horses, while these are ignored by sheep, which prefer legumes (Ginane et al., 2008). Chapter 8 will treat the effects of grazing in greater detail, and more generally those of some selected agricultural practices.

Table 4.2 This table shows a selection of studies in which trait values are compared across species found in different positions along various environmental gradients (e.g. species found in environments with high and low water availability). In these studies, measurements were conducted on individuals growing either in their natural environment or in controlled conditions. A list of abbreviations and their meaning is given in the Table of Abbreviations (p. xx). Subscripts 'area' and 'mass' refer to the area- and mass-based expressions of the trait, respectively.

Environmental Factor	Number and type of species	Traits studied	Reference
Water availability	20 shrubs from Californian chaparral	Water potential; 11 traits related to gas exchange and leaf function; 9 traits related to leaf and twig morphology; 3 traits related to wood and xylem properties; 3 traits related to canopy architecture; seed mass; 3 traits related to leaf phenology	Ackerly (2004)
	7 populations of 5 species (4 grasses, 1 legume)	SLA; LDMC; LA; plant height; aboveground biomass; rooting depth; leaf elongation rate; tissue water loss; survival	Volaire (2008)
	105 species across 50 sites in the Netherlands	SLA; LNC; LPC; LA; plant height; stem specific density	Ordoñez et al. (2010)
Temperature	52 herbaceous species from central England	DNA content; resistance to frost	MacGillivray & Grime (1995)
	4 vascular plant species dominant in Alaskan tundra	Above- and belowground biomass; shoot mass; tillering rate; maximum leaf elongation rate; number of active meristems; leaf and stem turnover; leaf length; shoot nitrogen and phosphorus contents; LNC; photosynthesis	Chapin & Shaver (1996)
	22 North American trees	Plant height growth; bud freezing tolerance	Loehle (1998)
	Between 9 and 16 herbaceous and woody species (depending on trait)	RGR; LAR; NAR; SLA; LMF; RMF; StMF; root/shoot ratio; carbon concentration	Atkin et al. (2006)
Salinity	Variable number and type (general synthesis)	Life form; growth form; photosynthetic pathway; seed size; seed germination; dispersal; pollination mode	Deil (2005)
	12 plant species from south-western USA salt marshes	Plant height; leaf number; leaf length; leaf width; leaf thickness; leaf volume; stem diameter; internode length; plant base diameter; number of primary branches; number of secondary branches	Richards et al. (2005)
	2 Mediterranean evergreen shrubs	7 traits related to ionic and water relations; 7 traits related to photosynthesis and growth; 8 morpho-anatomical and optical traits; 5 traits related to lipid peroxidation and polyphenol metabolism	Tattini et al. (2006)
	3 species from marshes of northern Belize	Above- (leaf or shoot) and below-ground biomass; leaf length; LA; leaf or shoot length RGR; rhizome length; total leaf or shoot nitrogen and phosphorus content	Macek & Rejmankova (2007)
Fire	Variable number and type (general synthesis)	Sprouting capacity; biomass allocation; (relative) growth rate; carbon allocation to reserves; seed production; plant height; seedling establishment; time to first sexual reproduction	Bond & Midgley (2003)
	1801 species from various databases	Sprouting capacity; persistence of propagules; growth rate; plant height; age at maturity; plant longevity; diaspore mass; seed production	Pausas et al. (2004)
	28 woody evergreen species from the Mediterranean area of north-eastern Catalonia	SLA; LDMC; LNC; LPC; LCC; relative leaf water content; relative shoot water content	Saura-Mas et al. (2009)

continued

Table 4.2 (*Continued*)

Environmental Factor	Number and type of species	Traits studied	Reference
Altitude	Variable number and type (general synthesis)	Life form; growth form; life cycle; seed dormancy and germination; seedling establishment; vegetative growth; photosynthesis; respiration; carbohydrate and lipid storage; drought resistance; phenology of dormancy; reproductive traits	Billings (1974)
	20 to 30 herbaceous perennial species from low and high altitude in the Central Alps (Austria)	Plant height; lateral spread; total shoot area; number of leaves; leaf angle; LA; degree of succulence; SLA; leaf thickness; length and dry matter fraction of petioles; anatomical characteristics of leaf tissues; cell sizes of leaves and roots; thickness of cell walls; repartition of stomata; stomatal density	Körner et al. (1989)
	1340 species from alpine and lowland sites in New Zealand, South America, and Austria	Leaf length, width and area; petiole length; number of leaves per 10 mm length of stem; length of green support; number of live leaves on a shoot; distance between stem apices; margin (entire vs dentate, serrate, undulate); division (morphological leaf simple vs compound); shape of leaf section; leaf hairiness; presence/absence of petiole	Halloy & Mark (1996)
	82 species from the French Alps	Plant height; vegetative spread; leaf angle; LA; SLA; leaf thickness; LNC; seed size	Choler (2005)
Climate	558 broad-leaved and 39 needle-leaved shrubs and trees from 182 geographical locations	LMA; leaf thickness; leaf density, elasticity modulus; osmotic potentials at full and zero turgor	Niinemets (2001)
	88 *Leucadendron* taxa (broad-leaved shrubs) from the Cape Floristic region (South Africa)	Plant height; LA; dispersal mode; beginning and end of flowering	Thuiller et al. (2004)
	> 2500 species from 175 sites around the world	LLS; LMA; LNC_{mass}; LNC_{area}; A_{mass}; A_{area}	Wright et al. (2005)
	2004 species from 90 sites distributed worldwide	Mechanical properties of leaves: work to shear, force to punch, force to tear; LMA; Leaf thickness; leaf tissue density	Onoda et al. (2011)
	Large number of species depending on trait data availability, from sites distributed worldwide	Plant height; LA; seed mass; LLS; spinescence; plant life span; LNC_{mass}; LNC_{area}; LPC_{mass}; LPC_{area}; SLA; LCC_{mass}; LCC_{area}; woodiness; ability to fix nitrogen; clonality; dispersal syndrome; evergreenness; presence of hairs on mature leaves; photosynthetic pathway; leaf form (compoundness)	Moles et al. (2014)

Variations in trait values have also been observed in response to many other environmental factors. The short overview given in Table 4.2 shows the wealth of traits that can be involved in plant responses to different environmental factors (see also Gillison, 2013). As shown in this table, the variety of traits measured as well as the range of species numbers screened in different studies make literature syntheses challenging, even to understand the response of plants to a single environmental gradient.

4.4.2 Accounting for species abundance in communities

The identification of response traits to different gradients of environmental factors presented in Section 4.4.1 is based on trait measures taken from individuals belonging to species representative of different positions along a given gradient: species originating from nutrient-limited environments vs species originating from environments rich in mineral nutrients, species from early stages of a succession as opposed to later stages of succession, etc. In these original environments, individuals are not growing as isolated plants, but are components of the larger plant community (see Chapter 5) found in these environments. In the majority of these plant communities, the various component species are present in differing abundances (Figure 4.8, and see Box 5.1 in Chapter 5 for more details). Selecting a given species in a particular habitat to identify response traits to an environmental factor does not generally explicitly take into account its local abundance in the sampled community (but see Grime, 1974, 1977; Käfer & Witte, 2004, and references therein).

Here we postulate that the structure of species abundances within a community is a reflection of interspecific differences in responses to both the abiotic environment and the biotic interactions, and that this constitutes an important element for understanding the relationships between traits and environmental factors. We assume that the dominant species are those which are the best adapted to local environmental conditions, resulting in a higher performance (Cingolani et al.,

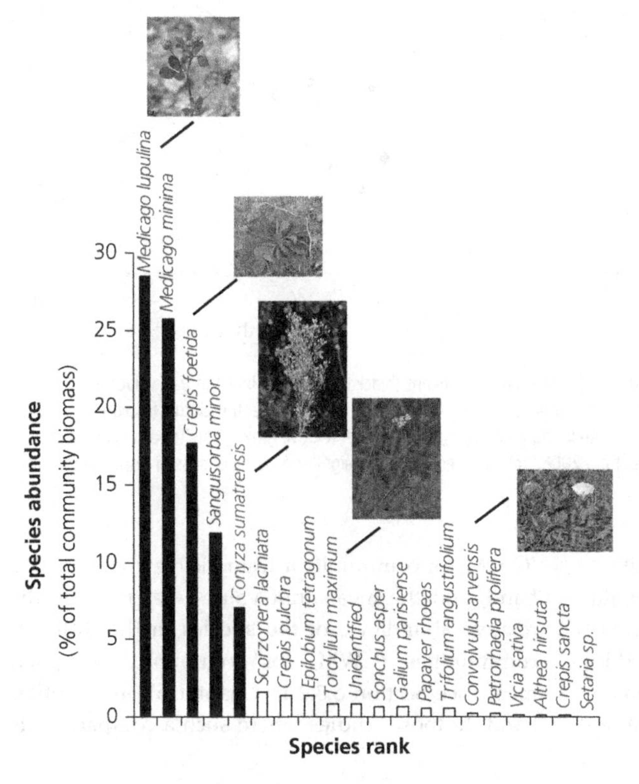

Figure 4.8 Example of a rank-abundance diagram of species in an abandoned field two years after the uprooting of vines in the French Mediterranean region, assessed from the measurement of the above-ground dry biomass of the species in this community. This field represents an early stage of secondary succession (data from Garnier et al., 2004). The black bars represent the 5 most abundant species ('dominant' in the terminology of Grime, 1998): photos for *Medicago lupulina, Crepis foetida*, and *Conyza sumatrensis*; the white bars represent species present in lower abundances ('subordinate' and/or 'transitory' in the terminology of Grime 1998): photos for *Tordylium maximum* and *Convolvulus arvensis*. For more details regarding rank-abundance diagrams, see Box 5.1 in Chapter 5.

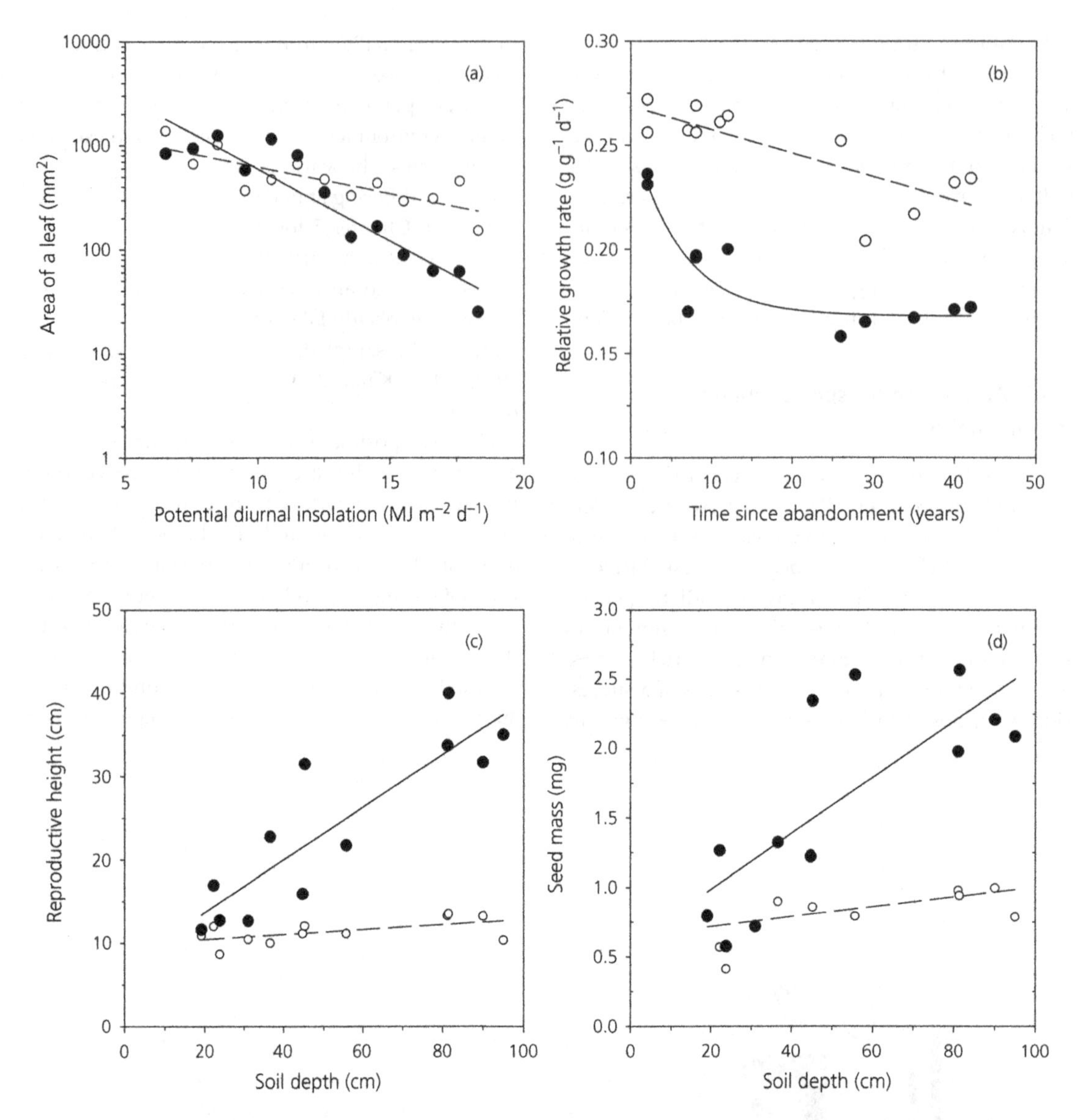

Figure 4.9 Comparison of arithmetic means (white circles, dotted lines) and weighted means (black circles, solid lines) of trait values at the community level along different types of gradients. (a) area of a leaf in response to light (Ackerly et al., 2002); potential diurnal insolation was predicted from an insolation model based on slope, aspect, and neighbouring topography; (b) potential relative growth rate in response to time since abandonment in a post-cultural succession (data from Vile et al., 2006); (c) plant reproductive height and (d) seed mass in response to soil depth (data from Bernard-Verdier et al., 2012).

2007; Vile et al., 2006). If this hypothesis is verified, the relationship between traits and environmental factors should be stronger when the abundance of species is taken into account. This can be tested by comparing the relationships between environmental factors and the arithmetic means of trait values (CArM, for community arithmetic mean) on the one hand, and the mean trait values weighted by the relative abundance of the species in the different communities (CWM, for community weighted mean: see Section 5.3.2 in Chapter 5) on the other hand. In those studies where such a comparison is possible (Ackerly et al., 2002; Cingolani et al., 2007; Dominguez et al., 2012; Garnier et al., 2004; Vile et al., 2006), the slope of the relationship between the CWM and the quantitative variable describing the gradient is indeed greater than that using the CArM (illustrated for four examples in Figure 4.9), which is in agreement with our hypothesis (see also the discussions in Castro-Díez, 2012; Pakeman et al., 2008).

As a consequence, we suggest that to improve the identification of response traits to environmental gradients, the abundance structure of communities should be taken into account. When this structure is not precisely known, the choice of species for trait measurements should be concentrated on the most abundant species, e.g. 'dominant' species *sensu* Grime (1998): for example, Garnier et al. (2004) showed that the signal shown when using the two most abundant species of communities representing different stages of secondary succession (cf. Figure 4.9b) was practically as strong as that shown when species composing 80% of the biomass of these communities were taken into account (see also Ansquer et al, 2009).

4.5 Ecological strategies

Up to now, this chapter has focused on the response of individual traits to different environmental factors, whether this is within a single species or among species. However, Table 4.1 shows that plant response to environmental variability always involves a combination of traits. Explicitly taking into account these combinations leads us to consider the concept of ecological strategy, which provides a more global approach to the understanding of functional diversity than simply the response of individual traits. The following section presents the basic concepts required to understand this notion of ecological strategy.

4.5.1 What is an ecological strategy?

The concept of ecological strategy is founded on the principle that similar environments exert similar selective forces on different species, leading to convergent evolution and adaptive patterns irrespective of their taxonomic identities (Silvertown et al., 1993). The underlying hypothesis is that extant organisms are the product of a small number of major selective forces which have operated throughout their evolutionary history and which have promoted their fitness in particular environments. Natural selection is then the result of the differences in fitness between individuals. It operates on the entire organisms, and the resulting patterns correspond to coordinated changes in multiple traits ('suites of traits'; see Figure 2.2 in Chapter 2). These combinations of traits, formed by natural selection, represent 'ecological strategies' which allow individuals to grow, survive, and reproduce in a given environment. Grime (1979) defines these strategies as 'groupings of similar or analogous genetic characteristics which recur widely among species or populations and cause them to exhibit similarities in ecology'. Craine (2009) proposed another definition in which the process of natural selection appears more explicitly: 'a strategy is a set of interlinked adaptations that arose as a consequence of natural selection and that promotes growth and successful reproduction in a given environment'. These two definitions, which emphasize slightly different aspects, can be considered jointly so as to provide a better understanding of the concept.

But why do these different strategies exist? Why does there not exist a single combination of traits that would allow one plant to dominate in all of the planet's habitats? The most obvious answer is that a combination of traits which leads to success in one type of environment does not necessarily lead to success in a different environment: there are trade-offs between traits, such that the presence of a trait value that is favourable in one environment may prove to be unfavourable in another. Examples of such functional compromises have been presented in the preceding chapter (see Figures 3.2 and 3.10), and Table 4.3 illustrates this idea for four examples drawn from the flora of the British Isles. The identification of ecological strategies, which correspond mostly to combinations of trait values of individuals of different species, constitutes an indirect demonstration of the limits of intraspecific trait variability (or of plasticity): as no species is sufficiently plastic to adapt to all types of environments, these are generally occupied by individuals of different species.

Table 4.3 Four examples of functional trade-offs identified for vascular plant species of the flora of the British Isles. From Grime & Pierce (2012). Reproduced with permission from Wiley.

Species	Trade-off
Betula populifolia	Production of numerous small winged seeds allows long-distance colonization of disturbed soil but limits seedling establishment in shaded conditions
Deschampsia flexuosa	Capacity to detoxify heavy metals on acidic soils is associated with susceptibility to iron deficiency on calcareous soils
Pilosella officinarum	Flattened growth-form of the shoot reduces water loss in dry habitats but prevents colonization of closed grassland
Tsuga canadensis	Slow growth and low respiratory losses permit seedling persistence in deep shade but restrict competitive ability early in vegetation succession

Different approaches to the identifications of strategies have been developed since the beginning of the twentieth century (cf. Table 1.2 in Chapter 1). A useful distinction can be made between 'life-history' and 'ecological' strategy models, which have mostly—but not always—developed in parallel. Although both ultimately aim at an understanding of the variation of, and controls on, the fitness of organisms, the approaches taken between these models differ. Life-history models place the emphasis on life histories, which can be defined as the probabilities of survival and the rates of reproduction at each age in the life cycle (Partridge & Harvey, 1988; Stearns, 1992). Ecological models tend to focus more on traits closely associated with productivity/stress and disturbance regimes as defined in Chapter 2 (for plants), be they vegetative or reproductive (cf. Table 1.2 in Chapter 1 and Table 4.5; Chapin et al., 1993; Craine, 2009; Grime, 1977). In the context of this book, we will therefore concentrate more on this second type, while borrowing elements from the first type where useful (e.g. identification of relevant environmental variables: see Section 4.5.2.1).

For plants, the historical development of ideas about ecological strategies is summarized with personal and complementary perspectives by e.g. Westoby (1998), Grime (2001), Craine (2009), and Gillison (2013). Here, we will briefly review the arguments that have led to the development of two- and three-strategy models, which has been a matter of intense debate in the scientific literature. We will then give particular emphasis to the model developed by Grime during the period 1970–1980, which is one of the most complete and influential strategy models developed in plant ecology so far (see e.g. Wilson & Lee, 2000).

4.5.2 Models of ecological strategies

The two basic elements required to define a strategy model—the characteristics of the environment and the characteristics of organisms in relation to the environment—are presented in the next two sections.

4.5.2.1 Environmental factors as selective forces

To develop a strategy model, the first stage consists of identifying the principal selective forces which have shaped the organisms and led to the observed combinations of traits. Central to the development of several influential strategy models is the idea that habitat characteristics are key selective forces shaping the strategies of organisms (Grime & Pierce, 2012; Southwood, 1988, for reviews). The *r-K* (life-history) selection model proposed by MacArthur and Wilson (1967) and expanded by Pianka (1970) is thought to be one of the first rigorous attempts to formalize this idea (Sibly & Calow, 1985). This model postulates that the characteristics of organisms and populations thereof are dependent on population density effects: one extreme consists of a 'perfect ecological vacuum with no density and no competition' (Pianka, 1970), while at the other extreme, density effects are maximal, the environment is saturated by organisms, and competition is strong (cf. Boyce, 1984; Gadgil & Solbrig, 1972; Pianka, 1970). Several authors have challenged this model, suggesting that an important dimension was missing from the *r-K* dichotomy: *K*-selection did not make an explicit distinction between relatively favourable and fully occupied environments and those which are predominantly and predictably unfavourable: selection operating in such environments has been defined as 'beyond *K*' (Greenslade, 1972) or 'adversity' selection (Greenslade, 1983; Grime, 1977; Southwood, 1977; Whittaker, 1975).

The combination of habitat features thought to be essential to shape the characteristics of organisms has been termed the 'habitat template' (Southwood, 1977). Reviewing such templates as identified in several strategy models, including those presented in the previous paragraph, Southwood (1988) concludes that key axes identified in most models relate to two main environmental factors considered either individually (e.g. in the *r*-*K* model) or in combination (e.g. in the *r*-*K*-adversity model). These are: i) disturbance, inducing the partial or complete destruction of individuals; examples include damage induced by fires, storms, landslides, or human activities (mowing, ploughing, etc.); and ii) adversity, representing increasingly unfavourable conditions for plant production; examples include low light, water or nutrient availability, suboptimal temperature, etc. (see Taylor et al., 1990, for a different perspective on this axis). Southwood (1988) further stresses the importance of a biotic interactions axis, which includes interactions such as predation, parasitism, and competition. This axis tends to run as a diagonal across the two abiotic axes corresponding to disturbance and adversity (i.e. stronger impact of biotic agents in favourable and undisturbed habitats). Interestingly, the dynamic equilibrium model designed by Huston (1979, 1994) also identifies disturbance (frequency or intensity) and productivity (expressing the inverse of adversity) as two major axes controlling species diversity in communities.

It should be noted here that these axes are defined using the same factors as those corresponding to the different types of gradients identified in Section 4.2, the 'stress' axis regrouping both resource and direct gradients.

4.5.2.2 Strategies: the match between organisms and the environment

The second stage of the process consists of identifying the characteristics of organisms which allow them to grow and reproduce in the different types of environments defined by this 'habitat template'. A number of predictions concerning the characteristics of organisms and populations thereof have been made in the context of life-history and ecological strategy models (e.g. Craine, 2009; Grime, 1977; Pianka, 1970; Southwood, 1988). For example, under extreme *r*-selection, the optimal strategy is to allocate resources to reproduction and produce as many offspring as possible. By contrast, when under *K*-selection, competition is strong, and the optimal strategy is to channel resources to maintenance and the production of a few extremely fit offspring (Gadgil & Solbrig, 1972; Pianka, 1970). This corresponds to the so-called colonization–competition trade-off. Detailed reviews of life-history features in relation to the environment are available in e.g. Reznick et al. (2002), and these will not be further discussed here. Below, we concentrate on plant ecological strategies, taking the elements of Grime's model as a framework for presentation.

Grime (1979) postulated that, for a given habitat, the pressures exerted upon individuals during the vegetative and reproductive phases of their lifecycle can be radically different. Additionally, as traits corresponding to these two phases are generally unrelated in plants (Grime et al., 1987; Shipley et al., 1989; Silvertown et al., 1992), he proposed the definition of a strategy model with two distinct components: one corresponding to the vegetative phase ('established phase') of plants, and the other based on their reproductive phase ('regeneration phase'). Grime's strategy model is probably the only one that has explicitly taken into account these two phases of plant functioning.

Strategies in the established phase

This is the three-strategy 'CSR' model initially proposed by Grime (1974, 1977), which is by far the best known and most recognized. C stands for 'Competitors', S for 'Stress-tolerators', and R for 'Ruderals'. These different strategies, termed 'primary', correspond respectively to plants that are found in environments: (1) where resource availability is high and level of disturbance is low; (2) where both resource availability and the level of disturbance are low; or (3) where both resource availability and the level of disturbance are high; in situations where resource availability is low and the level of disturbance is

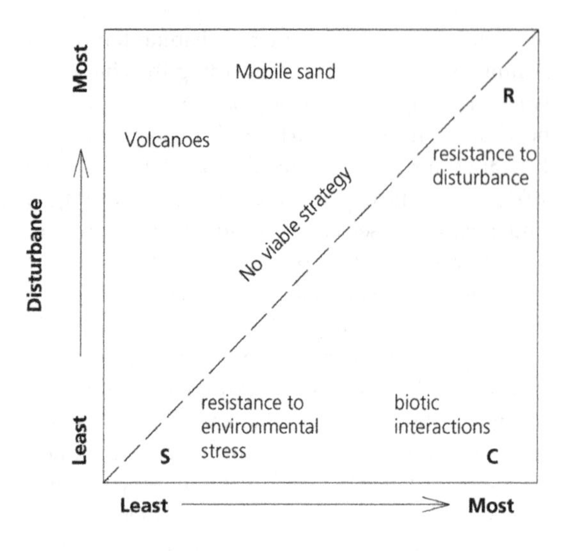

Figure 4.10 Ordination of strategies in the 'CSR' model of Grime (1977, 1979). The principal selective forces are represented in lower case characters, while the three primary strategies (C: Competitors; S: Stress-tolerators; R: ruderals) are represented in bold capital letters. Representation proposed by Fitter & Hay (2002), after Loehle (1988). Reproduced with permission from Elsevier.

high, no viable strategies can exist (Figure 4.10). In fact, these three primary strategies correspond to those organisms that are found at the extremities of the different gradients presented in Section 4.4 of this chapter.

Due to the form of the possible environmental space that can be occupied by plants, this model is called 'triangular' (see Grime, 1977, 1979). The initial triangular representation, which differs graphically from that presented in Figure 4.10, included obligatory compromises between the intensities of competition, stress, and ruderality of habitats, as the normalized sum of these intensities had to equal 1. In arguing that these environmental factors could be independent, Loehle (1988) proposed a rectangular representation which does not impose this construction constraint: the levels of competition, stress, and ruderality are defined by the combination of environmental favourability and disturbance, as shown in Figure 4.10.

Some important characteristics of plants belonging to the strategies C, S, and R are listed in Table 4.4. We also find, logically, some response traits presented in Table 4.1, such as the potential relative growth rate (see Box 3.3 in Chapter 3), leaf life span, the flowering period, or certain components related to seed production. Other traits, such as stem morphology, the quantity of litter produced, or the phenology of leaf production also appear important in determining the adaptation of plants to their environment, but they are less studied and relevant data is much less common in the literature. The responses of traits to variations in environmental factors (their 'plasticity' in response to one or several factors: see Section 4.3) are also part of the characteristics which potentially differentiate plants between the three primary strategies (Table 3 in Grime, 1977).

It is quite obvious that the combination of extreme habitats associated with the occurrence of the C, S, and R strategies only represents a subset of the range of possible plant habitats. In habitats containing less extreme combinations of resource availability and disturbance, Grime (1979) defined four 'secondary' strategies: competitive ruderals (C-R), stress-tolerant ruderals (S-R), stress-tolerant competitors (S-R), and finally 'C-S-R strategists', adapted to habitats in which the level of competition is limited by moderate intensities of stress and disturbance (see Grime, 1979, for further details of these secondary strategies).

Initially, assigning a species to a strategy was carried out using a combination of the following traits: the potential relative growth rate and a 'competitiveness index' calculated from vegetative height, lateral expansion, and an estimation of the maximal amount of accumulated litter (Grime, 1974). More recently, Hodgson et al. (1999) and Pierce et al. (2013) have developed algorithms based on respectively seven and three traits. The method developed by Hodgson et al. (1999) applies only to herbaceous species, and involves the following traits: canopy height, lateral expansion, the onset and duration of flowering, leaf dry matter content, specific leaf area, and the mass of an individual leaf. Pierce et al. (2013) developed a more general algorithm applicable to herbaceous and woody species, based on three leaf traits: specific leaf area, leaf dry matter content, and the area of a leaf. Whether these algorithms adequately capture CSR strategies for all types of vegetation remains to be established, and

Table 4.4 Some characteristics of competitive (C), stress-tolerant (S), and ruderal (R) plants (Grime, 1977; see Grime, 1979, pp. 48–49 for a more complete list of characteristics). Reproduced with permission from the University of Chicago Press.

	Competitive	Stress-tolerant	Ruderal
Morphology of shoot	High dense canopy of leaves; extensive lateral spread above and below ground	Extremely wide range of growth forms	Small stature, limited lateral spread
Leaf form	Robust, often mesomorphic	Often small or leathery, or needle-like	Various, often mesomorphic
Litter	Copious, often persistent	Sparse, sometimes persistent	Sparse, not usually persistent
Maximum potential relative growth rate	Rapid	Slow	Rapid
Life forms	Perennial herbs, shrubs, and trees	Lichens, perennial herbs, shrubs, and trees (often very long lived)	Annual herbs
Longevity of leaves	Relatively short	Long	Short
Phenology of leaf production	Well defined peaks of leaf production coinciding with period(s) of maximum potential productivity	Evergreens with various patterns of leaf production	Short period of leaf production in period of high potential productivity
Phenology of flowering	Flowers produced after (or more rarely before) periods of potential maximal productivity	No general relationship between time of flowering and season	Flowers produced at the end of temporarily favourable period
Proportion of annual production devoted to seeds	Small	Small	Large

identifying the minimum number of functional axes required to describe adequately plant structure and function (the intrinsic dimensionality of plant traits discussed by Laughlin, 2014) would be very useful in this context.

One of the major criticisms directed at this model is that the axes of stress and disturbance aggregate environmental factors that can in reality be very different. For the first axis, Grime's approach supposes that the major stress which determines the differing strategies is a nutrient stress, whether this be direct (an effective limitation due to low availability of mineral elements) or induced by other environmental factors. Craine (2009) and Craine et al. (2012) dispute this hypothesis, and propose different strategies depending on whether the stress is induced by low availability of mineral nutrients, by water or by light. Similarly, the disturbance axis can theoretically be subdivided as a function of the different variables necessary to effectively describe a disturbance regime (see Section 4.2, and Table 1 in White & Pickett, 1985). Another criticism suggests that there exist different types of adaptations to a single stress or the same disturbance. Grubb (1998) argues, for example, that plants can cope with a nutritional stress via three major types of strategies and not only one as suggested by the CSR model. This same type of argument has been presented for the case of disturbances by Steneck and Dethier (1994). A further argument put forward by Schulze et al. (2005) is that the strategy to which a species belongs may depend on the ontogenetic stage of the individual, which is certainly true in particular for trees (see also Cornelissen et al., 2003; Grubb, 1998). It is thus useful to state the life stage being considered for the assigned strategy. Overall, these criticisms, around which considerable debate continues to this day (see notably Austin, 2013; Craine, 2009), have not yet been completely addressed.

Strategies in the regeneration phase

Grime (1979) postulated that the selective forces leading to the development of different regeneration strategies were linked essentially with the

Table 4.5 The five major types of regeneration strategies identified for terrestrial plants, as well as their principal characteristics (after Grime, 1979, 1989). Reproduced with permission from Wiley.

Strategy	Functional characteristics	Conditions under which strategy appears to confer a selective advantage
Vegetative expansion	New shoots vegetative in origin and remaining attached to parent plant until well established	Productive or unproductive habitats subject to low intensities of disturbance
Seasonal regeneration in vegetation gaps	Independent offspring (seeds or propagules) produced in a single cohort	Habitats subjected to seasonally predictable disturbance by climate or biotic factors
Persistent seed or spore bank	Viable but dormant seeds or spores present throughout the year, some persisting more than 12 months	Habitats subjected to temporarily unpredictable disturbance
Numerous widely dispersed seeds or spores	Offspring numerous and exceedingly buoyant in air, widely dispersed and often of limited persistence	Habitats subjected to spatially unpredictable disturbance or relatively inaccessible (cliffs, walls, tree trunks, etc.)
Persistent juveniles	Offspring derived from an independent propagule but seedling or sporeling capable of long-term persistence in a juvenile state	Unproductive habitats subjected to low intensities of disturbance

spatial and temporal components of the disturbance regime, while the availability of resources appears as a less important selective factor (Table 4.5). Five major types of regeneration strategies have been identified in relation to different combinations of environmental factors: (1) vegetative expansion, (2) seasonal regeneration in vegetation gaps (occurring mostly in temperate climates), (3) regeneration from a persistent seed bank, (4) regeneration via the production of seeds or spores with a high capacity for dispersal, and (5) regeneration linked with the existence of persistent juveniles. The principal characteristics of these different strategies are presented in Table 4.5. It should be noted here that these traits relate more to those characteristics taken into account in life-history models.

While the ideas that form the basis of these strategies of the regeneration phase were published in 1979, they received much less attention than those focused on the adult stage, and also remain much more speculative. This results probably from two principal types of difficulty:

- the regeneration phase is complex to handle quantitatively, as it requires taking into account a temporal perspective that is seldom adequately addressed in current ecological studies;
- this phase appears to be essentially controlled by the disturbance regime, the components of which are also very difficult to quantify accurately (cf. Section 4.2, and Table 1 in White & Pickett, 1985).

4.5.3 Limits to strategy models

There have been numerous criticisms of strategy models such as those presented earlier in Section 4.5. For those concerning the *r*-*K* model and its further developments, detailed accounts are given by e.g. Stearns (1976, 1992), Parry (1981), and Reznick et al. (2002). This is also the case for the CSR model, some of which have been presented in Section 4.5.2.2 (see also Wilson & Lee, 2000). Further limitations include the way axes of environmental variation are defined, as 'stress' and 'disturbance' are concepts rather than quantitative variables (Westoby, 1998).

The fact that several of the concepts of the CSR model are still controversial and semantically problematic (Westoby, 1998) has resulted in the reluctance of some ecologists to adhere to the approach. Additionally, no formal method exists to place species within the CSR triangle (Westoby, 1998). As explained above, certain procedures have been developed to address this issue (Grime, 1974; Hodgson et al., 1999; Pierce et al., 2013), but the major problem of quantifying the environmental factors still remains. In fact, there is no current consensus on which environmental variables need to be measured in order to accurately and consistently describe the axes of stress and disturbance which form the basis of the CSR model. This is a major issue, as in the absence of such a quantification it is practically impossible to compare studies carried out in different environments: a situation considered as an extreme stress (or an extreme disturbance) in one study could be in the context of another study only a situation of moderate stress (or disturbance). This is probably a reason why this model has not been often used operationally in situations outside of the context in which it was developed, in contrast to its considerable and widespread use as a conceptual model. This problem of quantification corresponds to the comments made by Austin (1980), and mentioned in Section 4.2 of this chapter, regarding the lack of precision in the identification and description of gradients. This actually appears to be an important obstacle which needs to be overcome in order to be able to answer fundamental questions in plant ecology.

Some of the criticisms mentioned above are certainly valid. It seems obvious that any model whose aim is to propose generalizations will not be able to take into account all of the possible patterns and outcomes in fine detail. This is certainly true not only of the CSR model but also of the *r-K* life-history model. These types of models have however contributed, and continue to contribute, to our understanding of the interactions between organisms and environmental factors by proposing a coherent framework relating the characteristics of organisms with plant communities and the environment.

As stated by Stearns (1976), these strategy models are qualitative rather than quantitative. They should be recognized as general frameworks allowing for the development of questions which need further refinements for more specific cases.

4.6 Conclusions

Along environmental gradients, changes in trait values result from a combination of intraspecific variation and species' intrinsic differences, with the relative importance of each remaining to be determined. This chapter has shown that systematic changes in trait values in response to variation in different environmental factors allow us to identify 'response traits', which are specific to each particular factor. The concept of ecological strategies presents a more global perspective to the functioning of organisms by emphasizing the integrated responses of organisms to environmental factors rather than that of individual traits. Understanding the ecological and adaptive significance of these traits and strategies requires the explicit identification of the variables underlying these gradients and which lead to variation in attributes, whether these be direct gradients or gradients of resources or disturbance.

Chapters 3 and 4 cover levels of organization that extend essentially from the organ to the whole organism. They have allowed us to show how to characterize plants from a functional perspective, and to understand how variation in their functioning—or traits as markers of this functioning—allows the adaptation of plants to their environment. The following chapter addresses the level of communities (cf. Figure 1.2), at which the functions of individuals are modulated by the interactions with other individuals in the local space that constitutes the habitat.

4.7 Key points

1. Variation in environmental factors is an essential determinant of the structure and functioning of ecological systems and their components. This variation can be classified into four major categories: direct and indirect gradients, resource gradients, and disturbance gradients.
2. Traits whose attributes show systematic and predictable variation in response to gradients of environmental factors are called 'response traits'. Such response traits are different depending on the gradients considered. Understanding the ecological and adaptive significance of this response requires explicit identification of the variables underlying the original gradient and which give rise to variation in trait values, whether these be direct gradients, or gradients of resources or disturbance.
3. Along environmental gradients, changes in trait values result from a combination of intraspecific variation and species' intrinsic differences. The relative importance of these two components remains to be determined, and whether they need to be taken into account explicitly depends on the question being addressed and the scales of

space and time being considered. Additionally, a limited number of studies have shown that the responses of dominant species to different environmental factors are stronger than those of less abundant species.

4. The concept of ecological strategies provides a global perspective on the functioning of organisms, emphasizing their integrated response to environmental factors rather than that of single traits. These strategies result from fundamental trade-offs in resource use, with the underlying mechanism being that a combination of attributes which is favourable in one environment will be unfavourable in another environment characterized by different environmental conditions.
5. Grime's model of plant strategies, highly influential in plant ecology, proposes three principal plant strategies during the adult phase of their life cycle, based on suites of traits matching the different combinations of favourability and disturbance of the habitat. The model also identifies five major regeneration strategies, for which the prevailing disturbance regime is the primary determining factor. Strategy models, which are more qualitative than quantitative, are important elements of our understanding of the interactions between organisms and environmental factors.

4.8 References

Ackerly, D. (2004). Functional strategies of chaparral shrubs in relation to seasonal water deficit and disturbance. *Ecological Monographs, 74*, 25–44.

Ackerly, D. D., & Monson, R. K. (2003). Waking the sleeping giant: the evolutionary foundations of plant functions. *International Journal of Plant Sciences, 164*(3 Suppl.), S1–S6.

Ackerly, D. D., Knight, C. A., Weiss, S. B., Barton, K., & Starmer, K. P. (2002). Leaf size, specific leaf area, and microhabitat distribution of chaparral woody plants: contrasting patterns in species level and community level analyses. *Oecologia (Berlin), 130*, 449–457.

Aerts, R., & Chapin, F. S., III. (2000). The mineral nutrition of wild plants revisited: a re-evaluation of processes and patterns. *Advances in Ecological Research, 30*, 1–67.

Albert, C. H., Grassein, F., Schurr, F. M., Vieilledent, G., & Violle, C. (2011). When and how should intraspecific variability be considered in trait-based plant ecology? *Perspectives in Plant Ecology, Evolution and Systematics, 13*(3), 217–225. doi:10.1016/j.ppees.2011.04.003

Ansquer, P., Duru, M., Theau, J. P., & Cruz, P. (2009). Functional traits as indicators of fodder provision over a short time scale in species-rich grasslands. *Annals of Botany, 103*(1), 117–126. doi:10.1093/aob/mcn215

Atkin, O. K., Loveys, B. R., Atkinson, L. J., & Pons, T. L. (2006). Phenotypic plasticity and growth temperature: understanding interspecific variability. [; Proceedings Paper]. *Journal of Experimental Botany, 57*(2), 267–281. doi:10.1093/jxb/erj029

Austin, M. P. (1980). Searching for a model for use in vegetation analysis. *Vegetatio, 42*(1–3), 11–21. doi:10.1007/bf00048865

Austin, M. P. (2013). Inconsistencies between theory and methodology: a recurrent problem in ordination studies. *Journal of Vegetation Science, 24*(2), 251–268. doi:10.1111/j.1654-1103.2012.01467.x

Austin, M. P., & Smith, T. M. (1989). A new model for the continuum concept. *Vegetatio, 83*(1–2), 35–47. doi:10.1007/bf00031679

Bazzaz, F. A. (1996). *Plants in Changing Environments. Linking Physiological, Population, and Community Ecology*. Cambridge: Cambridge University Press.

Bernard-Verdier, M., Navas, M.-L., Vellend, M., Violle, C., Fayolle, A., & Garnier, E. (2012). Community assembly along a soil depth gradient: contrasting patterns of plant trait convergence and divergence in a Mediterranean rangeland. *Journal of Ecology, 100*(6), 1422–1433. doi:10.1111/1365-2745.12003

Billings, W. D. (1974). Adaptations and origins of alpine plants. *Arctic and Alpine Research, 6*(2), 129–142.

Björkman, O. (1981). Responses to different quantum flux densities. In O. L. Lange, P. S. Nobel, C. B. Osmond, H. Ziegler, & M. H. Zimmermann (eds), *Physiological Plant Ecology I. Responses to the Physical Environment* (pp. 57–107). Berlin: Springer-Verlag.

Bond, W. J., & Midgley, J. J. (2003). The evolutionary ecology of sprouting in woody plants. *International Journal of Plant Sciences, 164*(3), S103–S114. doi:10.1086/374191

Boomsma, C. R., Santini, J. B., Tollenaar, M., & Vyn, T. J. (2009). Maize morphophysiological responses to intense crowding and low nitrogen availability: an analysis and review. *Agronomy Journal, 101*(6), 1426–1452. doi:10.2134/agronj2009.0082

Boyce, M. S. (1984). Restitution of *r*- and *K*-selection as a model of density-dependent natural selection. *Annual Review of Ecology and Systematics, 15*(1), 427–447. doi:10.1146/annurev.es.15.110184.002235

Briske, D., & Richards, J. (1995). Plant responses to defoliation: a physiological, morphological, and demographic evaluation. In D. Bedunah & R. Sosebee (eds), *Wildland Plants: Physiological Ecology and Developmental Biology* (pp. 635–710). Denver: Society for Range Management.

Castro-Díez, P. (2012). Functional traits analyses: scaling-up from species to community level. *Plant and Soil, 357*, 9–12.

Chapin, F. S., III (1980). The mineral nutrition of wild plants. *Annual Review of Ecology and Systematics, 11*, 233–260.

Chapin, F. S., III, Autumn, K., & Pugnaire, F. (1993). Evolution of suites of traits in response to environmental stress. *The American Naturalist, 142*, S78–92.

Chapin, F. S., III, & Shaver, G. R. (1996). Physiological and growth responses of arctic plants to a field experiment simulating climatic changes. *Ecology, 77*, 822–840.

Choler, P. (2005). Consistent shifts in Alpine plant traits along a mesotopographical gradient. *Arctic, Antarctic, and Alpine Research, 37*(4), 444–453. doi:10.1657/1523-0430

Cingolani, A. M., Cabido, M., Gurvich, D. E., Renison, D., & Díaz, S. (2007). Filtering processes in the assembly of plant communities: are species presence and abundance driven by the same traits? *Journal of Vegetation Science, 18*, 911–920.

Cornelissen, J. H. C., Cerabolini, B., Castro-Díez, P., Villar-Salvador, P., Montserrat-Martí, G., Puyravaud, J. P., ... Aerts, R. (2003). Functional traits of woody plants: correspondence of species rankings between field adults and laboratory-grown seedlings? *Journal of Vegetation Science, 14*, 311–322.

Craine, J. M. (2009). *Resource Strategies of Wild Plants*. Princeton and Oxford: Princeton University Press.

Craine, J. M., Engelbrecht, B. M. J., Lusk, C. H., McDowell, N., & Poorter, H. (2012). Resource limitation, tolerance, and the future of ecological plant classification. *Frontiers in Plant Science, 3*. doi:10.3389/fpls.2012.00246

Deil, U. (2005). A review on habitats, plant traits, and vegetation of ephemeral wetlands: a global perspective. *Phytocoenologia, 35*(2–3),533–705. doi:10.1127/0340-269x/2005/0035-0533

Díaz, S., Lavorel, S., McIntyre, S., Falczuk, V., Casanoves, F., Milchunas, D. G., ... Campbell, B. D. (2007). Plant trait responses to grazing: a global synthesis. *Global Change Biology, 13*, 313–341.

Dominguez, M. T., Aponte, C., Perez-Ramos, I. M., Garcia, L. V., Villar, R., & Maranon, T. (2012). Relationships between leaf morphological traits, nutrient concentrations, and isotopic signatures for Mediterranean woody plant species and communities. *Plant and Soil, 357*(1–2), 407–424. doi:10.1007/s11104-012-1214-7

Duru, M., Balent, G., Gibon, A., Magda, D., Theau, J.-P., Cruz, P., & Jouany, C. (1998). Fonctionnement et dynamique des prairies permanentes. Exemple des Pyrénées centrales. *Fourrages, 153*, 97–113.

Eckstein, L. R., & Otte, A. (2004). Evidence for consistent trait-habitat relations in two closely related violets of contiguous habitat types from a fertilisation experiment. *Flora - Morphology, Distribution, Functional Ecology of Plants, 199*(3), 234–246. doi:10.1078/0367-2530-00151

Epstein, E., & Bloom, A. J. (2005). *Mineral Nutrition of Plants: Principles and Perspectives* (2nd ed.). Sunderland: Sinauer Associates, Inc.

Fitter, A. H., & Hay, R. K. M. (2002). *Environmental Physiology of Plants* (3rd ed.). London: Academic Press.

Fortunel, C., Violle, C., Roumet, C., Buatois, B., Navas, M.-L., & Garnier, E. (2009). Allocation strategies and seed traits are hardly affected by nitrogen supply in 18 species differing in successional status. *Perspectives in Plant Ecology, Evolution and Systematics, 11*(4), 267–283. doi:10.1016/j.ppees.2009.04.003

Gadgil, M., & Solbrig, O. T. (1972). The concept of r- and K-selection: evidence from wild flowers and some theoretical considerations. *The American Naturalist, 106*, 14–31.

Galloway, L. F. (2001). The effect of maternal and paternal environments on seed characters in the herbaceous plant *Campanula americana* (Campanulaceae). *American Journal of Botany, 88*(5), 832–840. doi:10.2307/2657035

Garnier, E., Cortez, J., Billès, G., Navas, M.-L., Roumet, C., Debussche, M., ... Toussaint, J.-P. (2004). Plant functional markers capture ecosystem properties during secondary succession. *Ecology, 85*(9), 2630–2637.

Garnier, E., Laurent, G., Bellmann, A., Debain, S., Berthelier, P., Ducout, B., ... Navas, M.-L. (2001). Consistency of species ranking based on functional leaf traits. *New Phytologist, 152*, 69–83.

Gillison, A. N. (2013). Plant functional types and traits at the community, ecosystem, and world level. In E. van der Maarel & J. Franklin (eds), *Vegetation Ecology* (2nd ed., pp. 347–386). Chichester, United Kingdom: Wiley-Blackwell (John Wiley & Sons, Ltd).

Ginane, C., Dumont, B., Baumont, R., Prache, S., Fleurance, G., & Farruggia, A. (2008). *Comprendre le comportement alimentaire des herbivores au pâturage: Intérêts pour l'élevage et l'environnement*. Paper presented at the 15ème journées de Rencontres Recherches Ruminants (3 R).

Givnish, T. J. (1988). Adaptation to sun and shade: a whole-plant perspective. *Australian Journal of Plant Physiology, 15*(1–2), 63–92.

Greenslade, P. J. M. (1972). Evolution in the staphylinid genus *Priochirus* (Coleoptera). *Evolution, 26*(2), 203–220. doi:10.2307/2407032

Greenslade, P. J. M. (1983). Adversity selection and the habitat templet. *The American Naturalist, 122*(3), 352–365. doi:10.2307/2461021

Grime, J. P. (1974). Vegetation classification by reference to strategies. *Nature* (London), *250*, 26–31.

Grime, J. P. (1977). Evidence for the existence of three primary strategies in plants and its relevance to ecological and evolutionary theory. *The American Naturalist, 111*, 1169–1194.

Grime, J. P. (1979). *Plant Strategies and Vegetation Processes*. Chichester: John Wiley & Sons.

Grime, J. P. (1989). The stress debate: symptom of impending synthesis? *Biological Journal of The Linnean Society, 37*(1–2), 3–17. doi:10.1111/j.1095-8312.1989.tb02002.x

Grime, J. P. (1998). Benefits of plant diversity to ecosystems: immediate, filter, and founder effects. *Journal of Ecology, 86*, 902–910.

Grime, J. P. (2001). *Plant Strategies, Vegetation Processes, and Ecosystem Properties* (2nd ed.). Chichester: John Wiley & Sons.

Grime, J. P., & Pierce, S. (2012). *The Evolutionary Strategies that Shape Ecosystems*. Oxford: Wiley-Blackwell.

Grime, J. P., Hunt, R., & Krzanowski, W. J. (1987). Evolutionary physiological ecology of plants. In P. Calow (ed.), *Evolutionary Physiological Ecology* (pp. 105–125). Cambridge: Cambridge University Press.

Grubb, P. J. (1998). A reassessment of the strategies of plants which cope with shortages of resources. *Perspectives in Plant Ecology, Evolution and Systematics, 1*, 3–31.

Guisan, A., & Zimmermann, N. E. (2000). Predictive habitat distribution models in ecology. *Ecological Modelling, 135*(2–3), 147–186. doi:10.1016/s0304-3800(00)00354–9

Gurevitch, J., Curtis, P. S., & Jones, M. H. (2001). Meta-analysis in ecology. *Advances in Ecological Research, 32*, 199–247. doi:10.1016/s0065-2504(01)32013-5

Halloy, S. R. P., & Mark, A. F. (1996). Comparative leaf morphology spectra of plant communities in New Zealand, the Andes, and the European Alps. *Journal of the Royal Society of New Zealand, 26*(1), 41–78.

Hejcman, M., Kristalova, V., Cervena, K., Hrdlickova, J., & Pavlu, V. (2012). Effect of nitrogen, phosphorus, and potassium availability on mother plant size, seed production, and germination ability of *Rumex crispus*. *Weed Research, 52*(3), 260–268. doi:10.1111/j.1365-3180.2012.00914.x

Hodgson, J. G., Montserrat-Martí, G., Cerabolini, B., Ceriani, R. M., Maestro-Martí, M., Peco, B., ... Villar-Salvador, P. (2005). A functional method for classifying European grasslands for use in joint ecological and economic studies. *Basic and Applied Ecology, 6*, 119–131.

Hodgson, J. G., Wilson, P. J., Hunt, R., Grime, J. P., & Thompson, K. (1999). Allocating C-S-R plant functional types: a soft approach to a hard problem. *Oikos, 85*, 282–294.

Huston, M. A. (1979). A general hypothesis of species diversity. *The American Naturalist, 113*, 81–101.

Huston, M. A. (1994). *Biological Diversity. The Coexistence of Species on Changing Landscapes*. Cambridge: Cambridge University Press.

Huston, M. A., & Smith, T. (1987). Plant succession: life history and competition. *The American Naturalist, 130*, 168–198.

Hutchinson, G. E. (1957). Concluding remarks. *Cold Spring Harb.Symp. Quant. Biol., 22*, 415–427.

Käfer, J., & Witte, J. -P.M. (2004). Cover-weighted averaging of indicator values in vegetation analyses. *Journal of Vegetation Science, 15*(5), 647–652. doi:10.1111/j.1654-1103.2004.tb02306.x

Kazakou, E., Garnier, E., Navas, M.-L., Roumet, C., Collin, C., & Laurent, G. (2007). Components of nutrient residence time and the leaf economics spectrum in species from Mediterranean old-fields differing in successional status. *Functional Ecology, 21*, 235–245.

Kazakou, E., Violle, C., Roumet, C., Navas, M.-L., Vile, D., Kattge, J., & Garnier, E. (2014). Are trait-based species rankings consistent across data sets and spatial scales? *Journal of Vegetation Science, 25*(1), 235–247. doi:10.1111/jvs.12066

Kitajima, K. (1994). Relative importance of photosynthetic traits and allocation patterns as correlates of seedling shade tolerance of 13 tropical trees. *Oecologia, 98* (3–4), 419–428. doi:10.1007/bf00324232

Körner, C., Neumayer, M., Pelaez Menendez-Riedl, S., & Smeets-Scheel, A. (1989). Functional morphology of mountain plants. *Flora (Jena), 182*, 353–383.

Lambers, H., Chapin, F. S., III, & Pons, T. L. (1998). *Plant Physiological Ecology* (1st ed.). New-York: Springer-Verlag.

Laughlin, D. C. (2014). The intrinsic dimensionality of plant traits and its relevance to community assembly. *Journal of Ecology, 102*(1), 186–193. doi:10.1111/1365-2745.12187

Lavorel, S., & Garnier, E. (2002). Predicting changes in community composition and ecosystem functioning from plant traits: revisiting the Holy Grail. *Functional Ecology, 16*, 545–556.

Leibold, M. A., Holyoak, M., Mouquet, N., Amarasekare, P., Chase, J. M., Hoopes, M. F., ... Gonzalez, A. (2004). The metacommunity concept: a framework for multi-scale community ecology. *Ecology Letters, 7*, 601–613.

Lepart, J., & Escarré, J. (1983). La succession végétale, mécanismes et modèles: analyse bibliographique. *Bulletin d'Ecologie, 14*(3), 133–178.

Loehle, C. (1988). Problems with the triangular model for representing plant strategies. *Ecology, 69*(1), 284–286.

Loehle, C. (1998). Height growth rate tradeoffs determine northern and southern range limits for trees. *Journal of Biogeography, 25*(4), 735–742. doi:10.1046/j.1365-2699.1998.2540735.x

MacArthur, R. H., & Wilson, E. O. (1967). *The Theory of Island Biogeography*. Princeton: Princeton University Press.

Macek, P., & Rejmankova, E. (2007). Response of emergent macrophytes to experimental nutrient and salinity additions. *Functional Ecology*, *21*(3), 478–488. doi:10.1111/j.1365-2435.2007.01266.x

MacGillivray, C. W., & Grime, J. P. (1995). Genome size predicts frost-resistance in British herbaceous plants: implications for rates of vegetation response to global warming. *Functional Ecology*, *9*(2), 320–325. doi:10.2307/2390580

McGill, B. J., Enquist, B. J., Weiher, E., & Westoby, M. (2006). Rebuilding community ecology from functional traits. *Trends in Ecology and Evolution*, *21*, 178–185.

McIntyre, S., Lavorel, S., Landsberg, J., & Forbes, T. D. A. (1999). Disturbance response in vegetation: towards a global perspective on functional traits. *Journal of Vegetation Science*, *10*, 621–630.

Messier, J., McGill, B. J., & Lechowicz, M. J. (2010). How do traits vary across ecological scales? A case for trait-based ecology. *Ecology Letters*, *13*, 838–848.

Moles, A. T., Perkins, S. E., Laffan, S. W., Flores-Moreno, H., Awasthy, M., Tindall, M. L., . . . Bonser, S. P. (2014). Which is a better predictor of plant traits: temperature or precipitation? *Journal of Vegetation Science*, *25*(5), 1167-1180. doi:10.1111/jvs.12190

Navas, M.-L., & Garnier, E. (2002). Plasticity of whole plant and leaf traits in *Rubia peregrina* in response to light, nutrient, and water availability. *Acta Oecologica*, *23*, 375–383.

Navas, M.-L., Roumet, C., Bellmann, A., Laurent, G., & Garnier, E. (2010). Suites of plant traits in species from different stages of a Mediterranean secondary succession. *Plant Biology*, *12*, 183–196.

Niinemets, Ü. (2001). Global-scale climatic controls of leaf dry mass per area, density, and thickness in trees and shrubs. *Ecology*, *82*, 453–469.

Onoda, Y., Westoby, M., Adler, P. B., Choong, A. M. F., Clissold, F. J., Cornelissen, J. H. C., . . . Yamashita, N. (2011). Global patterns of leaf mechanical properties. *Ecology Letters*, *14*(3), 301–312. doi:10.1111/j.1461-0248.2010.01582.x

Ordoñez, J. C., van Bodegom, P. M., Witte, J. P. M., Bartholomeus, R. P., van Hal, J. R., & Aerts, R. (2010). Plant strategies in relation to resource supply in mesic to wet environments: does theory mirror nature? *The American Naturalist*, *175*(2), 225–239. doi:10.1086/649582

Pakeman, R. J., Garnier, E., Lavorel, S., Ansquer, P., Castro, H., Cruz, P., . . . Vile, D. (2008). Impact of abundance weighting on the response of seed traits to climate and land use change. *Journal of Ecology*, *96*, 355–366.

Parry, G. D. (1981). The meanings of r- and K-selection. *Oecologia (Berlin)*, *48*, 260–264.

Partridge, L., & Harvey, P. H. (1988). The ecological context of life history evolution. *Science*, *241*, 1449–1455.

Pausas, J. G., Bradstock, R. A., Keith, D. A., Keeley, J. E., & Network, G. F. (2004). Plant functional traits in relation to fire in crown-fire ecosystems. *Ecology*, *85*(4), 1085–1100. doi:10.1890/02-4094

Pianka, E. R. (1970). On *r*- and K-selection. *The American Naturalist*, *104*, 592–597.

Pierce, S., Brusa, G., Vagge, I., & Cerabolini, B. E. L. (2013). Allocating CSR plant functional types: the use of leaf economics and size traits to classify woody and herbaceous vascular plants. *Functional Ecology*, *27*(4), 1002–1010. doi:10.1111/1365-2435.12095

Poorter, H., Niinemets, Ü., Poorter, L., Wright, I. J., & Villar, R. (2009). Causes and consequences of variation in leaf mass per area (LMA): a meta-analysis. *New Phytologist*, *182*, 565–588.

Poorter, H., Niinemets, Ü., Walter, A., Fiorani, F., & Schurr, U. (2010). A method to construct dose-response curves for a wide range of environmental factors and plant traits by means of a meta-analysis of phenotypic data. *Journal of Experimental Botany*, *61*(8), 2043–2055. doi:10.1093/jxb/erp358

Poorter, H., Niklas, K. J., Reich, P. B., Oleksyn, J., Poot, P., & Mommer, L. (2012). Biomass allocation to leaves, stems, and roots: meta-analyses of interspecific variation and environmental control. *New Phytologist*, *193*(1), 30–50. doi:10.1111/j.1469-8137.2011.03952.x

Poorter, L. (1999). Growth responses of 15 rain-forest tree species to a light gradient: the relative importance of morphological and physiological traits. *Functional Ecology*, *13*(3), 396–410. doi:10.1046/j.1365–2435.1999.00332.x

Prach, K., Pysek, P., & Smilauer, P. (1997). Changes in species traits during succession: a search for pattern. *Oikos*, *79*, 201–205.

Reznick, D., Bryant, M. J., & Bashey, F. (2002). *r*- and *K*-selection revisited: the role of population regulation in life-history evolution. *Ecology*, *83*(6), 1509–1520. doi:10.1890/0012-9658

Rice, S. A., & Bazzaz, F. A. (1989). Quantification of plasticity of plant traits in response to light intensity: comparing phenotypes at a common weight. *Oecologia (Berlin)*, *78*, 502–507.

Richards, C. L., Pennings, S. C., & Donovan, L. A. (2005). Habitat range and phenotypic variation in salt marsh plants. *Plant Ecology*, *176*(2), 263–273. doi:10.1007/s11258-004-0841-3

Robinson, D., & Rorison, I. H. (1988). Plasticity in grass species in relation to nitrogen supply. *Functional Ecology*, *2*, 249–257.

Saura-Mas, S., Shipley, B., & Lloret, F. (2009). Relationship between post-fire regeneration and leaf economics spectrum in Mediterranean woody species. *Functional Ecology, 23*(1), 103–110. doi:10.1111/j.1365-2435.2008.01474.x

Schilchting, C. D. (2002). Phenotypic plasticity in plants. *Plant Species Biology, 17*, 85–88.

Schulze, E.-D., Beck, E., & Müller-Hohenstein, K. (2005). *Plant Ecology*. Berlin: Springer-Verlag.

Shipley, B. (2010). *From Plant Traits to Vegetation Structure. Chance and Selection in the Assembly of Ecological Communities*. Cambridge: Cambridge University Press.

Shipley, B., Keddy, P. A., Moore, D. R. J., & Lemky, K. (1989). Regeneration and establishment strategies of emergent macrophytes. *Journal of Ecology, 77*, 1093–1110.

Sibly, R., & Calow, P. (1985). Classification of habitats by selection pressures: a synthesis of life-cycle and r/K theory. In R. M. Sibly & R. H. Smith (eds), *Behavioural Ecology. Ecological Consequences of Adaptive Behaviour* (pp. 75–90). Oxford: Blackwell Scientific Publications.

Silvertown, J., Franco, M., & McConway, K. (1992). A demographic interpretation of Grime's triangle. *Functional Ecology, 6*, 130–136.

Silvertown, J., Franco, M., Pisanty, I., & Mendoza, A. (1993). Comparative plant demography: relative importance of life-cycle components to the finite rate of increase in woody and herbaceous perennials. *Journal of Ecology, 81*, 465–476.

Southwood, T. R. E. (1977). Habitat, the templet for ecological strategies? Presidential address to the British Ecological Society, 5 January 1977. *Journal of Animal Ecology, 46*(2), 337–365.

Southwood, T. R. E. (1988). Tactics, strategies, and templets. *Oikos, 52*, 3–18.

Stearns, S. C. (1976). Life-history tactics: a review of the ideas. *Quarterly Review of Biology, 51*, 3–47.

Stearns, S. C. (1992). *The Evolution of Life Histories*. New York: Oxford University Press.

Steneck, R. S., & Dethier, M. N. (1994). A functional group approach to the structure of algal-dominated communities. *Oikos, 69*(3), 476–498. doi:10.2307/3545860

Sultan, S. E. (2001). Phenotypic plasticity for fitness components in *Polygonum* species of contrasting ecological breadth. *Ecology, 82*(2), 328–343. doi:10.1890/0012-9658

Tattini, M., Remorini, D., Pinelli, P., Agati, G., Saracini, E., Traversi, M. L., & Massai, R. (2006). Morpho-anatomical, physiological, and biochemical adjustments in response to root zone salinity stress and high solar radiation in two Mediterranean evergreen shrubs, *Myrtus communis* and *Pistacia lentiscus*. *New Phytologist, 170*(4), 779–794. doi:10.1111/j.1469-8137.2006.01723.x

Taylor, D. R., Aarssen, L. W., & Loehle, C. (1990). On the relationship between r/K selection and environmental carrying capacity: a new habitat templet for plant life-history strategies. *Oikos, 58*(2), 239–250. doi:10.2307/3545432

Thuiller, W., Lavorel, S., Midgley, G., Lavergne, S., & Rebelo, T. (2004). Relating plant traits and species distributions along bioclimatic gradients for 88 *Leucadendron* taxa. *Ecology, 85*(6), 1688–1699. doi:10.1890/03-0148

Tilman, D. (1982). *Resource Competition and Community Structure*. Princeton, New Jersey: Princeton University Press.

Valladares, F., & Niinemets, Ü. (2008). Shade tolerance, a key plant feature of complex nature and consequences. *Annual Review of Ecology, Evolution, and Systematics, 39*(1), 237–257. doi:10.1146/annurev.ecolsys.39.110707.173506

Vile, D., Shipley, B., & Garnier, E. (2006). Ecosystem productivity can be predicted from potential relative growth rate and species abundance. *Ecology Letters, 9*, 1061–1067.

Violle, C., & Jiang, L. (2009). Towards a trait-based quantification of species niche. *Journal of Plant Ecology, 2*(2), 87–93. doi:10.1093/jpe/rtp007

Violle, C., Enquist, B. J., McGill, B. J., Jiang, L., Albert, C. H., Hulshof, C., ... Messier, J. (2012). The return of the variance: intraspecific variability in community ecology. *Trends in Ecology and Evolution, 27*(4), 244–252. doi:10.1016/j.tree.2011.11.014

Violle, C., Garnier, E., Lecœur, J., Roumet, C., Podeur, C., Blanchard, A., & Navas, M.-L. (2009). Competition, traits, and resource depletion in plant communities. *Oecologia, 160*, 747–755.

Violle, C., Lecoeur, J., & Navas, M. -L. (2007). How relevant are instantaneous measurements for assessing resource depletion under plant cover? A test on light and soil water availability in 18 herbaceous communities. *Functional Ecology, 21*, 185–190.

Volaire, F. (2008). Plant traits and functional types to characterise drought survival of pluri-specific perennial herbaceous swards in Mediterranean areas. *European Journal of Agronomy, 29*(2–3), 116–124. doi:10.1016/j.eja.2008.04.008

Walters, M. B., & Reich, P. B. (1996). Are shade tolerance, survival, and growth linked? Low light and nitrogen effects on hardwood seedlings. *Ecology, 77*(3), 841–853. doi:10.2307/2265505

Westoby, M. (1998). A leaf-height-seed (LHS) plant ecology strategy scheme. *Plant and Soil, 199*, 213–227.

Westoby, M., & Wright, I. J. (2006). Land-plant ecology on the basis of functional traits. *Trends in Ecology and Evolution, 21*, 261–268.

Westoby, M., Falster, D. S., Moles, A. T., Vesk, P. A., & Wright, I. J. (2002). Plant ecological strategies: some leading dimensions of variation between species. *Annual Review of Ecology and Systematics*, *33*, 125–159.

White, P. S., & Pickett, S. T. A. (1985). Natural disturbance and patch dynamics, an introduction. In S. T. A. Pickett & P. S. White (eds), *The Ecology of Natural Disturbance and Patch Dynamics* (pp. 3–13). New York: Academic Press.

Whittaker, R. H. (1951). A criticism of the plant association and climatic climax concepts. *Northwest Science*, *25*, 17–31.

Whittaker, R. H. (1967). Gradient analysis of vegetation. *Biological Reviews*, *49*, 207–264.

Whittaker, R. H. (1975). The design and stability of plant communities. In W. H. van Dobben & R. H. Lowe-McConnell (eds), *Unifying Concepts in Ecology* (pp. 169–181). The Hague: Dr. W. Junk B. V. Publishers.

Whittaker, R. H. (1978). Direct gradient analysis. In R. H. Whittaker (ed.), *Ordination of Plant Communities* (2nd ed., pp. 7–51). The Hague: Junk.

Wilson, J. B., & Lee, W. G. (2000). C-S-R triangle theory: community level predictions, tests, evaluation of criticisms, and relation to other theories. *Oikos*, *91*, 77–96.

Wright, I. J., Reich, P. B., Cornelissen, J. H. C., Falster, D. S., Groom, P. K., Hikosaka, K., . . . Westoby, M. (2005). Modulation of leaf economic traits and trait relationships by climate. *Global Ecology and Biogeography*, *14*(5), 411–421.

CHAPTER 5

A functional approach to plant community structure

5.1 Introduction

Understanding how the sets of species which form a community are assembled requires first the description, analysis, and comparison of the patterns of community structure between and within different communities. Traditionally, the structure of a community has been investigated in a 'taxonomic' perspective by defining the number of species making up the community, as well as by the distribution of their abundances. Most often, a community is made up of a few dominant species which represent a large proportion of the total biomass or ground cover, and a large number of subordinate or minor species having a minor contribution to biomass or ground cover, and whose presence can vary greatly between seasons or years (Grime, 1998; Whittaker, 1965). A variety of indices and graphical representations have been proposed to compare communities which are found under differing conditions and which are composed of various numbers of species (see Box 5.1). These representations, while commonly used, remain highly descriptive and do not generally allow for a causal understanding of the observed patterns. To investigate the factors and mechanisms which control the composition of these assemblages, it is necessary to use other, more explanatory, approaches. The aim of this chapter is to review the utility of a functional approach to diversity, community structure, and assembly, which moves beyond the traditional approach to community structure based on taxonomy. After giving the definition of community assembly rules and the functional structure of a community, we will explain how the latter can be described using different indices. Subsequently, we will address the factors controlling variation in community functional structure between different environments, in particular when these are due to modifications of species abundances or intraspecific variation in traits, depending on the relevant spatial and temporal scales. We conclude with a discussion of the relationships between the functional and phylogenetic structures of communities, an area of current active research.

5.2 Community assembly rules and environmental filters

Diamond (1975) was one of the first to identify the rules controlling the assembly of communities, so as to determine what subset of species existing within a geographic region could be found living together in a given habitat. His research, based on more than 10 years of observations of bird communities in New Guinea, is based on the hypothesis of the dominant role of competition as the process determining the assemblage of species. One of the fundamentals of this hypothesis is the observation that certain species never coexist and thus have separated local distributions, the so-called 'checkerboard' pattern. Nevertheless, the first assembly rules established on the basis of observed associations between coexisting animals or plants in a given region appeared to be highly dependent on the local context, and did not lead to the establishment of general principles allowing the understanding of the assembly of other communities. The development of the theory of island biogeography, which estimates the specific

Box 5.1 Community structure: principal models of the distribution and abundance of species

(After Magurran, 2004; McGill, 2006, 2011; McGill et al., 2007.)

The simplest way to describe the structure of a community is by using synthetic indices representing the various facets of diversity (McGill, 2011). Abundance can be estimated by the number of individuals in the community, the percentage of vegetation cover, the biomass, and global energy usage. Specific richness corresponds to the number of species in the community. Diversity can be described, among different indices, by the Shannon–Weaver $\left(H = -\sum_{i=1}^{s} p_i \ln p_i\right)$ or Simpson indices $\left(E = 1 - \sum_{i=1}^{n} p_i^2\right)$, which take into account both s the number of species and p_i, the probability that an individual drawn at random belongs to species i; the first is more sensitive to the presence of rare species because of log transformation, and the second to the presence of dominant species. The evenness of the distribution of a community, reflecting the uniformity of abundance among species, can be established by comparing its actual distribution to a theoretical regular distribution by using, for example, the Pielou index ($E_H = H/H_{max}$ with $H_{max} = \ln S$). Dominance can be evaluated in an absolute manner by the abundance of the most common species, or relatively by dividing the species abundance by the total abundance. Rarity is the opposite of dominance; it is evaluated by the number of species representing less than 1–5% of the total number of individuals.

Another manner in which to define this structure consists of graphically presenting the distribution of species abundances (SAD for Species Abundance Distribution), or in other words defining the rules giving the expected abundance of each species within a community. A SAD can also be integrated into abundance vectors, which can be analysed in the same manner for all of the species in a community.

This distribution can be represented as a histogram of species frequency or a rank-abundance diagram, also called a Whittaker diagram after the name of the ecologist who first proposed it (Box 5.1, Figure 1). These representations are often used as they allow for the easy visual comparison

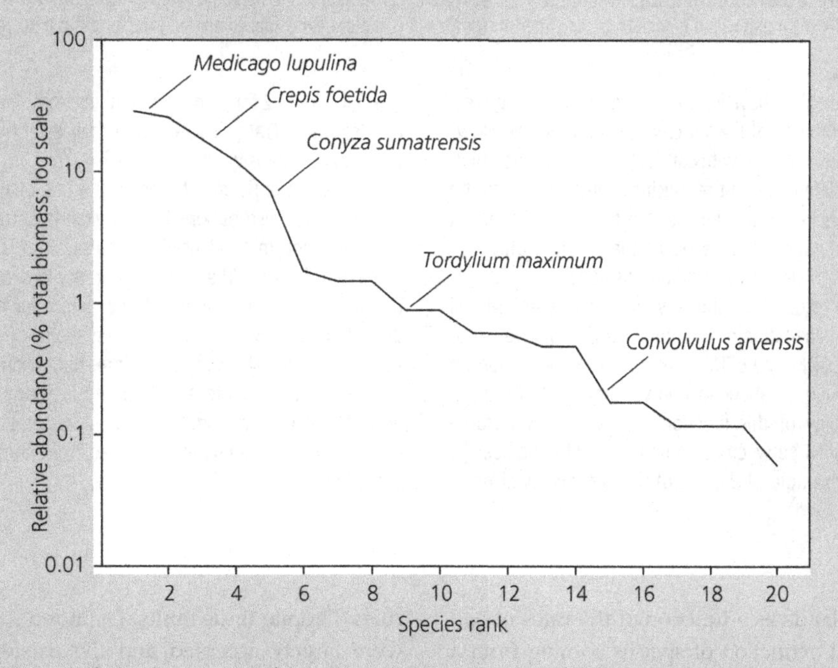

Box 5.1, Figure 1 A rank-abundance diagram of the species present in an abandoned field two years after its conversion from a vineyard in the Mediterranean region (Figure 4.8 redrawn using a logarithmic scale for abundance; the species named here are those presented as photographs in Figure 4.8).

continued

Box 5.1 *Continued*

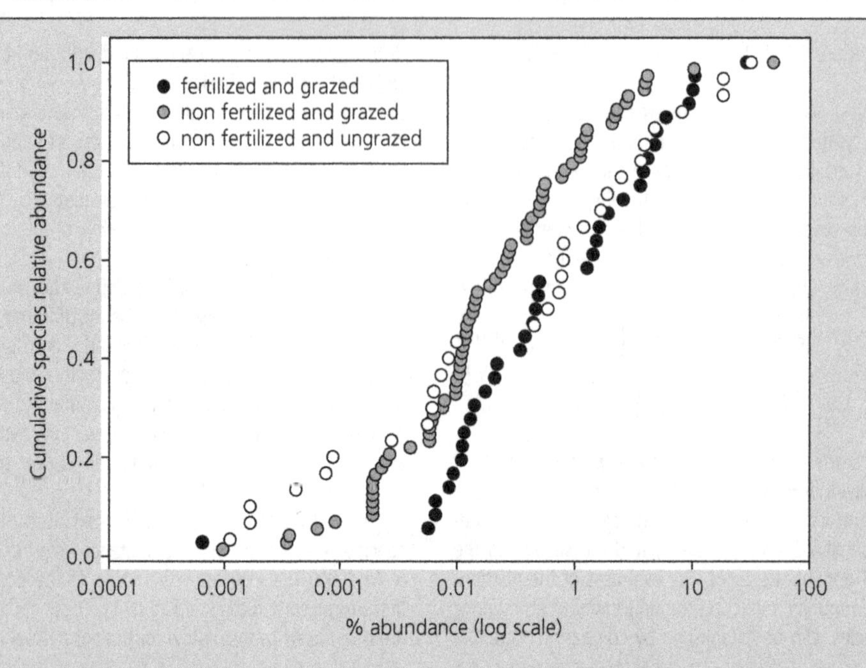

Box 5.2, Figure 2 Distribution of species abundances in three types of communities: a fertilized plot with a high grazing intensity (black circles), a non-fertilized plot with a moderate grazing intensity (grey circles), a plot neither grazed nor fertilized for the last 30 years (white circles).

of communities. They have led to mathematical descriptions, based on hypotheses of the structure of communities. However, they should be used with caution, because of their high sensitivity to differences in the specific richness between the communities being compared (i.e. the occupied proportion of the x axis), which partly explains their limited utility for validating or verifying theoretical propositions.

In order to resolve this difficulty, amongst others, McGill proposed the calculation of the Empirical Cumulative Density Function (ECDF): the ECDF corresponds to the proportion of species having an abundance lower than that of a given value. The utility of this function is that it allows statistical comparisons between communities of differing specific richness. An example of the use of this function is given in Box 5.1, Figure 2 for plant communities from the Larzac plateau (southern France), differing in their level of fertilization and grazing intensity.

One of the difficulties linked to the use of these models of abundance distributions is both experimental and statistical, with the method used to assess abundances having a major influence. The sampling area plays an important role as overly small samples do not allow for the inclusion of rare species, which influences the shape of the distribution. Finally, a SAD provides no information on the identity or type of species present, and thus cannot be used to account for the filtering effect of the habitat, explain the causes of rarity or dominance, or predict the functioning of a community.

richness of islands as a function of the rates of colonization and extinction of species coming from a reservoir of species on the mainland, provided a considerable advance in the description and prediction of observed assemblages (Diamond, 1975), but it did not establish explicit or mechanistic assembly rules. Despite these limits, Diamond's propositions were largely accepted and the existence of non-random checkerboard patterns of species distributions was confirmed by meta-analyses (Gotelli & McCabe, 2002). However, questions were also asked of the relative roles of competition and chance in the assembly of communities (Connor & Simberloff, 1979). This led to the emergence of hypotheses proposing the existence of new factors involved in the structuring of communities, such as other density-dependent interactions (predation, parasitism, etc.) and environmental fluctuations or disturbances, hypotheses which now have considerable support from current data (Wilson, 2011).

These new approaches to community assembly are based on the recognition of the impact of factors external to the community, such as factors linked to the dispersal of species and individuals and to local and regional environmental constraints, and of factors internal to the community, such as interactions between organisms and environmental micro-heterogeneity (Belyea & Lancaster, 1999; Lortie et al., 2004). In this view, assembly rules operate in a hierarchical fashion as a series of filters selecting species from the regional pool: (i) a first filter of species dispersal is controlled by stochastic biogeographical events, storage effects, or landscape structure; it determines the group of colonizing species potentially available in a given location at a given time, the composition of which is influenced by the order and timing of the arrival of species; (ii) an abiotic filter which corresponds to the influence of local environmental conditions (pH, humidity, soil temperature, etc.), the availability of resources, and the disturbance regime; it determines which species can establish locally; (iii) a biotic filter corresponding to the positive and negative interactions existing between living organisms within communities and whose impact is modulated by their density; it determines the set of neighbouring coexisting species within the community (Figure 5.1).

While generally accepted, this framework can be presented with differing hierarchical orders depending on authors. Some consider that the habitat effect *sensu stricto* is only due to local abiotic effects, and separate it from the filter due to biotic interactions. This could, however, be considered incorrect as the habitat is defined as the totality of the components of the local environment. For other authors, only competition and (more rarely) facilitation are considered as interactions filtering species or individuals. This restriction is also inaccurate as other types of biotic interactions, modified by density, such as host–parasite or predator–prey relationships, can also have a filtering effect (Cavender-Bares et al., 2009; Kraft et al., 2008). It should be clarified that biotic interactions whose effects are not modified by density (for example certain types of facilitation or competition by interference due to the accumulation of litter or modifications of the oxygenation of soils) appear to be *a priori* of similar

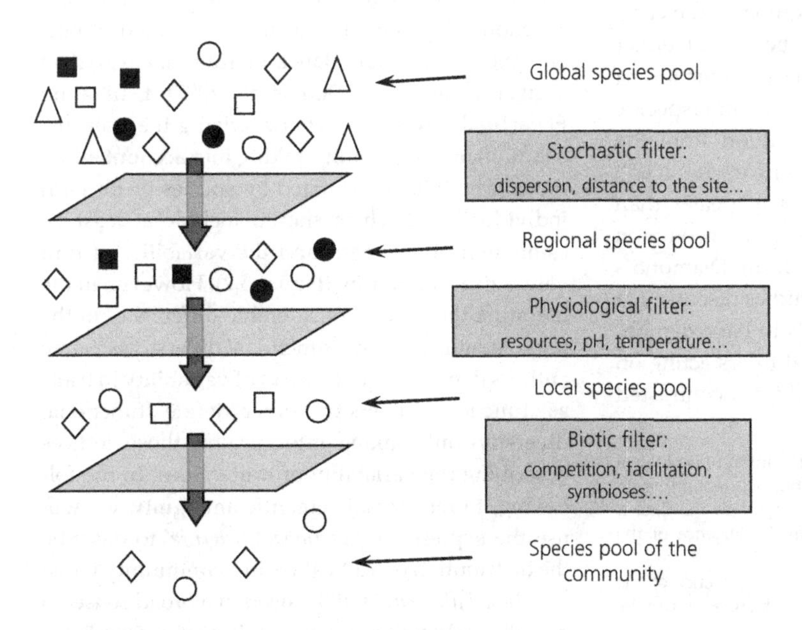

Figure 5.1 Community assembly rules represented as a series of nested filters (Lortie et al., 2004). The global species pool is the result of large-scale biogeographic events and the processes of speciation and extinction, and is subject to the action of three filters which define successively the regional pool, made up of species arriving in a location, the local pool, made up of species tolerant of local environmental conditions, and the community pool, made up of species not excluded by local interactions.

nature to environmental abiotic factors. In the following, we distinguish between the two filters linked to the habitat by classifying them as either biotic or abiotic, although we acknowledge that this distinction can sometimes be seen as subjective.

The relative effects of the different filters on the structuring of communities have been tested by new experimental and statistical methods, in particular with the generalization of the use of statistical null models[1]. These methods have shown that these filters have spatially nested effects (de Bello et al., 2013b) and that the effect of the local habitat on the assembly of communities is considerably stronger than that of stochastic effects, notably such as that of dispersal (Cornwell et al., 2006; Gilbert & Lechowicz, 2004; Götzenberger et al., 2012; Kraft et al., 2008): species coexisting in a community often show greater functional similarity than expected under a theoretical neutral model run with species from different habitats[2], while their identity, related to their taxonomic relationships, depends strongly on historical stochastic events, such as priority effects due to the order and timing of arrival at a site. This fact has been well established by studies of the temporal dynamics of the functional and floristic composition of artificially constructed grasslands. These studies show a degree of homogeneity of the functional composition of communities subject to the same environmental treatments, despite differences in the floristic composition of the communities. The latter appears to be linked either to differences in the dates of the introduction of species into the communities, or to interspecific differences in tolerance to competition from the local neighbouring species (Fukami et al., 2005). Opposite results, i.e. documenting a greater than expected functional diversity, are very uncommon although they can be expected from Diamond's ideas. A notable result from the studies discussed at the beginning of this paragraph is to have demonstrated the effect of environmental filters acting on traits (and not only on species) in highly contrasted systems such as temperate or tropical forests, temperate grasslands, and Mediterranean ecosystems.

[1] A null model is based on the randomization of ecological data or resampling, drawn from a distribution; depending on the randomized and not randomized portions of the data it should produce the expected pattern in the absence of the influence of ecological mechanisms.

[2] A theoretical neutral model assumes the absence of the structuring of information by ecological factors; it can be tested by the use of null models.

5.3 Characterizing the functional structure of a community

Demonstrating the major effects of filters on the distribution of traits within a community requires the development of new approaches for the characterization of communities via the description of their functional structure and the quantification of this by various indices.

5.3.1 Defining the functional structure of a community

The functional structure of a community was originally described by the abundance and diversity of functional groups, established on the basis of differences between species in life form, morphology, or strategies of resource acquisition (e.g. nitrogen fixing or non-nitrogen fixing species, species with C_4 or C_3 photosynthesis, etc.) (Lavorel et al., 1997). However, this type of *a priori* classification is not highly informative, and it is context dependent, for the majority of functions carried out by plants in a given environment. The generalized use of continuous traits has led to the estimation of various components of what was initially called 'community functional diversity' (FD) using the value, the range, and the relative abundance of traits measured in a given community (Díaz et al., 2007). Estimating FD actually consists of characterizing the trait distribution of a community, taking into account either mean trait values calculated by species or for each individual, and which should include at least an estimation of the mean and the variability of trait values in a community (Figure 5.2). However, in the literature there exist considerable differences in the understanding of what functional diversity is. Some authors define indices of mean of variability in traits as 'functional diversity', other define 'functional diversity' only taking into account those indices describing the variability of trait values. In the following, to remove all semantic ambiguity, we will use the expression *'functional structure'* to describe the distribution of trait values in a community while *'functional diversity'* will be used in a broad sense, to describe only trait value variability (cf. Figure 5.2).

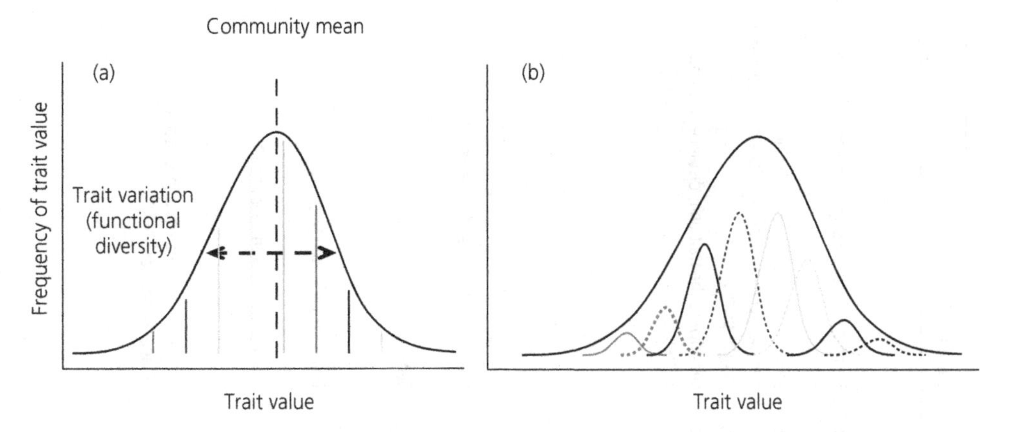

Figure 5.2 The functional structure of a community represented by the distribution of trait values in the community. Depending on the case, this distribution can be established using a mean value (a) for each species or (b) for each individual. In this second case, intraspecific variability is therefore taken into account. Levels of grey correspond to different species. Inspired by Andrew Siefert (unpublished).

5.3.2 Indices describing the functional structure of a community

Two types of indices have been developed, corresponding to the description of either the mean value or the variability of the traits in a community (Table 5.1).

A first way of estimating the mean value of the traits in a community can be formulated as follows:

$$CFP = \sum_{i=1}^{n} p_i * trait_i \qquad (5.1)$$

where CFP represents the community functional parameter (Violle et al., 2007), p_i and $trait_i$ being respectively a weighting factor and the value of the trait of the species i, with n representing the total number of species in the community. When only presence/absence data is available, all of the values of p_i are fixed at the same value (1/n), and the CFP is thus a non-weighted mean (this is the community arithmetic mean used in section 4.4.2 of Chapter 4). When species abundances are known, a standard procedure is to use the relative abundances of the species as values for p_i, which then allows the calculation of community weighted means (CWM) (Garnier et al., 2004; see also Hodgson et al., 2005, for different weighting methods). The CWM represents the most probable attribute that an individual would have if drawn at random from the community. It has proved to be a robust statistic which shows very little sensitivity to the experimental methods used for the estimation of the relative abundances of species and for trait values (Lavorel et al., 2008), which justifies its wide use in comparative studies.

A second group of indices is used to describe the functional variability which exists between the individuals or species of a given community (see Table 5.1, and the reviews by Aiba et al., 2013; Mouchet et al., 2010; Pavoine & Bonsall, 2011; Petchey & Gaston, 2006; Pla et al., 2012; Schleuter et al., 2010; and Weiher, 2011). These indices can be divided into three groups, depending on the component of diversity which they describe, the first two being derived from indices used to describe taxonomic diversity. Functional richness is defined as the volume of functional space occupied by the species (or individuals) in the community, functional evenness corresponds to the regularity of the distribution of trait abundances within this volume, and functional divergence represents the spread of the distribution of trait abundances within this volume (Figure 5.3) (definitions taken from Villéger et al., 2008). Functional diversity can also be described using other components such as specialization or functional originality (see Mouillot et al., 2013).

These components were initially estimated using single trait indices (Mason et al., 2005), but new methods including multiple traits have been proposed more recently (Villéger et al., 2008). The majority of currently used indices describe functional

Table 5.1 Comparison of the principal indices used to describe the functional structure of communities. The table indicates those indices linked to the mean (M) or the variability (V) of the trait distributions, or the three components of functional diversity (FD) originally defined by Mason et al. (2005): FR, functional richness; FE, functional evenness; FDv, functional divergence. Also indicated is whether species abundance is taken into account (Ab), whether a correlation has been demonstrated with species richness (SR), and also the principal documented limitations to their use (Limits). Compiled from Mouchet et al. (2010), Schleuter et al. (2010), Pavoine & Bonsall (2011); Weiher (2011); Aiba et al. (2013).

Index	Reference	Description	Mean/Var	FD	Ab	SR	Limits
Community-level arithmetic mean (CArM)	–	The arithmetic mean of species trait values	M	–	Yes	No	
Community-level weighted mean (CWM)	Garnier et al. (2004)	The mean of species trait values weighted by their relative abundance	M	–	Yes	No	Hump-back relationship with functional diversity
Functional diversity (FR_D)	Petchey et al. (2002)	The sum of the lengths of the branches of a functional classification	V	FR	No	Yes	Sensitive to trait unit and classification methods; creates distortions due to averaging at linkage points
Convex hull	Cornwell et al. (2006)	The volume included in the minimal convex envelope including all of the species in the functional space	V	FR	No	Yes	Sensitive to the trait unit, distance type and clustering
Functional richness (Fric)	Villéger et al. (2008)	Idem previous	V	FR	No	Yes	
Functional evenness (FEve)	Villéger et al. (2008)	The sum of the lengths of the branches of the minimal tree of functional classification, weighted by the relative abundance of species	V	FE	Yes	No	
Mean nearest neighbour distance (MNND)	Ricklefs & Travis (1980)	Mean of the functional distance to the closest neighbour in trait space	V	FE	Yes	No	Sensitive to the type of distance
Standard deviation nearest-neighbour distance (SDNND)	Ricklefs & Travis (1980)	The standard deviation of the functional distance to the closest neighbour in trait space	V	FE	No	No	Sensitive to the type of distribution
Functional attribute diversity (FAD)	Walker et al. (1999)	The sum of the distances between species pairs (in functional units)	V	FDv	No	Yes	Trivially dependent on species richness
Rao quadratic entropy (Q)	Rao (1982)	The sum of the distances between species pairs weighted by their relative abundance	V	FDv	Yes	No	Not monotonic with species richness
Functional divergence (FDiv)	Villéger et al. (2008)	The deviation of species from the minimal distance from the centre of gravity, weighted by their relative abundance	V	FDv	Yes	No*	*if the number of traits is large compared to the number of species
Functional dispersion (FDis)	Laliberté & Legendre (2010)	The mean distance of a species from the centroid of all species in a multi-dimensional space	V	FDv	Yes	No	
Community-level weighted variance (CWV)	Sonnier et al. (2010)	Variance of the trait values of species weighted by their relative abundance	V	FDv	Yes	No	Only for single traits

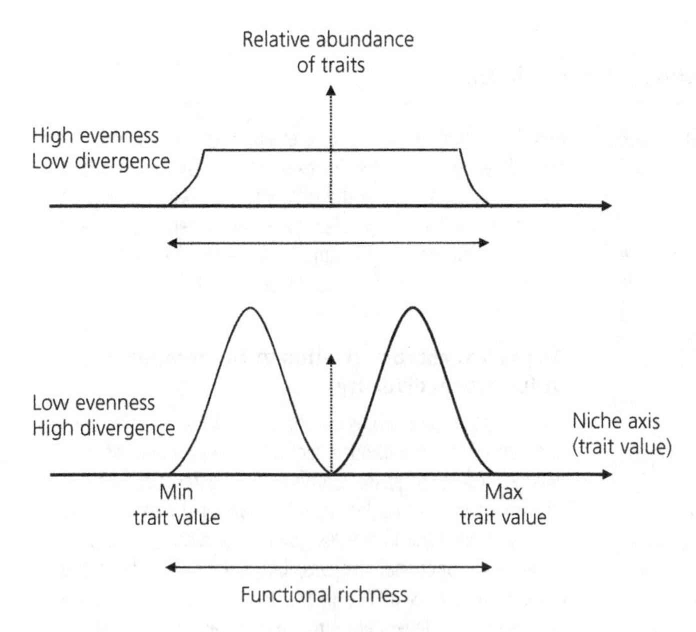

Figure 5.3 The three components of functional diversity defined by Mason et al. (2005) along an axis of trait values. The functional richness corresponds to the portion of the axis occupied by the trait values present in the community. The evenness and the functional divergence represent respectively the regularity and the inequalities in the distribution of the trait values along the axis. The figure shows two theoretical cases of patterns characterized by contrasts in evenness and functional divergence for the same values of functional richness and mean trait values.

divergence, and the most recent ones take into account the abundance of species (Table 5.1). Some use the sum (functional attribute diversity: Walker et al., 1999) or the mean (quadratic entropy: Lepš et al., 2006) of the functional distance between pairs of species localized in the same functional space; others use the distances between species established along a hierarchical classification (FD: Petchey et al., 2004) or the abundance distributions along functional axes (FD_{var}: Mason et al., 2003). All of these methods are available as packages for the statistical software R (see for example the presentation and discussion of the R package FDiversity in Pla et al., 2012).

The question of which index to use is crucial, especially given the recent increase in their number, and some criteria have been established for choosing an appropriate index (Box 5.2). The criteria for choosing an index should include one or more of the following: use one or a number of the traits linked to the targeted processes, explicitly characterize one of the components of the functional structure, take into account the abundance of species, cover different dimensions of functional structure, and have well-known relationships with specific richness as well as well-established mathematical properties (Box 5.2).

The majority of these indices have been compared according to these criteria in an effort to arrive at the most simple solution for the analysis of trait distributions. The conclusion from this work is simple: no universal index exists which can respond to all questions and which is robust to differing experimental conditions, due to the intrinsic diversity of the notion of functional diversity. Having first eliminated those indices with poor mathematical stability, indices should be chosen on the basis of links with processes which mainly explain the assembly of the studied communities, the distribution of trait values in the community, and the supposed relationships between functional diversity and niche (Aiba et al., 2013; de Bello et al., 2013a; Pavoine & Bonsall, 2011; Weiher, 2011). As a first approach, it has been proposed to bring together into the same framework the simple to use and robust indices of the CWM and Rao index (Lepš et al., 2006) or the community weighted variance (Sonnier et al., 2010), which would allow for the description of the mean and functional divergence of a community (Ricotta & Moretti, 2011). Other than these non-negligible advantages, the interest of these indices is that they can include numerous traits (Botta-Dukát, 2005), have low sensitivity to not taking into account intraspecific

Box 5.2 Five criteria for choosing a functional diversity index

Based on Garnier & Navas, 2012; Lepš et al., 2006; Pavoine & Bonsall, 2011; and Weiher, 2011.[3]

The choice of one or multiple traits

The choice of the type and number of traits enabling the precise quantification of the functional diversity of a community (FD) is crucial. Depending on the scientific question being addressed, the traits should allow for the identification of the mechanisms of coexistence between the species as well as the effects of the species on ecosystem functioning. Taking into account a number of traits in the calculation of an index allows for, in theory, greater precision in the estimation of FD, but can also lead to the inclusion of trivial relationships linked with species richness. Additionally, multi-trait indices may not vary between environments if they include traits with contrasting responses to environmental factors (Spasojevic & Suding, 2012). Finally, multi-trait indices can be sensitive to the calculation methods used, in particular the estimation of mathematical distances. These issues have led to Lepš et al. (2006) proposing that the traits used should be representative of relevant ecological strategies or have ad hoc responses to the tested environmental gradients. An additional stage consists of identifying the traits under evolutionary selection in the environment considered to have a significant impact on the targeted ecosystem functions, so as to limit the number of traits included. New methods of quantitative trait selection taking into account these aspects (Bernhardt-Romermann et al., 2008; Laliberté et al., 2012; Sonnier et al., 2012) have shown that the minimum number of traits selected can be highly variable (from six to ten traits on average) depending on the experimental or statistical method used.

Taking into account species abundance

Taking into account the abundance of species in the calculation of indices is based on the hypothesis that it is the dominant species that are most adapted to local environmental conditions (see Section 4.4.2 in Chapter 4) and have trait values different to those which explain the presence of species in an environment (see Section 5.4.4 and Laliberté et al., 2012; Shipley et al., 2006). This criterion is satisfied by the most recent indices (Table 5.1). It should, however, be noted that the method used to evaluate species abundance can influence the results. For example, an estimation based on biomass accentuates the differences between dominant species and other species, leading to an underestimation of diversity compared to an estimation based on the percentage cover or frequencies of species (Weiher, 2011).

An explicit characterization of the components of functional diversity

Choosing to quantify only one or other of the three components of FD is not without effect as richness, evenness, and divergence[4] each possess different characteristics. A comparison of these calculated indices with field data shows that: i) an increase in species richness generally induces an increase in functional richness, but not necessarily in the other components of FD; ii) the three components of FD are generally independent; and iii) they can be complementary in evaluations of the processes leading to community assembly—for example, functional richness allows for an evaluation of the effect of an environmental filter while functional divergence reflects the local equilibrium between the influences of biotic and abiotic factors (Bernard-Verdier et al., 2012; Mouchet et al., 2010; Pakeman, 2011).

Relationships with species richness

The relationships of evenness indices and functional divergence with species richness remain poorly understood, largely due to the effects of mathematical artefacts. This is an issue that requires urgent resolution (Pavoine & Bonsall, 2011). In reality, some researchers consider that species richness is the absolute measure of ecological diversity and that any 'good index' of FD should increase when a new species is introduced into the community. On the other hand, researchers for whom the functional dimension of diversity is important seek to eliminate the effect of species richness: a functional diversity index should only be sensitive to the introduction or loss of a species with trait values different to those otherwise represented. In order to provide indices appropriate for each type of study, it is urgent to characterize the behaviour of each index of FD in relation to variation in species richness.

[3] Some of these criteria, and in particular the first two, are also useful for characterizing the other component of functional structure, being the community weighted mean trait value. In this last case, the question of a choice of index is not relevant as the CWM is the only index that has been used up to now.

[4] Functional richness is the volume of functional space occupied by the species (or individuals) present in a community, functional evenness corresponds to the regularity of the abundance distribution of trait values within this volume, and functional divergence is represented by the inequalities in the abundance distribution of traits in this volume.

Mathematical status of the index

The last aspect to take into account, but not the least important, is the mathematical status of the chosen index. Many indices are mathematically complex, statistically poorly understood, and difficult to interpret from an ecological point of view, in particular as they are often drawn from fields other than functional ecology. Their calculation and conditions of use are the subject of much current discussion regarding their sensitivity to experimental and computing methods (see for example Aiba et al., 2013; Mouchet et al., 2010; Pavoine & Bonsall, 2011; Poos et al., 2009; Schleuter et al., 2010; Weiher, 2011). These publications have shown that there exists considerable variation in sensitivity to variable transformations, distance measures, and variation in the relative importance of the processes involved in community assembly. A recent proposal to improve their relevance and facilitate the choice of index by users would be to favour simpler solutions. In particular it would be useful to develop a series of harmonized indices for taxonomic, phylogenetic, and functional diversity possessing similar and well understood mathematical properties, based on clear ecological hypotheses, to allow for their comparison (Pavoine & Bonsall, 2011). One alternative would be to return to a simpler description of the shapes of trait distributions by calculating four principal central moments: mean, standard-deviation, skewness, and kurtosis (e.g. Enquist et al., 2015).

variability (Pakeman, 2014), and have well understood sensitivities to the choice of experimental and calculation methods (Lavorel et al., 2008; Lepš et al., 2006). Concerning this last point, the recent proposal of a framework allowing for the experimental decoupling of the CWM and the Rao[5] is a notable advance in the characterization of the functional structure of communities and its consequences in terms of ecosystem functioning (Dias et al., 2013).

Considering that all these calculations of functional diversity are approximations because they are based on a subset of traits, another option is to assess it by calculating phylogenetic diversity, defined as the amount of evolutionary history represented in the species of a particular community (Mouquet et al., 2012). Following the hypothesis that species closely related in the phylogeny have more similar trait values than distantly related species (Cavender-Bares et al., 2009; Srivastava et al., 2012; and see Section 3.4 in Chapter 3), some authors have suggested that phylogenetic diversity probably represents the most synthetic estimate of community trait space. However, experimental assessment of such relationships has yielded mixed results.

[5] The CWM is linked to functional diversity indices by a bell-shaped relationship, as the extremes of the distribution, areas corresponding to either very low or high mean trait values, always correspond to situations with very low functional divergence (Dias et al., 2013).

5.4 Factors controlling community functional structure

Considering the functional structure of community questions the assemblage of species as proposed by Belyea and Lancaster (1999). It shifts the central focus away from the diversity, abundance, and identity of species towards the functioning of species. The community is viewed as the result of the effect of environmental filters, which exclude those phenotypes which do not possess appropriate values of response traits. The actual community is thus composed of those phenotypes possessing adequate trait values (Keddy, 1992; McGill et al., 2006; Weiher & Keddy, 1999). This reformulation has important consequences, which are explored in the following section. It theoretically allows access to the processes underlying the effects of environmental factors within and among spatial scales, and thus to explain the differences in abundance observed between species. It also brings the focus onto the individual, the level at which traits are measured and at which environmental filters operate, leading to questions on the importance of intraspecific variability in the functional structure of a community.

5.4.1 Processes linked to biotic and abiotic filters

The processes controlling the effects of abiotic and biotic filters are by nature very different. The first filter selects species depending on their capacity to survive and reproduce according to the local abiotic conditions, this being their fundamental niche. Keddy (1992) illustrated this idea for wetlands by showing that the lower limit of the distribution of species in relation to the distance to the riverbank is determined by the ability of species to survive seasonal soil flooding.

The second filter is related to the realized niche of species, as it selects species depending on their capacity to survive and reproduce in interaction with other living organisms. As indicated elsewhere, the effect of this filter is often interpreted as a function of the principle of competitive exclusion first developed by Gause (1937), ignoring other types of density-dependent interactions. This principle states that in a stable and homogenous habitat, two species using the same limiting resource cannot coexist, with one of the species eventually excluding the other depending on their competitive ability. It has been integrated into a number of other theoretical hypotheses, and in particular is the foundation of the principle of limiting functional similarity. This principle postulates that species cannot coexist unless they occupy different ecological niches: they are differentiated along at least one axis of their niches, thus limiting the negative effects of biotic interactions on their performance (MacArthur & Levins, 1967). This second principle, while having been verified by a number of experiments (Götzenberger et al., 2012), can be invalidated, at least for some traits, under certain conditions. This is the case over time under highly competitive conditions, when the less competitive species are excluded and the species-poor community is formed of functionally similar high-competitive species (Mayfield & Levine, 2010). These complex interactions between processes linked to competition have direct consequences for the structure of communities over time. They lead to concomitant modifications of species richness and functional diversity: both are large under moderate to high competition and get lower when competitive exclusion is operating. Moreover, traits linked to niche differentiation are not the same as those which take into account competitive hierarchies (Herben & Goldberg, 2014). It is thus difficult to infer the influences of processes of biotic interactions on the basis of the distribution of traits in a community, given that the influence of other types of interactions, such as facilitation, are only beginning to be demonstrated (Schöb et al., 2012).

It should always be remembered that the processes linked to environmental filters have a major but not exclusive effect on the functional structure of communities, as they are influenced by the priority effect created by the order of arrival of species, which depends on the local history and capacities for dispersal of organisms. The following section describes the currently known consequences of interactions between these processes for the functional structure of communities.

5.4.2 The role of filtering processes on trait variability

These biotic and abiotic filtering processes result in very different consequences for the distribution of traits within a community, by modifying the different components of functional structure. They can act on functional richness by restricting the range of trait values in comparison to that found in the regional pool of species or individuals; this exclusion of individuals presenting inadequate trait values corresponds *sensu stricto* to a filtering effect. In addition, these processes can also act on the distribution of trait values for a given range of trait values. When the variability of trait values is less than that expected under a null model, the distribution of traits is said to be convergent, aggregated, or underdispersed. In the opposite case, when the variability of trait values is greater than that expected under a null model, the trait distribution is said to be divergent or over-dispersed (Figure 5.4a) (Bennett et al., 2013; de Bello et al., 2013b; McGill et al., 2006; Weiher & Keddy, 1995, 1999).

To disentangle the effects of filters on the local distribution of traits, the following hypotheses have been proposed based on a two-step approach (Figure 5.5). As has been already indicated, the availability of resources and the disturbance regime have a preponderant role amongst the local abiotic factors and act upon differing types of traits. When the abiotic filter depends on the availability of resources, the species within the community share equivalent physiological tolerances to local environmental conditions, which results in a restricted range of trait values linked to resource use as compared to the regional species pool

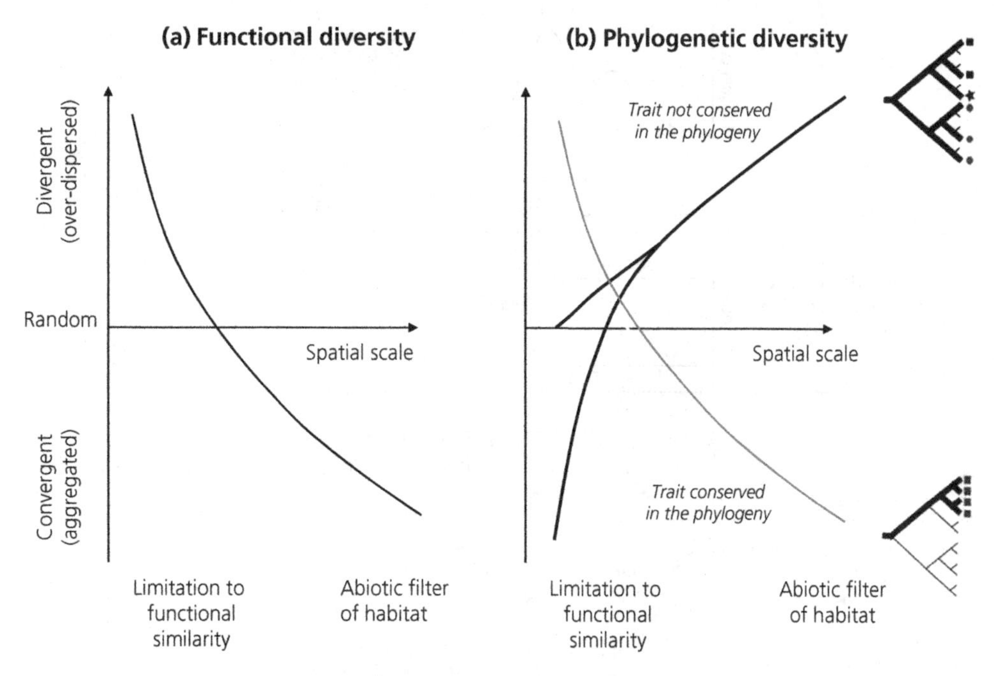

Figure 5.4 Comparison of the trait distribution in a community and in the corresponding phylogenetic tree of species found in the community as a function of the spatial scale considered. The effects of biotic interactions, which lead to a limitation in functional similarity ('*limiting similarity*') dominate at the scale of neighbouring individuals, and become progressively diluted with increasing distance as the effects of the abiotic conditions of the habitat become predominant. a) The distribution of trait values in a community can be divergent at the scale of neighbours in response to biotic interactions, and then convergent at larger scales as abiotic factors of the habitat become more influential. b) The distribution of trait values in the phylogenetic tree depends both on the relative importance of biotic and abiotic factors depending on the spatial scale, but also on the degree of trait conservatism in the phylogeny. When traits are conserved in the phylogeny, phylogenetic diversity responds in the same manner as functional diversity with changes in spatial scale, passing from divergent (over-dispersed) to convergent (aggregated) with increasing distance. In contrast, when traits are not conserved within the phylogeny, functional and phylogenetic diversity vary in opposite ways with increasing distance. The right-hand side of the figure illustrates the types of extreme distributions found in phylogenies, when traits are conserved within the phylogeny; the differences in levels of grey correspond to different types of traits linked to the same function.

(Bernard-Verdier et al., 2012; Cornwell & Ackerly, 2009; de Bello et al., 2009). In the vast majority of cases, the distribution of these traits is also convergent, which corresponds to high similarity in functioning between the selected species (Grime, 2006). However, strong environmental constraints, such as those linked to low soil water availability, can at the same time restrict the range of trait values linked to resource use, and also lead to divergence in the distribution of some traits in relation to the local coexistence of multiple response strategies to drought (Bernard-Verdier et al., 2012; Cornwell & Ackerly, 2009; Gross et al., 2013). Additionally, when the abiotic filter depends primarily on the disturbance regime, it is essentially the traits linked to the regeneration of species which will respond. There is generally a diversity of strategies allowing the establishment of new individuals after disturbance, which results in a divergent distribution for the corresponding traits (Grime, 2006; Grime & Pierce, 2012), and a stronger effect of disturbance on the functional structure of seed banks than on established vegetation (Pakeman & Eastwood, 2013).

The biotic filter also affects the distribution of traits. The hypothesis of limiting functional similarity between interacting individuals (MacArthur & Levins, 1967) results in a regular spacing of individuals in niche space, which corresponds to a divergent distribution of traits in the community (Cornwell & Ackerly, 2009; Stubbs & Wilson, 2004; Wilson, 2007). However, as already discussed, this filter can also lead to a convergence in trait distribution in cases where, the less competitive

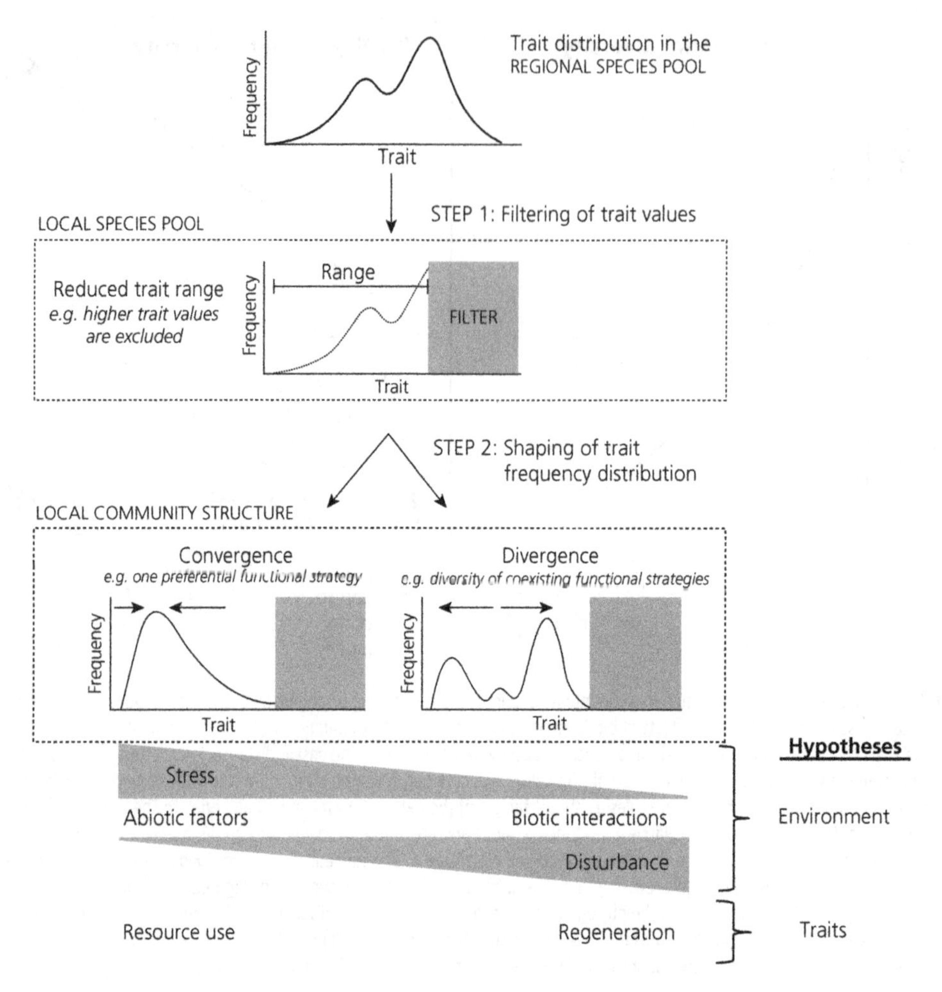

Figure 5.5 Representation of how assembly rules determine the functional structure of communities. At the first stage, the range of trait values which make up the regional pool is restricted by the effect of the habitat filter. The range of resulting traits can be distributed in either a convergent or divergent manner depending on whether the variability of values is high or low in comparison to the distribution obtained according to a null model (absence of environmental effects). Two types of hypotheses are generally proposed (and discussed in the text): i) trait distributions are more variable in favourable environments as the strong biotic interactions present lead to limitations in functional similarity; ii) traits linked to the tolerance to abiotic environmental factors, in particular with the utilization of resources, are more convergent than traits associated with regeneration. Taken from Bernard-Verdier et al. (2012).

species having been excluded from the community over the course of time, the remaining species are all tolerant to competition (Chesson, 2000; Gerhold et al., 2013; Mayfield & Levine, 2010). It is possible that these two effects of biotic filters on the functional structure of communities can occur concomitantly as they do not act on the same traits (Herben & Goldberg, 2014; Weiher et al., 1998). Traits linked to architecture and descriptive of the spatial extension of plants, such as height or rooting depth, are linked to processes leading to niche differentiation such as limiting functional similarity. In contrast, traits linked to the acquisition of resources, such as the 'fast traits' defined by Reich (2014, cf. Chapter 3), are linked to competitive hierarchies and thus can be indicative of competitive exclusion. The analysis of the structure of functional diversity of either one or the other of these types of traits would thus lead to different conclusions as to the role of biotic interactions.

The distribution of traits within a given community is thus the result of the combination of local abiotic and biotic factors: the influence of the first is expected to be greater in environments with major constraints and lead to convergent distributions, and that of the second being predominant in environments with few constraints and with increasing competition in relation to increasing vegetation cover and productivity (Grime, 2006; Weiher et al., 1998). Results of recent experiments have allowed for the partial differentiation of these hypotheses. On the one hand, we have seen earlier that no direct link exists between the type of filter and the dispersion of traits. On the other hand, recent studies carried out under highly contrasted conditions have shown that the processes linked to abiotic and biotic filters generally interact, but with differing incidence depending on the trait considered and on environmental conditions (de Bello et al., 2013b; Gross et al., 2013).

These results illustrate that the question of separating the impacts of these different processes on community assembly *in natura* cannot be resolved as yet, despite the fact that their influences have been increasingly documented in a large range of environments (Götzenberger et al., 2012; Weiher et al., 2011). The heterogeneity of results is probably due to the high sensitivity of the indices and null models used to analyse the interactions between processes, including the species pool used for randomizations and priority effect, which has a strong impact on community assembly (Aiba et al., 2013; de Bello et al., 2012). A possible approach to improving our understanding of these issues would be, once again, a widespread standardization of protocols and of the statistical methods applied (choice of null models, the use of indices with mathematical distances and variable transformations compatible with the tested processes, etc.) and better accounting for the spatial scales of investigation, as well as an improved characterization of environmental conditions (cf. Chapter 4).

5.4.3 Variations in the effect of filters with spatial or temporal scale

The relative influence of the different processes linked to abiotic and biotic filters varies with spatial scale: that of biotic interactions will be strong at the scale of neighbouring plants, which corresponds to the zone of resource use and sharing, and then will become progressively weaker in comparison to the effects of abiotic factors as the scale of observation increases. These spatial variations in the relative effects of processes should result in changes in the distribution of traits as a function of increasing spatial scale: being divergent at the level of neighbouring plants, then becoming random when the relative effects of the two types of processes cancel each other out, before finally becoming convergent (Figure 5.4a) (Weiher et al., 2011; Weiher & Keddy, 1995). This general theoretical model has been validated in number of different types of communities (Cavender-Bares et al., 2009; Laliberté et al., 2013; Stubbs & Wilson, 2004; Weiher & Keddy, 1995).

The distribution of trait values can also vary over time. The variability of traits observed during the regeneration phase of herbaceous species or tropical trees (Fukami et al., 2005; Kraft et al., 2008) is due to the coexistence of a number of effective establishment strategies (Grime, 2006; Grime & Pierce, 2012), while the traits of adult individuals tend to converge as a consequence of the non-random mortality of seedlings and juveniles. In addition, the monitoring of artificial herbaceous communities over a number of years has shown that differences in the functional structure of communities decrease over the course of time, limitations of edaphic resources being the major factor, while functional divergence increases within each community, reflecting niche differentiation in response to increasing competition (Roscher et al., 2013).

5.4.4 Environmental filters and species abundances

An important question is whether the dominant species in an environment respond differently to environmental filters to other species. In general the answer seems to be yes, as experiments carried out in different environments have shown that dominant species have restricted ranges of trait values as compared to those of other species in the community (Cornwell & Ackerly, 2010; Laliberté et al., 2012; Sonnier et al., 2012; Yan et al., 2013). For example, dominant woody species in arid areas of California have a higher average wood density than other species while those woody species dominating in wetter sites have a lower average wood density than that of subordinate species (Cornwell & Ackerly, 2010). Similarly, the dominant species of subtropical forests growing along a moisture gradient show extreme trait values as compared to other species, with lower leaf nitrogen concentrations in dry areas, and with high specific leaf area and low leaf dry matter contents in wet environments (Yan et al., 2013). As a consequence, functional divergence appears often to be larger than chance expectation, subordinate species creating skewness in trait distributions. However, these abundances are equally influenced by stochastic processes (Laliberté et al., 2012), such as the priority effect linked to the order in which species arrived, which generates competitive or facilitative associations between species, influencing their local performance.

Traits explaining variations in species abundances often respond to changes in environmental factors which have been modified over time. This is the case for changes in the flora of the United Kingdom over the last 30 years, a period marked by increases in the availability of nitrogen, and which has been characterized by progressive increases in abundance of species with high SLA and of large size (Smart et al. 2005). In contrast, traits linked to abundances at a local scale may not be linked to abundances at the scale of the landscape (Cornwell & Ackerly, 2010; Yan et al., 2013), with the reasons for such differences being poorly understood.

All of these reasons help explain why the relationship between response traits and the environment is generally stronger when the abundance of species is taken into account (Cingolani et al., 2007; Garnier et al., 2004; Pakeman et al., 2008; Shipley et al., 2006; Vile et al., 2006; and see Section 4.4.2 in Chapter 4).

5.4.5 The role of trait intraspecific variability

The use of traits in comparative ecology is largely based on the hypothesis of a greater variation in trait values between species than within species (McGill et al., 2006; Weiher et al., 2011; Westoby et al., 2002): according to this hypothesis, the differences in traits observed between species are conserved between environments, even when their absolute values change, validating the use of mean species trait values to infer processes. The confirmation of this hypothesis is essential in order to generalize the use of trait values from databases (Cordlandwehr et al., 2013, and see Chapter 9; Kazakou et al., 2014). However, some traits can show non-negligible amounts of intraspecific variability (Albert et al., 2010; Garnier et al., 2001; Kazakou et al., 2014; and see Section 4.3 in Chapter 4), and it is important to specify those situations in which it needs to be taken into account.

A first issue is to determine the relative importance of inter- and intraspecific variability in the changes in communities observed between environments. When these changes are primarily due to species replacements, the effect of intraspecific variability becomes comparatively weak. This can be seen along a gradient of soil depth in southern France, where taking into account the intraspecific variability of traits linked to the leaf economic spectrum did not significantly change estimations of their weighted means (Pérez-Ramos et al., 2012). This has also been demonstrated along gradients of topography and aridity in California where intraspecific variability explained 20% of the variation in recorded leaf surface areas (Cornwell & Ackerly, 2009). In contrast, in cases where the amount of intraspecific variability becomes significant, such as where it reaches between 30% and 45% of the variation of height and SLA along a flooding gradient (Jung et al., 2010), not taking it into account can lead to an erroneous evaluation of functional diversity and of assembly processes (see also Pakeman, 2014, using simulated data).

The relative importance of intraspecific variability compared to that existing between species is nevertheless difficult to determine as it is contingent on local conditions. First, it varies greatly between traits, both in terms of its amplitude and the type of response to environmental conditions (see Chapter 4 for examples, and Garnier et al., 2001; Roche et al., 2004). Second, the relative importance of intraspecific variability depends largely on the scale at which it is evaluated (Albert et al., 2010, 2011; Messier et al., 2010). For example, in trees from tropical forests, respectively 21% and 35% of the total variance of SLA and LDMC is due to interspecific variability, while the intraspecific variability measured between leaves represented 10% and 15%, and that between individuals 22% and 17% (Messier et al., 2010). The relative importance of intra- and interspecific variability changes as a function of the local equilibrium between the effects of abiotic and biotic factors: strong biotic constraints limit intraspecific variability and increase interspecific variability, in agreement with the principle of limiting functional similarity, while the two types of variability have equivalent levels in environments where abiotic constraints are strong (Albert et al., 2011).

This duality of responses of intraspecific variability to environmental filters can be decomposed into two components (Ackerly & Cornwell, 2007) based on the differences established between the niches α and β (Pickett & Bazzaz, 1978). The niche α corresponds to the manner in which interacting plants differ in their use of local resources, while the niche β characterizes the differences in the requirements of plants growing in different environments. As an analogy, the value of trait α is the difference between the trait value of a species and the community weighted mean calculated for all the taxa present in the community, while the value β of a trait corresponds to the mean weighted value of a species along an environmental gradient. The values β of traits being controlled by habitat abiotic factors are generally less divergent than expected from null models, while α values of traits are generally highly variable, reflecting the low functional similarity between interacting individuals. The partitioning proposed by Ackerly & Cornwell (2007) is shown in Figure 5.6 for the example of Mediterranean rangelands located on the Larzac plateau in southern France (cf. Box 8.1 in Chapter 8), and can be applied to all traits.

The recent recognition of the importance of intraspecific trait variability in the structuring of communities has led to the reconsideration of the hypothesis that different environmental filters also act on the selection of genotypes in the context of community ecology (Violle et al., 2012). Abiotic factors favour those individuals with trait values close to the average, which corresponds to stabilizing selection, while biotic factors lead to local increases in intraspecific variability so as to limit negative interactions between individuals. Tests of this theory are currently underway; and its validation would confirm the strong impact of intraspecific variability on the determination of local diversity.

Despite this rather fragmentary understanding, it is possible to identify those questions for which intraspecific trait variability should be taken into account (Albert et al., 2011, and see Figure 4.7 in Chapter 4) and provide recommendations as to the ways in which this should be done. Intraspecific trait variability should be characterized in studies focused on traits subject to evolutionary changes (for example in long-term studies), where the extremes of trait values have a particularly important ecological significance (for example the phenomena of seed dispersal), or where the response functions to certain environmental factors need to be explicitly taken into account. While it may be ignored for studies carried out at scales greater than that of the ecosystem, it should be taken into account in regional studies focused on one or a number of species, as well as in local-scale studies aiming to understand the assembly rules of species based on estimations of functional diversity; lack of consideration could lead to erroneous conclusions.

Figure 5.6 The partitioning of species traits along an environmental gradient according to the concept of α and β niches (Pickett & Bazzaz, 1978). The α niche corresponds to the niche differences between co-existing species, while the β niche corresponds to niche differences observed among environments. (a) Trait values from eight species growing along all, or part, of a gradient of the availability of soil resources in Mediterranean rangelands, represented by the community weighted means of SLA from six communities spread along this gradient. The y axis corresponds to the mean SLA value of each species in each community (a different symbol and type of line for each species). The perennial grass *Bromus erectus* is the only species present in all six communities and is shown as black circles and dashed line. (b) The values of SLA of three species are extracted from the previous graph: *Bromus erectus* (circles) is present along the whole gradient; *Stipa pennata* (squares) is present only in communities with a low SLA; while *Poa bulbosa* (triangles) is found only in sites with high SLA. For each species, the β value of the SLA is indicated with an unfilled symbol; its position on the x axis is obtained by summing over all of the sites occupied by the species the product of the weighted mean of the SLA for the community and the local abundance of the species. The α value of the trait for each species is indicated by the vertical dashed line linking in a perpendicular manner the β value of the trait to the x = y line (dotted line). *Bromus erectus* is the dominant species across the gradient, and has a β value close to the centre of the gradient of the weighted mean of SLA and a very low α value. In contrast, the two other species have β values located at the extremities of the gradient where the species are most abundant, and very high α values. These calculations were carried out using the approach developed by Ackerly & Cornwell (2007) and Shipley (2010), using the data of Fayolle (2008).

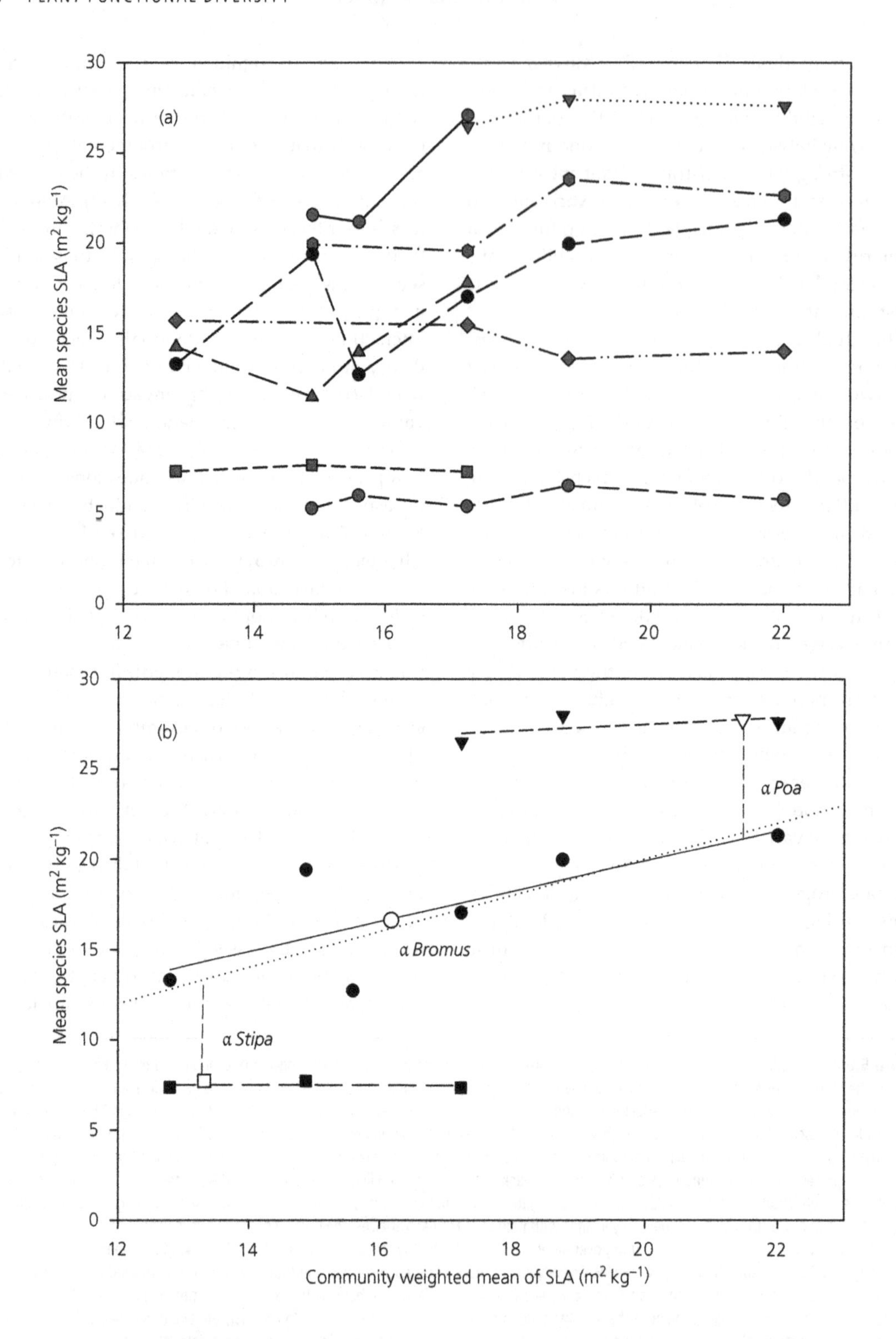
(a)
Mean species SLA (m² kg⁻¹)
30
25
20
15
10
5
0
12
14
16
18
20
22
(b)
α Poa
α Bromus
α Stipa
Mean species SLA (m² kg⁻¹)
30
25
20
15
10
5
0
12
14
16
18
20
22
Community weighted mean of SLA (m² kg⁻¹)

In this last case, a choice must also be made as to the index, i.e. community weighted mean or functional diversity index, to be used depending on its sensitivity to trait intraspecific variability (Albert et al., 2012; Pakeman, 2014). Finally, extracting data from databases appears not to be appropriate for estimating the functional structure of communities with highly variable traits such as chemical traits, for evaluations at infra-community scales and/or for studies in vegetation growing in extreme environments (Cordlandwehr et al., 2013; Kazakou et al., 2014).

5.4.6 Phylogeny, traits, and community assembly

In the preceding text we have placed the emphasis on two complementary approaches to diversity; taxonomic and, in particular, functional. There exists, however, a third manner in which to approach diversity through the phylogenetic relationships between species: in this case, it consists of identifying the structure of the evolutionary relationships between species coexisting in a community. A currently debated question is characterizing the relationships existing between these three types of diversity (Cadotte et al., 2013; Pavoine & Bonsall, 2011) and more precisely to understand if the distribution of traits in a community is dependent on the phylogenetic relationships between the species (Webb et al., 2002). This section will present the different hypotheses discussed in the literature regarding this topic, as well as the methods used to test them.

A frequently proposed hypothesis is that closely related species have more similar environmental requirements than more distantly related species; this is the hypothesis named 'phylogenetic niche conservatism' (Webb et al., 2002). According to this hypothesis, when an abiotic filter is strong, both the functional diversity and the phylogenetic diversity have a convergent distribution (Figure 5.4). Similarly, as closely related individuals will be subject to more intense competition than more distantly related individuals, a strong biotic filter will lead to a divergent distribution of the two types of diversity, corresponding to niche differentiation. In cases where the niche conservatism hypothesis is not valid, there will be no relationship between the distributions of the two types of diversity.

These predictions, summarized in Figure 5.4, have prompted a large number of experiments to determine the phylogenetic structure of communities (see Cavender-Bares et al., 2009; Pavoine & Bonsall, 2011, for reviews). A first conclusion resulting from this work is that the phylogenetic structure of communities is generally not random regardless of the spatial scale or phylogenetic level considered, which indicates a major role for historical and current environmental filters on the observed distribution of species. For example, the phylogenetic structure of Mediterranean forests can be largely explained by the distribution of traits controlling responses to fire, which in this context is a powerful past and present abiotic filter (Verdú & Pausas, 2007), while the distribution of traits linked to water use in a Central American tropical forest is directly explained by the ancestral climatic niche of the species (Sedio et al., 2013). The impact of biotic filters on community phylogenetic structure is much less clear. The hypothesis by which competition should generate an over-dispersion of phylogenetic structure, relative to niche differentiation (Webb et al., 2002), is becoming more and more called into question. On the one hand, theoretical studies have shown that the loss of species via competitive exclusion can limit the over-dispersion of phylogenetic structure, even when those traits linked with competition are conserved in the phylogeny (Mayfield & Levine, 2010). On the other hand, these effects have not been verified *in natura*, in communities where the interactions between species are more complex (Bennett et al., 2013). A second conclusion is that traits differ greatly in their level of phylogenetic conservatism. For example, seed mass is a highly conserved trait while vegetative height can differ greatly amongst individuals of these same species (Cavender-Bares et al., 2006). More generally, some authors have suggested that traits associated with the β niche, that is those linked to differences between habitats, are more phylogenetically conserved than those linked with the α niche, associated with the coexistence of species within a community (Prinzing et al., 2008; Silvertown et al., 2001).

The next step consists of testing the hypothesis of a relationship between functional diversity and phylogenetic diversity (Webb et al., 2002), because species that are closely related in the phylogeny tend to show more similar trait values than distantly related species (Cavender-Bares et al., 2009; Srivastava et al., 2012). The majority of currently completed experiments or modelling studies have refuted this hypothesis or have only found weak evidence of relationships between these types of diversity (Pavoine & Bonsall, 2011; Weiher et al., 2011). Amongst others, a study carried out in grasslands along a soil depth gradient in Mediterranean climatic conditions showed that, after having corrected for variation in taxonomic diversity, the phylogenetic diversity was not related to the functional diversity, regardless of which of the eight measured traits, or multi-trait indices, were used (Bernard-Verdier et al., 2013). The generalization of these results would lead to invalidating the relevance of phylogenetic diversity as a proxy for functional diversity, and its use, for example, to predict ecosystem processes (Cadotte et al., 2009, see also Chapter 6). However, it remains possible to use the phylogenetic structure of a community as an indicator of the 'dark' component of functional diversity, represented by traits which are poorly identified or difficult to measure (Cadotte et al., 2013; Weiher et al., 2011). This has been successfully done, for example, in a comparison of the assembly of arctic communities along an environmental gradient (Spasojevic & Suding, 2012).

In order to advance the understanding of the linkages between phylogenetic and functional diversity and community assembly, it is first necessary to overcome methodological blockages. The first of these is to improve the reliability of detecting phylogenetic signals in traits (see for example the phylogeny of specific leaf area, Figure 3.18 in Chapter 3). This is becoming more and more possible with the increasing availability of phylogenetic data, increases in the processing power of computers, and the development of relevant software tools, despite the persistence of some remaining statistical difficulties (Cavender-Bares et al., 2009). It will then be necessary to define 'good indices' for phylogenetic diversity taking into account the limits of those currently in use, as described in a number of reviews (Pausas & Verdú, 2010; Pavoine & Bonsall, 2011; Webb et al., 2002; Weiher et al., 2011). These indices, as for those used to describe functional diversity, are sensitive to the number of species sampled and the type of null model used, which makes their interpretation difficult and limits comparisons among different studies, as previously noted for some indices of functional diversity. Some authors have proposed coupling the analysis of phylogenetic and functional structures to better take into account the processes underlying community assembly (Cadotte et al., 2013; Pavoine & Bonsall, 2011). These complementary approaches should allow for a better understanding of the evolutionary history of traits and the processes underlying their past and present variability. Some results of this type currently exist, and have formed the basis for theoretical models for the interpretation of patterns of observed diversity, but they still require further confirmation (Pavoine & Bonsall, 2011).

5.5 Conclusions

The current scientific renaissance of community ecology is largely due to the strong explanatory potential of the functional component of diversity: identifying the links between identity, evolutionary history, and species functioning remains as a major challenge in advancing our understanding of the processes underlying community structure and dynamics. Further work is required to improve the methods used to answer these questions, in particular that of a harmonization and standardization of tools and approaches between scientific disciplines that until now have had little collaboration. Another challenge is to establish the relationships between communities and ecosystem functions, the subject of the following chapter.

5.6 Key points

1. The rules which determine the composition of communities, or assembly rules, are based on the hypothesis of the combined effects of different types of environmental filters on the regional pool of individuals: a first filter associated with stochastic dispersal events determines which species are potentially available; a second filter

is that of local abiotic conditions, which select for those species able to tolerate these conditions; finally, a third filter is that of biotic effects, which determine the coexistence of interacting species. The effect of these nested filters defines the distribution of traits in the community.

2. The functional structure of a community is defined by the distribution of trait values of its component individuals. It can be quantified using two major elements: (a) the community weighted mean of a trait, which is the sum for all of the species in a community of the proportion of each species in the community multiplied by its trait value, and (b) functional divergence, which can be expressed using a large number of indices. Amongst these, the Rao index and the community weighted variance (CWV) of a trait are widely used. They allow for the evaluation of functional divergence using either one or multiple traits, take into account species abundance, and are robust to variation in experimental and statistical methods.
3. The functional structure of a community varies as a function of the relative importance of the different filters and their underlying processes. The abiotic filter selects individuals on the basis of their tolerance of local conditions. When these are linked to resource availability, this leads to a restriction in the range of trait values relevant to resource use, and a convergent distribution with little dispersion around the mean. When the filter is linked to disturbance, it can lead to an over-dispersion or divergence of trait values, especially those linked to regeneration. Finally the biotic filter can have contrasting effects on trait distributions. It can lead to a divergence in trait values as expected from the hypothesis of limiting functional similarity, leading to a regular spacing of individuals within the functional space, or to a convergence in the distribution of trait values due to the competitive exclusion of less competitive individuals. The relative importance of these processes on functional structure varies between traits and as a function of the spatial and temporal scales considered.
4. The functional structure of a community depends on the intra- and interspecific variability of traits. However, their relative importance varies between traits and with the spatial scale considered: the biotic filter acts in a very local manner and tends to reduce intraspecific variability and increase interspecific variability consistent with the principle of limiting functional similarity. However, a strong abiotic filter tends to limit both types of variability in a comparable manner.
5. The structure of a community can also be described by combining traits with the evolutionary links between species (phylogenetic diversity). It is often postulated that closely related species have more similar trait values than species chosen at random. The hypothesis that follow is that, it should be possible to use the phylogenetic structure of a community to infer its functional structure. This has not been verified in a number of situations however, and therefore deserves more careful assessment.

5.7 References

Ackerly, D. D., & Cornwell, W. K. (2007). A trait-based approach to community assembly: partitioning of species trait values into within- and among-community components. *Ecology Letters, 10*, 135–145.

Aiba, M., Katabuchi, M., Takafumi, H., Matsuzaki, S.-i. S., Sasaki, T., & Hiura, T. (2013). Robustness of trait distribution metrics for community assembly studies under the uncertainties of assembly processes. *Ecology, 94*(12), 2873–2885. doi:10.1890/13-0269.1

Albert, C. H., de Bello, F., Boulangeat, I., Pellet, G., Lavorel, S., & Thuiller, W. (2012). On the importance of intraspecific variability for the quantification of functional diversity. *Oikos, 121*(1), 116–126. doi:10.1111/j.1600-0706.2011.19672.x

Albert, C. H., Grassein, F., Schurr, F. M., Vieilledent, G., & Violle, C. (2011). When and how should intraspecific variability be considered in trait-based plant ecology? *Perspectives in Plant Ecology, Evolution and Systematics, 13*(3), 217–225. doi:10.1016/j.ppees.2011.04.003

Albert, C. H., Thuiller, W., Yoccoz, N. G., Soudant, A., Boucher, F., Saccone, P., & Lavorel, S. (2010). Intraspecific functional variability: extent, structure, and sources of variation. *Journal of Ecology, 98*, 604–613.

Belyea, L. R., & Lancaster, J. (1999). Assembly rules within a contingent ecology. *Oikos, 86*, 402–416.

Bennett, J. A., Lamb, E. G., Hall, J. C., Cardinal-McTeague, W. M., & Cahill, J. F. (2013). Increased competition does not lead to increased phylogenetic overdispersion in a native grassland. *Ecology Letters, 16*(9), 1168–1176. doi:10.1111/ele.12153

Bernard-Verdier, M., Flores, O., Navas, M.-L., & Garnier, E. (2013). Partitioning phylogenetic and functional diversity into alpha and beta components along an environmental gradient in a Mediterranean rangeland. *Journal of Vegetation Science*, *24*(5), 877–889. doi:10.1111/jvs.12048

Bernard-Verdier, M., Navas, M.-L., Vellend, M., Violle, C., Fayolle, A., & Garnier, E. (2012). Community assembly along a soil depth gradient: contrasting patterns of plant trait convergence and divergence in a Mediterranean rangeland. *Journal of Ecology*, *100*(6), 1422–1433. doi:10.1111/1365-2745.12003

Bernhardt-Romermann, M., Romermann, C., Nuske, R., Parth, A., Klotz, S., Schmidt, W., & Stadler, J. (2008). On the identification of the most suitable traits for plant functional trait analyses. *Oikos*, *117*(10), 1533–1541. doi:10.1111/j.0030-1299.2008.16776.x

Botta-Dukát, Z. (2005). Rao's quadratic entropy as a measure of functional diversity based on multiple traits. *Journal of Vegetation Science*, *16*(5), 533–540. doi:10.1111/j.1654-1103.2005.tb02393.x

Cadotte, M., Albert, C. H., & Walker, S. C. (2013). The ecology of differences: assessing community assembly with trait and evolutionary distances. *Ecology Letters*, *16*(10), 1234–1244. doi:10.1111/ele.12161

Cadotte, M. W., Cavender-Bares, J., Tilman, D., & Oakley, T. H. (2009). Using phylogenetic, functional, and trait diversity to understand patterns of plant community productivity. *PLoS ONE*, *4*(5), e5695. doi:10.1371/journal.pone.0005695

Cavender-Bares, J., Keen, A., & Miles, B. (2006). Phylogenetic structure of Floridian plant communities depends on taxonomic and spatial scale. *Ecology*, *87*(7), S109–S122. doi:10.1890/0012-9658

Cavender-Bares, J., Kozak, K. H., Fine, P. V. A., & Kembel, S. W. (2009). The merging of community ecology and phylogenetic biology. *Ecology Letters*, *12*, 693–715.

Chesson, P. (2000). Mechanisms of maintenance of species diversity. *Annual Review of Ecology and Systematics*, *31*, 343–366.

Cingolani, A. M., Cabido, M., Gurvich, D. E., Renison, D., & Díaz, S. (2007). Filtering processes in the assembly of plant communities: are species presence and abundance driven by the same traits? *Journal of Vegetation Science*, *18*, 911–920.

Connor, E. F., & Simberloff, D. (1979). The assembly of species communities: chance or competition? *Ecology*, *60*(6), 1132–1140. doi:10.2307/1936961

Cordlandwehr, V., Meredith, R. L., Ozinga, W. A., Bekker, R. M., van Groenendael, J. M., & Bakker, J. P. (2013). Do plant traits retrieved from a database accurately predict on-site measurements? *Journal of Ecology*, *101*(3), 662–670. doi:10.1111/1365-2745.1

Cornwell, W. K., & Ackerly, D. D. (2009). Community assembly and shifts in plant trait distributions across an environmental gradient in coastal California. *Ecological Monographs*, *79*, 109–126.

Cornwell, W. K., & Ackerly, D. D. (2010). A link between plant traits and abundance: evidence from coastal California woody plants. *Journal of Ecology*, *98*(4), 814–821.

Cornwell, W. K., Schwilk, D. W., & Ackerly, D. D. (2006). A trait-based test for habitat filtering: convex hull volume. *Ecology*, *87*(6), 1465–1471. doi:10.1890/0012-9658

de Bello, F., Carmona, C. P., Mason, N. W. H., Sebastià, M.-T., & Lepš, J. (2013a). Which trait dissimilarity for functional diversity: trait means or trait overlap? *Journal of Vegetation Science*, *24*(5), 807–819. doi:10.1111/jvs.12008

de Bello, F., Price, J. N., Münkemüller, T., Liira, J., Zobel, M., Thuiller, W., . . . Pärtel, M. (2012). Functional species pool framework to test for biotic effects on community assembly. *Ecology*, *93*(10), 2263–2273. doi:10.1890/11-1394.1

de Bello, F., Thuiller, W., Lepš, J., Choler, P., Clement, J. C., Macek, P., . . . Lavorel, S. (2009). Partitioning of functional diversity reveals the scale and extent of trait convergence and divergence. *Journal of Vegetation Science*, *20*(3), 475–486. doi:10.1111/j.1654-1103.2009.01042.x

de Bello, F., Vandewalle, M., Reitalu, T., Lepš, J., Prentice, H. C., Lavorel, S., & Sykes, M. T. (2013b). Evidence for scale- and disturbance-dependent trait assembly patterns in dry semi-natural grasslands. *Journal of Ecology*, *101*(5), 1237–1244. doi:10.1111/1365-2745.12139

Diamond, J. M. (1975). Assembly of species communities. In M. L. Cody & J. M. Diamond (eds), *Ecology and evolution of communities* (pp. 342–444). Cambridge, MA, USA: Belknap Press.

Dias, A. T. C., Berg, M. P., de Bello, F., Van Oosten, A. R., Bílá, K., & Moretti, M. (2013). An experimental framework to identify community functional components driving ecosystem processes and services delivery. *Journal of Ecology*, *101*(1), 29–37. doi:10.1111/1365-2745.12024

Díaz, S., Lavorel, S., Chapin, F. S., III, Tecco, P. A., Gurvich, D. E., & Grigulis, K. (2007). Functional diversity—at the crossroads between ecosystem functioning and environmental filters. In J. G. Canadell, D. E. Pataki, & L. F. Pitelka (eds), *Terrestrial Ecosystems in a Changing World* (pp. 81–91). Berlin: Springer-Verlag.

Enquist, B. J., Norberg, J., Bonser, S. P., Violle, C., Webb, C. T., & Savage, V. M. (2015). Scaling from traits to ecosystems: developing a general Trait Driver Theory via integrating trait-based and metabolic scaling theories. *Advances in Ecological Research*, *52*, 249–318. doi:10.1016/bs.aecr.2015.02.001.

Fayolle, A. (2008). *Structure des communautés de plantes herbacées sur les Grands Causses: stratégies fonctionnelles des espèces et interactions interspécifiques*. PhD Dissertation, SupAgro Montpellier, Montpellier.

Fukami, T., Bezemer, T. M., Mortimer, S. R., & van der Putten, W. H. (2005). Species divergence and trait convergence in experimental plant community assembly. *Ecology Letters*, *8*(12), 1283–1290. doi:10.1111/j.1461-0248.2005.00829.x

Garnier, E., Cortez, J., Billès, G., Navas, M.-L., Roumet, C., Debussche, M., ... Toussaint, J.-P. (2004). Plant functional markers capture ecosystem properties during secondary succession. *Ecology*, *85*(9), 2630–2637.

Garnier, E., Laurent, G., Bellmann, A., Debain, S., Berthelier, P., Ducout, B., ... Navas, M.-L. (2001). Consistency of species ranking based on functional leaf traits. *New Phytologist*, *152*, 69–83.

Garnier, E., & Navas, M.-L. (2012). A trait-based approach to comparative functional plant ecology: concepts, methods, and applications for agroecology. A review. *Agronomy for Sustainable Development*, *32*, 365–399. doi:10.1007/s13593-011-0036-y

Gause, G. F. (1937). Experimental populations of microscopic organisms. *Ecology*, *18*(2), 173–179.

Gerhold, P., Price, J. N., Püssa, K., Kalamees, R., Aher, K., Kaasik, A., & Pärtel, M. (2013). Functional and phylogenetic community assembly linked to changes in species diversity in a long-term resource manipulation experiment. *Journal of Vegetation Science*, *24*(5), 843–852. doi:10.1111/jvs.12052

Gilbert, B., & Lechowicz, M. J. (2004). Neutrality, niches, and dispersal in a temperate forest understory. *Proc. Natl. Acad. Sci. USA*, *101*(20), 7651–7656. doi:10.1073/pnas.0400814101

Gotelli, N. J., & McCabe, D. J. (2002). Species co-occurrence: a meta-analysis of J. M. Diamond's assembly rules model. *Ecology*, *83*(8), 2091–2096. doi:10.2307/3072040

Götzenberger, L., de Bello, F., Bråthen, K. A., Davison, J., Dubuis, A., Guisan, A., ... Zobel, M. (2012). Ecological assembly rules in plant communities—approaches, patterns, and prospects. *Biological Reviews*, *87*(1), 111–127. doi:10.1111/j.1469-185X.2011.00187.x

Grime, J. P. (1998). Benefits of plant diversity to ecosystems: immediate, filter, and founder effects. *Journal of Ecology*, *86*, 902–910.

Grime, J. P. (2006). Trait convergence and trait divergence in herbaceous plant communities: mechanisms and consequences. *Journal of Vegetation Science*, *17*, 255–260.

Grime, J. P., & Pierce, S. (2012). *The Evolutionary Strategies that Shape Ecosystems*. Oxford: Wiley-Blackwell.

Gross, N., Börger, L., Duncan, R. P., & Hulme, P. E. (2013). Functional differences between alien and native species: do biotic interactions determine the functional structure of highly invaded grasslands? *Functional Ecology*, *27*(5), 1262–1272. doi:10.1111/1365-2435.12120

Herben, T., & Goldberg, D. E. (2014). Community assembly by limiting similarity vs. competitive hierarchies: testing the consequences of dispersion of individual traits. *Journal of Ecology*, *102*(1), 156–166. doi:10.1111/1365-2745.12181

Hodgson, J. G., Montserrat-Martí, G., Cerabolini, B., Ceriani, R. M., Maestro-Martí, M., Peco, B., ... Villar-Salvador, P. (2005). A functional method for classifying European grasslands for use in joint ecological and economic studies. *Basic and Applied Ecology*, *6*, 119–131.

Jung, V., Violle, C., Mondy, C., Hoffmann, L., & Muller, S. (2010). Intraspecific variability and trait-based community assembly. *Journal of Ecology*, *98*, 1134–1140.

Kazakou, E., Violle, C., Roumet, C., Navas, M.-L., Vile, D., Kattge, J., & Garnier, E. (2014). Are trait-based species rankings consistent across data sets and spatial scales? *Journal of Vegetation Science*, *25*(1), 235–247. doi:10.1111/jvs.12066

Keddy, P. A. (1992). Assembly and response rules: two goals for predictive community ecology. *Journal of Vegetation Science*, *3*, 157–164.

Kraft, N. J. B., Valencia, R., & Ackerly, D. (2008). Functional traits and niche-based tree community assembly in an Amazonian forest. *Science*, *322*, 580–582. doi:10.1126/science.1160662

Laliberté, E., & Legendre, P. (2010). A distance-based framework for measuring functional diversity from multiple traits. *Ecology*, *91*(1), 299–305.

Laliberté, E., Norton, D. A., & Scott, D. (2013). Contrasting effects of productivity and disturbance on plant functional diversity at local and metacommunity scales. *Journal of Vegetation Science*, *24*(5), 834–842. doi:10.1111/jvs.12044

Laliberté, E., Shipley, B., Norton, D. A., & Scott, D. (2012). Which plant traits determine abundance under long-term shifts in soil resource availability and grazing intensity? *Journal of Ecology*, *100*(3), 662–677. doi:10.1111/j.1365-2745.2011.01947.x

Lavorel, S., Grigulis, K., McIntyre, S., Williams, N. S. G., Garden, D., Dorrough, J., ... Bonis, A. (2008). Assessing functional diversity in the field—methodology matters! *Functional Ecology*, *22*, 134–147.

Lavorel, S., McIntyre, S., Landsberg, J., & Forbes, T. D. A. (1997). Plant functional classifications: from general groups to specific groups based on response to disturbance. *Trends in Ecology and Evolution*, *12*, 474–478.

Lepš, J., de Bello, F., Lavorel, S., & Berman, S. (2006). Quantifying and interpreting functional diversity of natural communities: practical considerations matter. *Preslia*, *78*, 481–501.

Lortie, C. J., Brooker, R. W., Choler, P., Kikvidze, Z., Michalet, R., Pugnaire, F. I., & Callaway, R. M. (2004). Rethinking plant community theory. *Oikos*, *107*, 433–438.

MacArthur, R. H., & Levins, R. (1967). The limiting similarity, convergence, and divergence of coexisting species *The American Naturalist, 101*, 377–385.

Magurran, A. E. (2004). *Measuring Biological Diversity* (2nd ed.). Oxford: Blackwell.

Mason, N. W. H., MacGillivray, K., Steel, J. B., & Wilson, J. B. (2003). An index of functional diversity. *Journal of Vegetation Science, 14*, 571–578.

Mason, N. W. H., Mouillot, D., Lee, W. G., & Wilson, J. B. (2005). Functional richness, functional evenness, and functional divergence: the primary components of functional diversity. *Oikos, 111*(1), 112–118.

Mayfield, M. M., & Levine, J. M. (2010). Opposing effects of competitive exclusion on the phylogenetic structure of communities. *Ecology Letters, 13*, 1085–1093.

McGill, B. J. (2006). A renaissance in the study of abundance. *Science, 314*, 770–772.

McGill, B. J. (2011). Species abundance distributions. In A. E. Magurran & B. J. McGill (eds), *Biological Diversity: Frontiers in Measurement and Assessment* (pp. 105–122). Oxford: Oxford University Press.

McGill, B. J., Enquist, B. J., Weiher, E., & Westoby, M. (2006). Rebuilding community ecology from functional traits. *Trends in Ecology and Evolution, 21*, 178–185.

McGill, B. J., Etienne, R. S., Gray, J. S., Alonso, D., Anderson, M. J., Benecha, H. K., ... White, E. P. (2007). Species abundance distributions: moving beyond single prediction theories to integration within an ecological framework. *Ecology Letters, 10*(10), 995–1015. doi:10.1111/j.1461-0248.2007.01094.x

Messier, J., McGill, B. J., & Lechowicz, M. J. (2010). How do traits vary across ecological scales? A case for trait-based ecology. *Ecology Letters, 13*, 838–848.

Mouchet, M. A., Villéger, S., Mason, N. W. H., & Mouillot, D. (2010). Functional diversity measures: an overview of their redundancy and their ability to discriminate community assembly rules. *Functional Ecology, 24*(4), 867–876.

Mouillot, D., Graham, N. A. J., Villéger, S., Mason, N. W. H., & Bellwood, D. R. (2013). A functional approach reveals community responses to disturbances. *Trends in Ecology and Evolution, 28*(3), 167–177. doi:10.1016/j.tree.2012.10.004

Mouquet, N., Devictor, V., Meynard, C. N., Munoz, F., Bersier, L. F., Chave, J., ... Thuiller, W. (2012). Ecophylogenetics: advances and perspectives. *Biological Reviews, 87*(4), 769–785. doi:10.1111/j.1469-185X.2012.00224.x

Pakeman, R. J. (2011). Functional diversity indices reveal the impacts of land use intensification on plant community assembly. *Journal of Ecology, 99*(5), 1143–1151. doi:10.1111/j.1365-2745.2011.01853.x

Pakeman, R. J. (2014). Functional trait metrics are sensitive to the completeness of the species' trait data? *Methods in Ecology and Evolution, 5*(1), 9–15. doi:10.1111/2041-210x.12136

Pakeman, R. J., & Eastwood, A. (2013). Shifts in functional traits and functional diversity between vegetation and seed bank. *Journal of Vegetation Science, 24*(5), 865–876. doi:10.1111/j.1654-1103.2012.01484.x

Pakeman, R. J., Garnier, E., Lavorel, S., Ansquer, P., Castro, H., Cruz, P., ... Vile, D. (2008). Impact of abundance weighing on the response of seed traits to climate and land use change. *Journal of Ecology, 96*, 355–366.

Pausas, J. G., & Verdú, M. (2010). The jungle of methods for evaluating phenotypic and phylogenetic structure of communities. *BioScience, 60*(8), 614–625. doi:10.1525/bio.2010.60.8.7

Pavoine, S., & Bonsall, M. B. (2011). Measuring biodiversity to explain community assembly: a unified approach. *Biological Reviews, 86*(4), 792–812. doi:10.1111/j.1469-185X.2010.00171.x

Pérez-Ramos, I. M., Roumet, C., Cruz, P., Blanchard, A., Autran, P., & Garnier, E. (2012). Evidence for a 'plant community economics spectrum' driven by nutrient and water limitations in a Mediterranean rangeland of southern France. *Journal of Ecology, 100*(6), 1315–1327. doi:10.1111/1365-2745.12000

Petchey, O. L., & Gaston, K. J. (2006). Functional diversity: back to basics and looking forward. *Ecology Letters, 9*, 741–758.

Petchey, O. L., Casey, T., Jiang, L., McPhearson, P. T., & Price, J. (2002). Species richness, environmental fluctuations, and temporal change in total community biomass. *Oikos, 99*(2), 231–240. doi:10.1034/j.1600-0706.2002.990203.x

Petchey, O. L., Hector, A., & Gaston, K. J. (2004). How do different measures of functional diversity perform? *Ecology, 85*, 847–857.

Pickett, S. T. A., & Bazzaz, F. A. (1978). Organization of an assemblage of early successional species on a soil moisture gradient. *Ecology, 59*, 1248–1255. doi:10.2307/1938238

Pla, L., Casanoves, F., & di Rienzo, J. (2012). *Quantifying Functional Biodiversity*. Dordrecht: Springer.

Poos, M. S., Walker, S. C., & Jackson, D. A. (2009). Functional-diversity indices can be driven by methodological choices and species richness. *Ecology, 90*(2), 341–347. doi:10.1890/08-1638.1

Prinzing, A., Reiffers, R., Braakhekke, W. G., Hennekens, S. M., Tackenberg, O., Ozinga, W. A., ... van Groenendael, J. M. (2008). Less lineages—more trait variation: phylogenetically clustered plant communities are functionally more diverse. *Ecology Letters, 11*, 809–819.

Rao, C. R. (1982). Diversity and dissimilarity coefficients—a unified approach. *Theoretical Population Biology, 21*, 24–43.

Reich, P. B. (2014). The world-wide 'fast–slow' plant economics spectrum: a traits manifesto. *Journal of Ecology, 102*(2), 275–301. doi:10.1111/1365-2745.12211

Ricklefs, R. E., & Travis, J. (1980). A morphological approach to the study of avian community organization. *The Auk, 97*(2), 321–338.

Ricotta, C., & Moretti, M. (2011). CWM and Rao's quadratic diversity: a unified framework for functional ecology. *Oecologia, 167*(1), 181–188. doi:10.1007/s00442-011-1965-5

Roche, P., Díaz Burlinson, N., & Gachet, S. (2004). Congruency analysis of species ranking based on leaf traits: which traits are the more reliable? *Plant Ecology, 174*, 37–48.

Roscher, C., Schumacher, J., Lipowsky, A., Gubsch, M., Weigelt, A., Pompe, S., ... Schulze, E.-D. (2013). A functional trait-based approach to understand community assembly and diversity–productivity relationships over 7 years in experimental grasslands. *Perspectives in Plant Ecology, Evolution and Systematics, 15*(3), 139–149. doi:10.1016/j.ppees.2013.02.004

Schleuter, D., Daufresne, M., Massol, F., & Argillier, C. (2010). A user's guide to functional diversity indices. *Ecological Monographs, 80* 469–484.

Schöb, C., Butterfield, B. J., & Pugnaire, F. I. (2012). Foundation species influence trait-based community assembly. *New Phytologist, 196*(3), 824–834. doi:10.1111/j.1469-8137.2012.04306.x

Sedio, B. E., Paul, J. R., Taylor, C. M., & Dick, C. W. (2013). Fine-scale niche structure of Neotropical forests reflects a legacy of the Great American Biotic Interchange. *Nature Communications, 4*, 2317. doi:10.1038/ncomms3317

Shipley, B. (2010). *From Plant Traits to Vegetation Structure. Chance and Selection in the Assembly of Ecological Communities*. Cambridge: Cambridge University Press.

Shipley, B., Vile, D., & Garnier, E. (2006). From plant traits to plant communities: a statistical mechanistic approach to biodiversity. *Science, 314*, 812–814.

Silvertown, J., Dodd, M., & Gowing, D. (2001). Phylogeny and the niche structure of meadow plant communities. *Journal of Ecology, 89*(3), 428–435.

Smart, S. M., Bunce, R. G. H., Marrs, R., LeDuc, M., Firbank, L. G., Maskell, L. C., ... Walker, K. J. (2005). Large-scale changes in the abundance of common higher plant species across Britain between 1978, 1990, and 1998 as a consequence of human activity: tests of hypothesised changes in trait representation. *Biological Conservation, 124*(3), 355–371.

Sonnier, G., Navas, M.-L., Fayolle, A., & Shipley, B. (2012). Quantifying trait selection driving community assembly: a test in herbaceous plant communities under contrasted land use regimes. *Oikos, 121*(7), 1103–1111. doi:10.1111/j.1600-0706.2011.19871.x

Sonnier, G., Shipley, B., & Navas, M.-L. (2010). Quantifying relationships between traits and explicitly measured gradients of stress and disturbance in early successional plant communities. *Journal of Vegetation Science, 21*(6), 1014–1024.

Spasojevic, M. J., & Suding, K. N. (2012). Inferring community assembly mechanisms from functional diversity patterns: the importance of multiple assembly processes. *Journal of Ecology, 100*(3), 652–661. doi:10.1111/j.1365-2745.2011.01945.x

Srivastava, D. S., Cadotte, M. W., MacDonald, A. A. M., Marushia, R. G., & Mirotchnick, N. (2012). Phylogenetic diversity and the functioning of ecosystems. *Ecology Letters, 15*(7), 637–648. doi:10.1111/j.1461-0248.2012.01795.x

Stubbs, W. J., & Wilson, J. B. (2004). Evidence for limiting similarity in a sand dune community. *Journal of Ecology, 92*(4), 557–567. doi:10.1111/j.0022-0477.2004.00898.x

Verdú, M., & Pausas, J. G. (2007). Fire drives phylogenetic clustering in Mediterranean Basin woody plant communities. *Journal of Ecology, 95*(6), 1316–1323. doi:10.1111/j.1365-2745.2007.01300.x

Vile, D., Shipley, B., & Garnier, E. (2006). Ecosystem productivity can be predicted from potential relative growth rate and species abundance. *Ecology Letters, 9*, 1061–1067.

Villéger, S., Mason, N. W. H., & Mouillot, D. (2008). New multidimensional functional diversity indices for a multifaceted framework in functional ecology. *Ecology, 89*(8), 2290–2301. doi:10.1890/07-1206.1

Violle, C., Enquist, B. J., McGill, B. J., Jiang, L., Albert, C. H., Hulshof, C., ... Messier, J. (2012). The return of the variance: intraspecific variability in community ecology. *Trends in Ecology and Evolution, 27*(4), 244–252. doi:10.1016/j.tree.2011.11.014

Violle, C., Navas, M.-L., Vile, D., Kazakou, E., Fortunel, C., Hummel, I., & Garnier, E. (2007). Let the concept of trait be functional! *Oikos, 116*, 882–892.

Walker, B., Kinzig, A., & Langridge, J. (1999). Plant attribute diversity, resilience, and ecosystem function: the nature and significance of dominant and minor species. *Ecosystems, 2*, 95–113.

Webb, C. O., Ackerly, D. D., McPeek, M. A., & Donoghue, M. J. (2002). Phylogenies and community ecology. *Annual Review of Ecology and Systematics, 33*, 475–505.

Weiher, E. (2011). A primer of trait and functional diversity. In A. E. Magurran & B. J. McGill (eds), *Biological Diversity: Frontiers in Measurement and Assessment* (pp. 175–193). Oxford: Oxford University Press.

Weiher, E., Clarke, G. D. P., & Keddy, P. A. (1998). Community assembly rules, morphological dispersion, and the coexistence of plant species. *Oikos, 81*(2), 309–322.

Weiher, E., Freund, D., Bunton, T., Stefanski, A., Lee, T., & Bentivenga, S. (2011). Advances, challenges, and a developing synthesis of ecological community assembly theory. *Philosophical Transactions of the Royal Society B: Biological Sciences, 366*, 2403–2413

Weiher, E., & Keddy, P. (1999). *Ecological Assembly Rules. Perspectives, Advances, Retreats.* Cambridge: Cambridge University Press.

Weiher, E., & Keddy, P. A. (1995). Assembly rules, null models, and trait dispersion: New questions front old patterns. *Oikos, 74*(1), 159–164.

Westoby, M., Falster, D. S., Moles, A. T., Vesk, P. A., & Wright, I. J. (2002). Plant ecological strategies: some leading dimensions of variation between species. *Annual Review of Ecology and Systematics, 33*, 125–159.

Whittaker, R. H. (1965). Dominance and diversity in land plant communities. *Science, 147*, 250–260.

Wilson, J. B. (2007). Trait-divergence assembly rules have been demonstrated: limiting similarity lives! A reply to Grime. *Journal of Vegetation Science, 18*(3), 451–452. doi:10.1658/1100-9233

Wilson, J. B. (2011). The twelve theories of co-existence in plant communities: the doubtful, the important, and the unexplored. *Journal of Vegetation Science, 22*(1), 184–195.

Yan, E.-R., Yang, X.-D., Chang, S. X., & Wang, X.-H. (2013). Plant trait-species abundance relationships vary with environmental properties in subtropical forests in eastern China. *PLoS ONE, 8*(4), e61113. doi:10.1371/journal.pone.0061113

CHAPTER 6

Plant traits and ecosystem properties

6.1 Introduction

The properties of an ecosystem depend on a large variety of factors, such as the climate, the availability of nutrients, or the composition of communities of organisms locally present. Chapin et al. (2011) proposed regrouping these factors into two major categories (Figure 6.1): i) independent control factors (corresponding to environmental states) such as climate, underlying geology, topography, potential biological diversity, and time, and ii) interactive control factors, which are those which act upon and which are themselves modulated by ecosystem properties; these include resource availability, environmental conditions (temperature, pH, etc.), disturbances, local biological diversity, and human activities. The existence of different controls on ecosystem properties at the global scale is illustrated by Figure 6.2, which shows the relationships between the Net Primary Productivity (cf. Box 6.1) of different biomes and annual precipitation on the one hand (Figure 6.2a) and average vegetation height on the other hand (Figure 6.2b).

The expression 'ecosystem properties' ('EP', hereafter) used in this chapter refers to both the pools (quantities) and fluxes (flows) of materials and energy in ecological systems, and ecosystem analysis seeks to understand the factors that regulate these properties (cf. Chapin et al., 2011). A detailed ecosystem analysis of the different factors involved is beyond the scope of this book, but the reader can find detailed treatments of these topics in e.g. Whittaker (1975), Schulze & Zwölfer (1987), Aber & Melillo (2001), and Chapin et al. (2011). The present chapter will be more particularly concerned with studies explicitly considering the controls that plants, being a major component of the living world (Figure 6.1), can exert over ecosystem properties, either on their own or in combination with other factors.

6.2 Plant communities and ecosystem properties

The recognition of the major role that plant communities play in determining EP is not recent, as revealed by a number of bibliographic reviews (Cardinale et al., 2012; Chapin et al., 2011; Eviner & Chapin, 2003; Grime, 2001; Naeem et al., 2012; Vitousek & Hooper, 1993; Wardle, 2002). Identification of the respective roles of species richness and the functional structure of plant communities in controlling EP has been the subject of an intense scientific debate over the last few decades (see e.g. Balvanera et al., 2006; Cardinale et al., 2006, 2007; Hooper et al., 2005; Loreau et al., 2001; Naeem et al., 2009). A growing consensus seems to be emerging, based on the idea that EP depend more on the functional characteristics of species than on their number (Cadotte et al., 2011; Chapin, 1993; Chapin et al., 2000; Díaz & Cabido, 2001; Díaz et al., 2007a; Grime, 1997; Lavorel, 2013; Lavorel & Garnier, 2002). These characteristics influence two distinct components of community functional structure: the mean value and the variability of the traits of individuals which make up these communities (discussed in Chapter 5). These components are implicated in two different, non-exclusive hypotheses concerning the relationship between diversity and EP. The first hypothesis, called the 'dominance' hypothesis, stipulates that the traits of the dominant species will influence EP: the so-called 'mass ratio effect' (Grime, 1998; Smith & Knapp, 2003).

Plant Functional Diversity. Eric Garnier, Marie-Laure Navas, and Karl Grigulis.

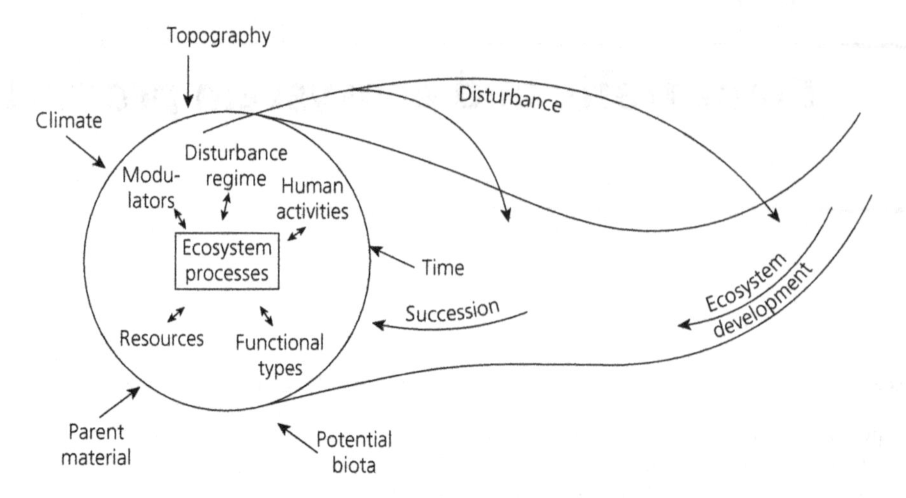

Figure 6.1 Relationships between independent control factors (outside the circle), interactive control factors (within the circle), and ecosystem properties (inside the rectangle). The circle represents the boundary of the ecosystem, whose structure and functioning respond to and affect interactive control factors, which are ultimately governed by the independent control factors. Ecosystem properties change through long-term development and short-term succession. Taken from Chapin et al. (2011). Reproduced with permission from Springer.

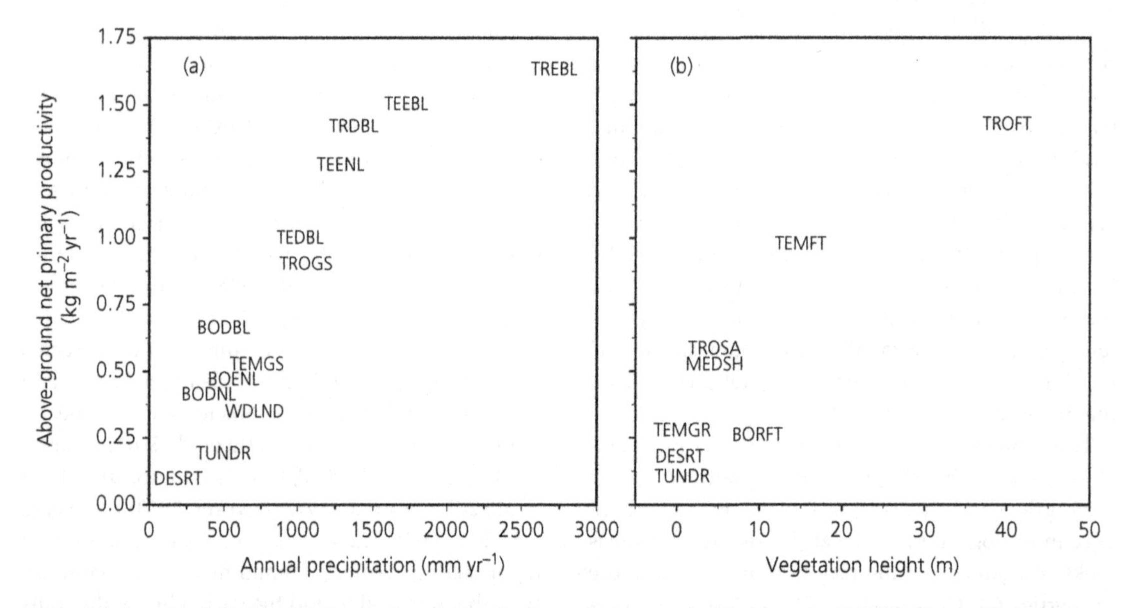

Figure 6.2 Examples of abiotic and biotic controls on the net above-ground primary productivity (above-ground NPP) of terrestrial ecosystems. Relationships between above-ground NPP and (a) annual precipitation (from Gower, 2002) and (b) mean vegetation height (data taken from Saugier et al., 2001) for different biomes. It should be noted that the classification used to define the different biomes may differ between the two figures. The correlation coefficients are (a): $r = 0.90$ ($P < 0.001$, $n = 13$) ; and (b): $r = 0.91$ ($P < 0.001$, $n = 8$). In (a): BODBL: boreal deciduous broadleaved forests; BODNL: boreal deciduous needle-leaved forests; BOENL: boreal evergreen needle-leaved forests; DESRT: deserts; TEMGS: temperate grasslands; TROGS: tropical grasslands; TEDBL: temperate deciduous broadleaved forests; TEEBL: temperate evergreen broadleaved forests; TEENL: temperate evergreen needle-leaved forests; TRDBL: tropical deciduous broadleaved forests; TREBL: tropical evergreen broadleaved forests; TUND: tundra; WDLND: woodlands. In (b): BORFT: boreal forests; DESRT: deserts; MEDSH: Mediterranean shrublands; TEMFT: temperate forests; TEMGR: temperate grasslands; TROFT: tropical forests; TROSA: savannahs and tropical grasslands; TUND: tundra.

The second hypothesis, called the 'niche complementarity' hypothesis (Petchey & Gaston, 2006; Tilman, 2001), stipulates that it is primarily the presence of different species which use environmental resources in a complementary manner that will influence ecosystem functioning.

Since ecosystem properties not only depend on the biotic components of ecosystems, but also on abiotic factors (cf. Figure 6.1), the dominance and complementarity hypotheses should be tested by controlling for these abiotic factors either experimentally (in common garden experiments for example) or statistically (see the approach using structural equation models used by e.g. Grace et al., 2007; Shipley, 2010). The framework proposed by Diaz et al. (2007b; cf. Figure 7.5 in Chapter 7), which aims at disentangling these different effects on ecosystem properties and services, takes these different aspects into account in a progressive, hierarchical manner. The effects of abiotic factors are evaluated first (this concerns experiments conducted *in natura*), then dominance and complementarity effects are assessed, and finally the potential effects of certain species or groups of species which cannot be taken into account by these two biotic components are analysed. All the steps of this framework are theoretically required to tease out the different factors controlling ecosystem properties, but it has been followed in a limited number of studies so far (see Sections 6.3 and 6.4 below). Studies in which one or several of these steps have been considered will therefore be discussed in this chapter, since significant conclusions can still be drawn from such studies. This is particularly the case for those in which the dominance and complementarity hypotheses have been tested separately—with or without taking into account the effects of abiotic factors—and which constitute a substantial piece of our current knowledge.

In the next sections, we will first present the functional basis of relationships between plant traits and ecosystem properties for selected components of biogeochemical cycles. To do so, simple additivity among species will be assumed, which forms the basis of the mass ratio effect at the core of the dominance hypothesis. Experimental tests of this hypothesis will therefore be presented in the sections following the description of each selected property. We will then review studies that have tested the complementarity hypothesis as a standalone hypothesis, before giving an overview of studies in which the dominance and complementarity hypotheses have been tested in combination. Finally, the new approach of ecosystem allometry will be briefly presented.

6.3 Functional bases for trait-ecosystem properties relationships and the dominance hypothesis

Since specific ecosystem properties are affected by a combination of traits while particular key traits are simultaneously involved in the control of multiple processes (de Bello et al., 2010; Díaz et al., 2006; Eviner & Chapin, 2003; Lavorel & Garnier, 2002; see Figure 7.3 in Chapter 7), understanding relationships between traits and ecosystem properties should be done for each property individually. This is done here for selected components of biogeochemical cycles, which have been particularly studied.

As a first step towards a better understanding of these relationships, simple additivity among species is assumed. This approach assumes that the effect of each species on the EP studied is proportional to its abundance in the community (see e.g. Equation 6.1 below for the case of net primary productivity), which means that potential effects of species interactions and/or complementarity are not taken into account. This assumption forms the basis of the dominance hypothesis initially proposed by Grime (1998) under the name of the 'mass ratio hypothesis' (or the 'mass ratio effect'), which postulates that certain traits of locally abundant species determine the magnitude of ecosystem processes, at least over time scales during which community composition is not modified. These traits are called 'effect traits' (Díaz & Cabido, 2001; Lavorel & Garnier, 2002; cf. Figure 2.6). According to the dominance hypothesis, ecosystem properties therefore depend on the weighted means of effect traits at the level of the community (Díaz et al., 2007a; Garnier et al., 2004; Violle et al., 2007b; see Chapter 5). Table 6.1 gives a list of studies that have tested the dominance hypothesis without considering potential complementarity effects.

Table 6.1 List of studies that have tested the dominance hypothesis without considering complementarity effects. Abbreviations: ANPP, above-ground net primary productivity; SANPP, specific above-ground net primary productivity; AGBmass, above-ground standing biomass; total soil C, total soil carbon content; total soil N, total soil nitrogen content. When the effects of environmental factors have been taken into account in addition to those of community structure, these are mentioned in the 'Environmental factors' column ('-' denotes that no environmental factor has been considered in the analyses).

Study	Some essential features of the study	Property	Environmental factors	Reference
1	Secondary succession in Southern France (*in natura*)	ANPP, SANPP, AGBmass, soil carbon, soil nitrogen, litter decomposition	–	Garnier et al. (2004)
2	Secondary succession in Southern France (*in natura*)	SANPP	–	Vile et al. (2006)
3	Secondary succession in Southern France (*in natura*)	Litter decomposition	–	Cortez et al. (2007)
4	11 grassland sites across Europe and Israel (*in natura*)	Standing litter	–	Garnier et al. (2007)
5	Secondary succession in Southern Sweden (*in natura*)	Litter decomposition	–	Quested et al. (2007)
6	Alpine grasslands in the French Alps (*in natura*)	AGBmass, ANPP, SANPP, litter decomposition, accumulated litter at snow melt	–	Quétier et al. (2007)
7	Alpine grasslands in the French Alps (*in natura*)	Soil water content	–	Gross et al. (2008)
8	18 grassland communities in the French Pyrénées (*in natura*)	Dry matter digestibility, rate of dry matter production, timing of production	Nutrient limitation	Ansquer et al. (2009)
9	60 forests plots across Amazonia (*in natura*)	ANPP	–	Baker et al. (2009)
10	11 grassland sites across Europe and Israel (*in natura*)	Litter decomposition	–	Fortunel et al. (2009)
11	Two permanent grasslands differing in management (*in natura*)	Dry matter digestibility	–	Andueza et al. (2010)
12	Mixed litters from pot-grown plants (common garden)	Litter decomposition	–	Pakeman et al. (2011)
13	Range of grasslands across England (*in natura*)	Soil microbial community composition	Climate, soil type, management, soil nutrient and carbon stocks	de Vries et al. (2012)
14	Alpine grasslands in the French Alps (*in natura*)	Total soil C, total soil N, green biomass, standing litter, digestibility	Altitude, nutrient limitation	Lavorel & Grigulis (2012)
15	Plant and invertebrate communities from a range of annual crops and uncropped land habitats in the UK (*in natura*)	Structure of invertebrate consumers community	–	Storkey et al. (2013)
16	Mixed litters from field-growing trees (microcosm experiment)	Litter decomposition	–	Tardif & Shipley (2013)
17	9 grassland sites across France (*in natura*)	Dry matter digestibility	Climate, management	Gardarin et al. (2014)
18	Mixed litters from field-growing herbaceous plants (common garden experiment)	Litter decomposition	–	Tardif et al. (2014)

For each component of biogeochemical cycles discussed below, we first present the rationale for the relationship between traits and ecosystem properties, before reviewing the studies in Table 6.1 which have formally tested the dominance hypothesis for this particular component.

6.3.1 Components of net primary productivity

Net Primary Productivity (NPP) represents the net carbon gain by the vegetation and determines the amount of energy available to sustain all organisms (Chapin et al., 2011). NPP can be considered as a variable integrating the functioning of the whole ecosystem, as it determines the magnitude of the fluxes of different mineral elements, as well as of secondary productivity (McNaughton et al., 1989). NPP represents the balance of gross primary productivity (the gross uptake of CO_2 by vegetation) minus autotrophic respiration, and can be assessed in numerous ways (see Luyssaert et al., 2007; Scurlock et al., 2002, for reviews). The approach developed below takes a biomass dynamics perspective on NPP, which allows one to scale up from the growth-related traits of organisms described in the previous chapters to the biomass production of ecosystems.

6.3.1.1 Functional bases of relationships with traits

Numerous traits linked to plant growth or the leaf economic spectrum described in Chapter 3 are known for their impacts on NPP of ecosystems or some of its components (Boxes 6.1 and 6.2). Building on the initial work by Chapin (1993) prolonged by Chapin et al. (1996), Lavorel and Garnier (2002) established a formal relationship between species traits and NPP, which takes into account the proportion of different species in the community. Starting from a simple biomass-based expression of NPP, and using a number of simplifying assumptions (Box. 6.1), NPP can be shown to depend on several

Box 6.1 Net primary productivity: components and estimations

NPP is defined as the total photosynthetic gain of the vegetation, minus the quantity of respired carbon, per unit area of soil surface and per unit time. In a biomass dynamics perspective, it has units of mass × (soil surface)$^{-1}$ × time^{-1}. As a coarse approximation (see e.g. Luyssaert et al., 2007, for discussion), it can be considered to correspond to the variation in living (ΔB_{Living}) and dead biomass (ΔB_{Litter}) over a given time interval (ΔT), to which is added the losses of dead tissue via the process of decomposition over this same interval (Long et al., 1989; Scurlock et al., 1999):

$$NPP = \sum_{i=1}^{n} \left[\frac{\Delta B_{Living_i} + \Delta B_{Litter_i} + [B_{Litter} \times r_D]_i}{\Delta T_i} \right] \quad (6.1.1)$$

where r_D is the relative rate of litter decomposition, and ΔT_i the time interval between two successive samples. The summation corresponds to consecutive production intervals over a given time period: these can be, for example, seasonal variations in production over a year. A precise characterization of NPP is extremely difficult and time consuming, which explains why it is often estimated using a number of simplifications and approximations (Scurlock et al., 2002).

The formulation presented in Equation 6.1.1 corresponds to those methods of measurement used in herbaceous systems; other formulations are possible, in particular in the case of forests (Clark et al., 2001; Scurlock et al., 1999).

Above- and below-ground components of NPP

NPP corresponds to the sum of the NPP of both above- and below-ground plant components. The latter is very difficult to measure *in situ*, and the majority of studies consider only the NPP of the above-ground component (above-ground NPP or ANPP in the following discussion). In herbaceous systems, the two most widely used methods for estimating ANPP are (Scurlock et al., 2002):

1) Calculating the difference between the living above-ground biomass (*BA*) of communities harvested over the course of successive time intervals, which results in the following simplification of Eq. E6.1.1:

$$ANPP = \sum_{i=1}^{n} \frac{\Delta BA_i}{\Delta T_i} \quad (6.1.2)$$

continued

Box 6.1 *Continued*

This approximation is acceptable when the time interval ΔT is short in comparison to the speed of organ renewal. In the majority of cases, only two harvests are taken: one at the beginning of the vegetation growth season (BA_0), and the other at the moment that the above-ground biomass is maximal (BA_{max}). The ANPP is therefore calculated as:

$$ANPP = \frac{BA_{max} - BA_0}{\Delta T} \quad (6.1.3)$$

2) By measuring the living above-ground biomass at the peak of biomass (BA_{max} above). This approach assumes, in addition to the above assumptions, that the biomass at the beginning of the season (B_0) is small in comparison to BA_{max}. While this is often the case in communities dominated by annuals, the widespread use of BA_{max} represents a fairly poor estimation of ANPP in vegetation dominated by perennials.Scurlock et al. (2002) noted, however, that this method can be successfully used for rough comparisons between temperate grassland ecosystems, but not to compare between temperate and tropical systems.

Specific primary productivity and production efficiency

NPP depends strongly on the amount of living plant biomass in an ecosystem (Saugier et al., 2001; see also Figure 6.9). An argument similar to that developed for comparisons between the absolute and relative growth rates of plants (Box 3.3) also applies here. To take into account these biomass effects (cf. Section 6.3.1 of the main text), it is possible to calculate a quantity corresponding to the ratio between the growth increment of living biomass and the standing biomass of the vegetation, such as:

$$SNPP = \sum_{i=1}^{n} \left[\frac{\Delta B_{Living\,i}}{\Delta T_i} * \frac{1}{B_{Living\,i}} \right] \quad (6.1.4)$$

called 'specific net primary productivity' (Garnier et al., 2004). It should be noted that the use of this equation requires the same assumptions as those outlined above in the case where NPP was calculated on the basis of living biomass only. When only one time interval is taken into account Equation 6.1.4 reduces to:

$$SNPP = \frac{\Delta B_{Living}}{\Delta T * B_{Living}} \quad (6.1.5)$$

This quantity has been referred to using several names: 'community relative growth rate' (Chapin, 1993), 'rate of biomass renewal' (Cebrián & Duarte, 1995) or 'biomass efficiency production' (Reich et al., 1992, 1997).

variables including plant traits, as (see Box 6.2 for the different steps leading to this equation):

$$NPP = \sum_{k=1}^{n} \frac{N_k * M_{Tkt_1} * \left(e^{RGR_k(t_2-t_1)_k} - 1\right)}{\Delta T} \quad (6.1)$$

In this equation, n is the total number of species in the community, N_k is the number of individuals of the species k per unit of soil surface area, M_{Tkt1} is the mean biomass of an individual of the species k at the time t_1, RGR_k and $(t_2 - t_1)_k$ correspond respectively to the RGR and the active growth period of the species k, and ΔT is the period of time over which the NPP is being estimated.

NPP is a variable expressed per unit of soil surface area (cf. Box 6.1). In Equation 6.1, trait values are weighted by the term $[N_k \times (M_{Tt1})_k]$, which corresponds to the biomass of species k in the community, per unit of soil surface area. Thus NPP depends not only on the traits of the species present, but also on their biomass per unit of area (cf. Table 6.2, Figure 6.9, and Saugier et al., 2001). To take into account this biomass effect and to evaluate more precisely the role of traits as components of the productivity of ecosystems, Reich et al. (1992, 1997) have proposed the calculation of an ecosystem 'production efficiency', this being the ratio between the above-ground NPP and the above-ground biomass of the vegetation (see Table 6.2). Garnier et al. (2004) proposed a reformulation of production efficiency based on Equation 6.1 (see Box 6.1), leading to the concept of the 'specific net primary productivity' of the ecosystem (cf. Box 6.2):

$$SNPP = log_e \left(\sum_{k=1}^{n} \frac{p_k * e^{RGR_k(t_2-t_1)_k}}{\Delta T} \right) \quad (6.2)$$

Box 6.2 Plant traits and components of net primary productivity

As described in Box 6.1, net primary productivity relates to variations in living biomass per unit area of soil surface. A supplementary stage consists of decomposing this biomass into the sum of biomasses of the constituent species of the community (Lavorel & Garnier, 2002). The total biomass of a community can thus be expressed:

$$B_{Living} = \sum_{k=1}^{n} N_k \times M_{Tk} \quad (6.2.1)$$

where N_k and M_{Tk} represent, respectively, the number of individuals and the average biomass of an individual of the species k, and n is the total number of species present in the community. By combining Equation 6.2.1 with Equation 6.1.2 for NPP (Box 6.1), applied to both the above- and below-ground parts for a given time interval ΔT, we obtain:

$$NPP = \sum_{k=1}^{n} \frac{N_k \times \left(M_{Tkt_2} - M_{Tkt_1}\right)}{\Delta T} \quad (6.2.2)$$

where M_{Tkt2} and M_{Tkt1} are the total biomasses of the species k at the times t_2 and t_1 respectively. For each species k, the relationship between M_{Tkt2} and M_{Tkt1} is written:

$$M_{Tkt2} = M_{Tkt1} \times e^{RGR_k(t_2 - t_1)_k} \quad (6.2.3)$$

where RGR_k and $(t_2 - t_1)_k$ are, respectively, the relative growth rate (cf. Box 3.3 in Chapter 3) and the active growth period of the species k; the latter is not necessarily identical to the time interval ΔT. Equation 6.2.2 can thus be written as:

$$NPP = \sum_{k=1}^{n} \frac{N_k \times M_{Tkt_1} \times \left(e^{RGR_k(t_2 - t_1)_k} - 1\right)}{\Delta T} \quad (6.2.4)$$

Which is the Equation 6.1 presented in the main text.

This same approach can be applied to the specific net primary productivity (Equation 6.1.4 and 6.1.5 from Box 6.1, and see Garnier et al., 2004). Beginning with Equation 6.1.5:

$$SNPP = \frac{\Delta B_{Living}}{\Delta T * B_{Living}} \approx \frac{log_e(B_{Living})_{t_2} - log_e(B_{Living})_{t_1}}{\Delta T} \quad (6.2.5)$$

where log_e represents the natural logarithm of the variables involved, and by combining this equation with the equations 6.2.1 and 6.2.3, we obtain:

$$SNPP = \frac{log_e\left(\sum_{i=1}^{k} N_k \times (M_{Tkt_1}) \times e^{RGR_k(t_2 - t_1)_k}\right) - log_e\left(B_{Living}\right)_{t_1}}{\Delta T} \quad (6.2.6)$$

Knowing that the difference between the logarithms of these two variables is equal to the logarithm of the ratio between these variables, and if we use p_k for the expression $\left[N_k \times (M_{Tkt1})/(B_{Living})_{t_1}\right]$, Equation 6.2.6 becomes:

$$SNPP = log_e\left(\sum_{k=1}^{n} \frac{p_k \times e^{RGR_k(t_2 - t_1)_k}}{\Delta T}\right) \quad (6.2.7)$$

In this expression, p_k represents the proportion (in biomass) of the species k in relation to the total biomass of the community. This equation is the same as Equation 6.2 in the main text.

in which p_k is the initial proportion of the species k in the community, which replaces the biomass factor used in the calculation of NPP (Equation 6.1). The SNPP, as well as the production efficiency, are both expressed in units of grams per gram of green biomass per unit of time, and corresponds to the relative growth rate of the whole community. It is thus likely that significant relationships exist between the SNPP and the weighted means of traits linked to the leaf economics spectrum, similar to those observed at the level of the whole plants.

6.3.1.2 Testing the dominance hypothesis

Net primary productivity

One of the first syntheses on the topic was conducted at a coarse scale, by compiling disparate data for four types of contrasted biomes from the temperate zone (Table 6.2; Chapin, 1993). This study showed a positive association between the arithmetic mean of relative growth rates of whole plants (RGR) calculated for all the species in the community with the efficiency of biomass production

Table 6.2 Biomass, net primary productivity, and production efficiency (above-ground productivity / above-ground biomass) of four major types of ecosystems from the temperate zone, as well as the means (not weighted by species abundance) of the maximum heights and relative growth rates (RGR) of species representative of these ecosystems. Taken from Chapin (1993). Reproduced with permission from Elsevier.

Variable	Grassland	Shrubland	Deciduous forest	Evergreen forest
Above-ground biomass (g m^{-2})	0.3±0.02	3.7±0.5	15±2	31±8
Above-ground net primary productivity (g m^{-2} $year^{-1}$)	0.3±0.02	0.4±0.07	1.0±0.08	0.8±0.08
Production efficiency (g g^{-1} $year^{-1}$)	1.0	0.1	0.07	0.03
Maximum height (cm)	100	400	2200	2200
Laboratory RGR (g g^{-1} $week^{-1}$)	1.3	0.8	0.7	0.4

(the ratio between the net primary productivity and above-ground biomass; cf. Box 6.1). Given that the RGR of species was not weighted by their local abundance, these results do not constitute a formal test of the dominance hypothesis. They still suggest an association between the growth of representative species of the different ecosystems and their productivity, however, in agreement with this hypothesis.

Given the difficulty in estimating the RGR in the field, Equation 6.1 was tested using either RGR data measured in the laboratory (RGR_{max}), or via measures of traits involved in the leaf economics spectrum and which are significantly associated with RGR (see Chapter 3). Leoni et al. (2009) showed, for example, that the interspecific differences in RGR_{max} measured under controlled conditions of the dominant species of grazed and ungrazed plots of grasslands in Uruguay were good estimators of the values of above-ground NPP measured in these same plots (Altesor et al., 2005). These results suggest that the hierarchy of species based on their values of RGR_{max} as determined under controlled conditions is conserved for the RGR achieved under field conditions, despite differences in ontogeny, plant environment interactions, and the effects of competition between plants grown in the laboratory or in the field (compare the efficiencies of production and growth rates in Table 6.2, and see the discussion in Vile et al., 2006). Additionally, relationships have been established between the mean of traits that are 'markers' of RGR and the field-estimated NPP: the above-ground NPP was linked to the community weighted mean (CWM) of the leaf area in subalpine grasslands in the French Alps (Gross et al., 2008), and to the unweighted mean of canopy leaf nitrogen concentration (in addition to canopy leaf area index) across 128 cold temperate and boreal forests (Reich, 2012).

As mentioned in Box 6.1, numerous studies use the maximum above-ground biomass produced during a vegetation season (AGB_{max}) as an estimation of the above-ground NPP of ecosystems (cf. Scurlock et al., 2002). Negative and positive relationships were respectively found between AGB_{max} and the community weighted mean of the specific leaf area and leaf dry matter content, in the herbaceous stages of secondary successions in Mediterranean old-fields (Garnier et al., 2004), while an increase in plant stature (maximum height) and a decrease in LDMC were accompanied by increased standing biomass in alpine grasslands (Quétier et al., 2007).

Relationships between traits and components of primary productivity have also been tested in common garden experiments. In the study by Pontes et al. (2007) involving 13 perennial C_3 forage species grown over two years at two levels of nitrogen availability and two defoliation regimes, above-ground NPP was negatively related to leaf nitrogen content per unit of fresh mass, suggesting that productive species dilute nitrogen in their tissues. In monocultures of species originating from Mediterranean fields grown at two levels of nitrogen availability, a multiple regression combining specific leaf area and leaf nitrogen concentration explained 59% of the variance in measured above-ground NPP during the first year of growth (Garnier et al., unpublished results).

A large-scale experiment carried out in Jena (Germany) during several years included monocultures of 61 common species from Central European grasslands growing under homogenous environmental conditions. The results obtained after four months of growth were relatively complex (Heisse et al., 2007). Once plant seedling density was accounted for, it appears that the differences in the above-ground biomass observed between the plots could be explained using different trait combinations. These include the date of seedling emergence, the vertical distribution of biomass along the stem, the size of individual stems, biomass allocation between roots and stems, and some of the traits involved in the leaf economics spectrum. In fact, high levels of biomass production could be achieved with multiple combinations of these different traits. In another experiment of a similar type, Craine et al. (2002) grew 33 grassland species over five years under conditions of low nitrogen availability. Amongst those species which were not nitrogen fixers, those which produced and accumulated large quantities of biomass were characterized by tissues of low nitrogen concentration, long-lived organs, and a high root/shoot ratio. These two last traits were identified in a model proposed by Berendse & Aerts (1987) as being those which allowed a species to dominate under conditions of low environmental nutrient availability (see Chapter 4). Under such conditions of low nitrogen availability, Craine et al. (2002) also showed that nitrogen fixation allowed leguminous species to produce more above-ground biomass than other species, in association with a higher production of stem biomass and a lower production of fine roots.

These few experiments carried out in experimental gardens with relatively limited sets of species have shown that combinations of traits influencing ecosystem properties can vary as a function of the chosen environmental conditions and the duration of the experiment. In particular, the results obtained over short time scales should not be extrapolated to the long term, especially when dealing with perennial species.

Specific net primary productivity

On the basis of a bibliographic review concerning forest ecosystems, Reich et al. (1992) showed that the efficiency of above-ground biomass production (Equation 6.2) was positively correlated with the specific leaf area and negatively with the leaf life span of the major species found in these ecosystems. In addition, Garnier et al. (2004) tested Equation 6.2 using data drawn from the study of a secondary succession in the south of France. These authors found strong positive relationships between the above-ground SNPP and the weighted means of the specific leaf area and the leaf nitrogen concentration. It is interesting to note that for the same successional series, Vile et al. (2006) found a positive relationship between the above-ground SNPP and the weighted mean of the potential relative growth rates—measured under optimal growth conditions in the laboratory—of the species making up the different communities. All of these three studies illustrate the strong effects of traits involved in the leaf economics spectrum on the specific productivity of ecosystems. These traits thus have demonstrated effects on functions at least at three levels of organization: the leaf, the whole plant, and the community. This is illustrated for the case of the specific leaf area in Figure 6.3, which shows the relationships between this trait and the leaf net photosynthetic rate, the relative growth rate of whole plants, and the SNPP of plant communities. These results tend to validate the dominance hypothesis for the different components of ecosystem productivity (see also Grime & Pierce, 2012).

The dominance hypothesis has not been verified in all of the contexts in which it has so far been studied, however. A study including 60 plots located in Amazonian forest, distributed from the coast of the Pacific Ocean to that of the Atlantic Ocean over a distance of 4000 km, did not find any evidence of a relationship between the gross wood production and the weighted mean of the maximum height or that of the wood density. However, this production varied considerably between the plots, as a function of environmental factors which were not able to be clearly identified in this study (Baker et al., 2009). It appears that some supplementary work will be required in a larger number of situations in order to unravel the relative importance of abiotic factors and the functional structure of communities on the components of NPP (cf. Figure 7.5).

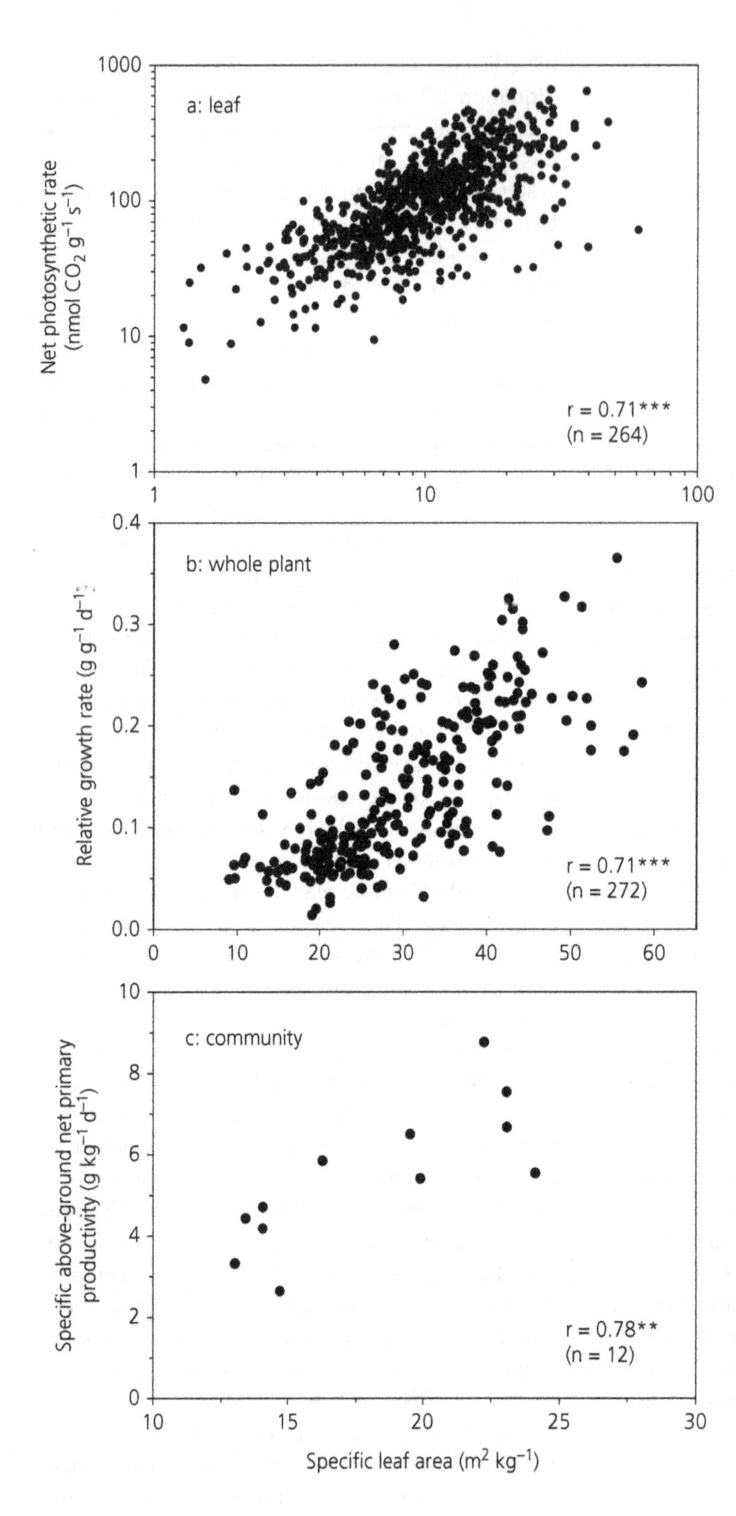

Figure 6.3 Relationships between specific leaf area and processes measured at three levels of organization. (a): leaf net photosynthetic rate; (b): whole plant relative growth rate; (c): above-ground specific net primary productivity of the plant community. For this last case, the specific leaf area is weighted by the relative abundance of the species in the community (cf. eq. 6.2). The Pearson correlation coefficient (r) and the number of replicates (in brackets) are indicated for each relationship. Significance levels: **, $P < 0.01$; ***, $P < 0.001$. It should be noted that the quantities represented on the x axes are equivalent and that they are expressed in units comparable between the three levels of organization. Taken from Violle et al. (2007b).

6.3.2 Litter decomposition and dry matter digestibility

6.3.2.1 Functional bases of relationships with traits

Processes such as the efficiency of production (or the SNPP) and litter decomposition are tightly linked in ecosystems (e.g. Cebrián & Duarte, 1995; Chapin et al., 2011). However, the development of a formalism linking traits to decomposition comparable to that developed for the SNPP above is nevertheless difficult, as the relationships to be taken into account are less direct in the case of decomposition. These would in particular require formalizing links between the traits of living leaves and those litter characteristics that influence their decomposition, such as the ratios between the concentrations of carbon and nitrogen and/or the ratios between lignin content and nitrogen concentration (see Figure 6.5, and the bibliographical syntheses in e.g. Aerts, 1997; Aerts & Chapin, 2000; Chapin et al., 2011; Wardle, 2002).

A number of studies also suggest that the rate of decomposition of plant litter should be related to the quality of the plant material as used by herbivores (Bardgett & Wardle, 2010; Lavorel & Grigulis, 2012; Wardle et al., 2004; Figure 6.4), a relationship discussed at the organism level in Chapter 3. Together with the amount and the timing of dry matter production, the quality of the plant material is a critical component of herbage nutritive value. It can be estimated by the measurement of dry matter digestibility of green material, which provides a synthetic measure of the amount of energy available for herbivores in plant components (Bruinenberg et al., 2002). Differences in digestibility result from differences in the structural/metabolic tissue ratio, which depends on plant development, and from the temporal variation in the digestibility of structural tissues as they age (Duru et al., 2008). The poorly digestible biomass fraction in plant tissues relates to high structural carbohydrates content in cell walls, together with a high degree of lignification and high fibre concentration (Bruinenberg et al., 2002; Choong et al., 1992).

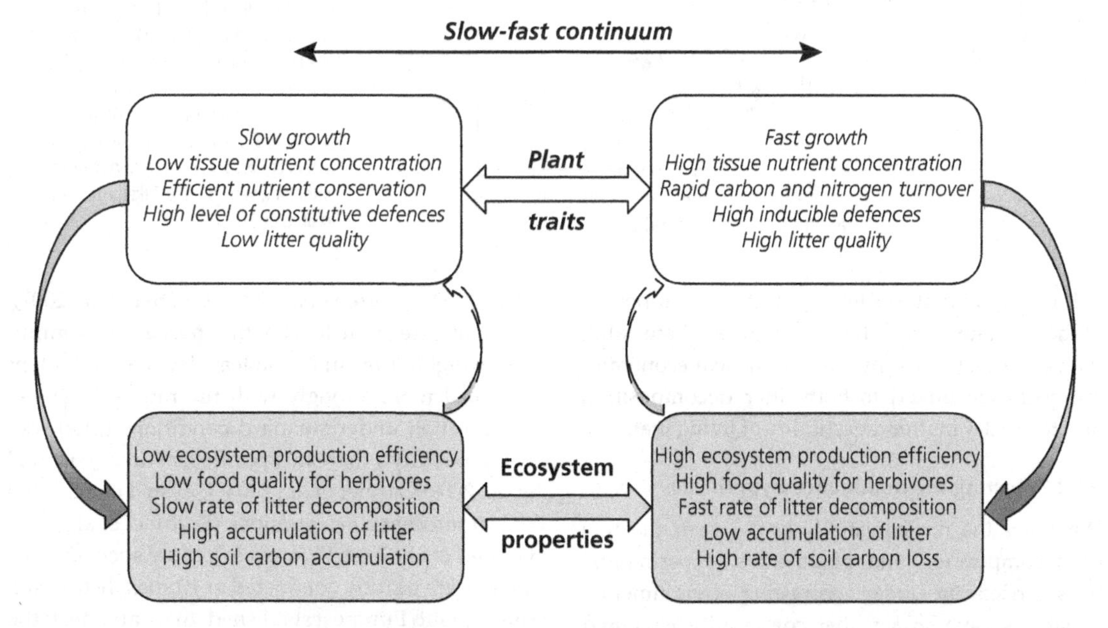

Figure 6.4 One major driver of ecosystem functioning is the difference in fundamental plant traits between slow and fast growing species (the 'slow–fast continuum': cf. Chapter 3). Plant traits serve as determinants of the quality, quantity, and fluxes of resources in the ecosystem. The effects of organism traits on ecosystem properties are represented by the large curved arrows on the two external sides of the boxes, and feedback effects from ecosystems to organisms are represented by the smaller curved arrows on the internal sides of the boxes (based on Wardle et al., 2004; Bardgett & Wardle, 2010; Lavorel & Grigulis, 2012; Reich, 2014).

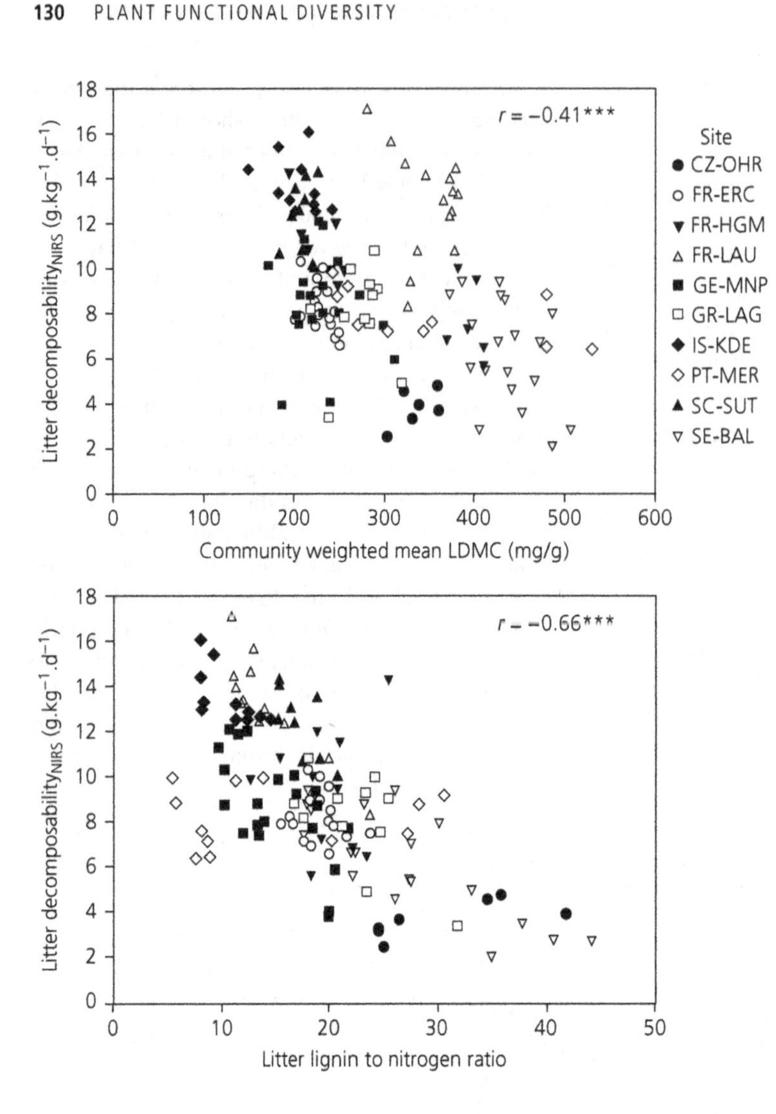

Figure 6.5 Correlations between the decomposability (i.e. decomposition estimated in the laboratory under standardized conditions) of litter from herbaceous communities predicted by near infrared spectroscopy and: (a) the community weighted mean of leaf dry matter content (LDMC), and (b) the initial lignin/nitrogen ratio of the litter. The correlation between the weighted mean of the LDMC and the lignin/nitrogen ratio is positive and significant (r = 0.34***, not shown). These relationships were established for 10 sites in Europe and Israel subjected to differences in land use. The Pearson correlation coefficients (r) and the level of significance of the relationships are indicated on the figure (***, $P < 0.001$). Site abbreviations: the first two letters are used to indicate the country with CZ: Czech Republic; FR: France; GE: Germany; GR: Greece; IS: Israel; PT: Portugal; SC: Scotland; SE: Sweden, and the three last letters corresponding to different sites within a given country. Taken from Fortunel et al. (2009).

If the broad patterns linking traits and properties at the ecosystem level shown in Figure 6.4 are valid, we expect that traits involved in the leaf economics spectrum are linked to both litter decomposition and to the dry matter digestibility of living material.

6.3.2.2 Testing the dominance hypothesis

Whatever the mechanistic links between the rate of decomposition and plant traits, experimental tests carried out under contrasting environmental situations have shown that community weighted means of a number of traits involved in the leaf economics spectrum do indeed have an impact on the rate of litter decomposition (Cortez et al., 2007; Fortunel et al., 2009; Garnier et al., 2004; Pakeman et al., 2011; Quested et al., 2007; Quétier et al., 2007). As is the case at the level of the species, the community weighted mean of the leaf dry matter content is linked more strongly with the rate of litter decomposition under standard conditions (litter 'decomposability') than are specific leaf area (Fortunel et al., 2009; Garnier et al., 2004; Quested et al., 2007) or leaf nutrient concentrations (Fortunel et al., 2009; Garnier et al., 2004; Figure 6.5a). More specifically, a comparative study conducted at 10 sites distributed throughout Europe established that, amongst the traits of living leaves, the weighted mean of the leaf dry matter content was the parameter best correlated with the ratio of lignin to nitrogen of fresh litter, with this latter trait explaining 44% of the variation in the rate of litter decomposition (Fortunel et al., 2009: Figure 6.5b).

Tardif and Shipley (2013) and Tardif et al. (2014) used a different approach to test the biomass ratio hypothesis as applied to litter decomposition: they compared the decomposition rates of mixed-species litter with the community weighted means of decomposition rates measured on single-species litters which composed the mixtures. They found close associations between observed and calculated decomposition rates, both for tree litter decomposing in controlled conditions (Tardif & Shipley, 2013) and for herbaceous litter decomposing at three grassland sites (Tardif et al., 2014).

As the quantity of standing litter depends on the equilibrium between its production and decomposition, plant traits linked to either of these two processes can potentially have effects on this quantity. For example, it has been shown that the community weighted mean of leaf dry matter content was in fact positively linked to the quantity of aboveground standing litter in communities established in the 10 European sites and compared in the previously cited study (Garnier et al., 2007). But other traits also seem to have influences on this property, such as plant height, nitrogen concentration, and the fracture resistance of leaves (see Díaz et al., 2007b).

The dominance hypothesis has been tested with respect to digestibility in few situations, and in somewhat different ways. Using a discrete approach based on the classification of grass species into different types (cf. Section 8.2.6 in Chapter 8), Duru et al. (2008) and Ansquer et al. (2009) showed that grasslands with a high proportion of grasses with low leaf dry matter content had a higher digestibility than grasslands dominated by grasses with a high leaf dry matter content. Ansquer et al. (2009) conducted a more formal test of the dominance hypothesis, and found a negative relationship between digestibility and the proportion of a plant functional type consisting of stress-tolerant species with high leaf dry matter content, calculated for grasses in 18 grassland plots. Andueza et al. (2010) found a negative relationship between the community weighted mean of leaf dry matter content and digestibility in a comparison of two meadows differing in management intensity, while Lavorel & Grigulis (2012) found a positive relationship between digestibility and CWM of leaf nitrogen concentration in a set of 63 montane grassland plots.

In a recent study conducted in nine grassland sites over a large range of climatic conditions, soil resource levels, and management regimes, Gardarin et al. (2014) found that community digestibility and most community weighted means of traits responded to climatic factors and management regimes, but that relationships were not always significant when each site was considered separately. Community digestibility was significantly related to one or more plant traits within each site, and to all of the measured traits when considering all the sites. Leaf dry matter content (LDMC) had the most consistent effect on digestibility across all sites, with a strikingly similar effect within each site. Potential evapotranspiration was negatively related to digestibility, and explained a large part of the between-site variance. In addition, the intensity of disturbance (biomass removal) was associated with a high digestibility.

6.3.3 Other components of biogeochemical cycles

Numerous other ecosystem properties involved in biogeochemical cycles are certainly affected by traits (see the reviews by Eviner & Chapin, 2003; Wardle, 2002). Amongst these, soil carbon storage has been particularly intensively studied (e.g. De Deyn et al., 2008; Wardle et al., 2004).

6.3.3.1 Functional bases of relationships with traits

In contrast to the processes presented above, a certain number of complex intermediate stages need to be taken into account to understand the links between plant traits and soil carbon storage. Soil carbon storage is in fact the result of carbon inputs and exports, and both of these two processes can be influenced by a large number of traits as well as by the potential interactions between species (Figure 6.6). Traits linked to plant growth and the production and quality of litter certainly play an important role in soil carbon storage, but their relative importance in relation to other factors, as presented

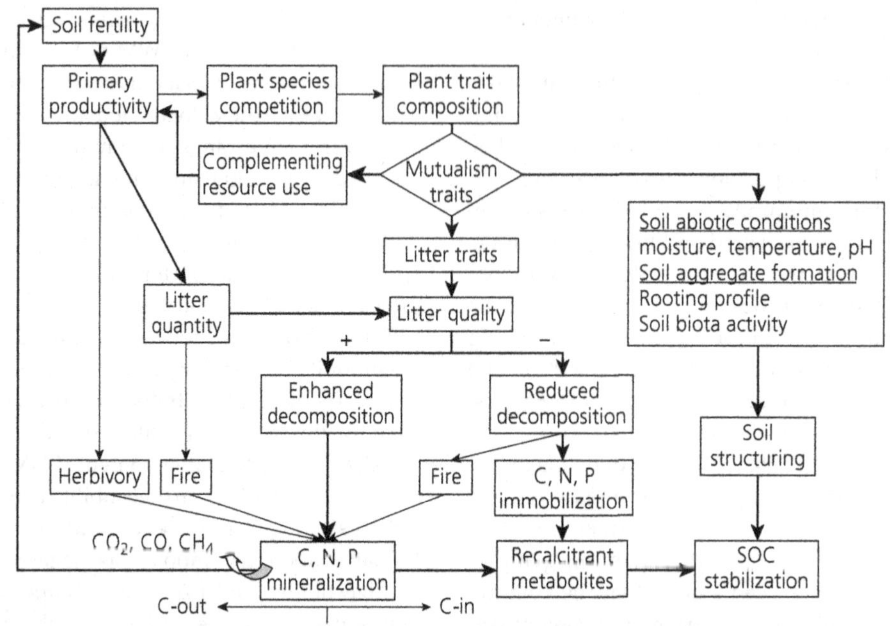

Figure 6.6 Potential plant trait composition effects on soil carbon sequestration, through their influence on the ratio between carbon gains (C-in) and losses (C-out). The trait composition of a community, which reflects the nature and abundance of the principal plant functional groups but also the biotic interactions in which they are involved, has a major influence on the fate of soil carbon. It should also be noted that that the impact of external events such as fire or herbivory can modify the storage of soil carbon by affecting the quantity of material which is mineralized. Taken from De Deyn et al. (2008). Reproduced with permission from Wiley.

in Figure 6.6, has not previously been evaluated in a systematic manner.

6.3.3.2 Testing the dominance hypothesis

Given the number of steps and processes involved (Figure 6.6), a high number of traits can potentially influence soil carbon content (De Deyn et al., 2008; Eviner & Chapin, 2003), and some above-ground traits have the potential to influence edaphic processes. For example, in a post-cultural succession in the Mediterranean region, Garnier et al. (2004) found a negative correlation between the total soil carbon content and the weighted mean of the specific leaf area and leaf nitrogen concentration, and a positive correlation with leaf dry matter content. These relationships result from the fact that these traits of living leaves lead to slow rates of decomposition (Cortez et al., 2007; Quested et al., 2007) and high rates of litter accumulation (Garnier et al., 2007), two variables which have a major effect on the flux of carbon entering the soil (Figure 6.6).

6.3.4 Soil water content

The various fluxes and compartments involved in the water balance of ecosystems are an aspect of their functioning which is not necessarily influenced by the same traits as those involved in the storage and transfer of carbon and soil mineral elements discussed previously. This section will describe how the traits of species present in a community influence the quantity of water present in the soil.

6.3.4.1 Functional bases of relationships with traits

At any given moment, the quantity of water present in the soil is a function of the balance between precipitation, evapotranspiration, runoff of surface water, and the infiltration of water into the soil (Figure 6.7, and see e.g. Larcher, 2003). To our knowledge, no clear information exists as to the manner in which plant species can affect the last two of

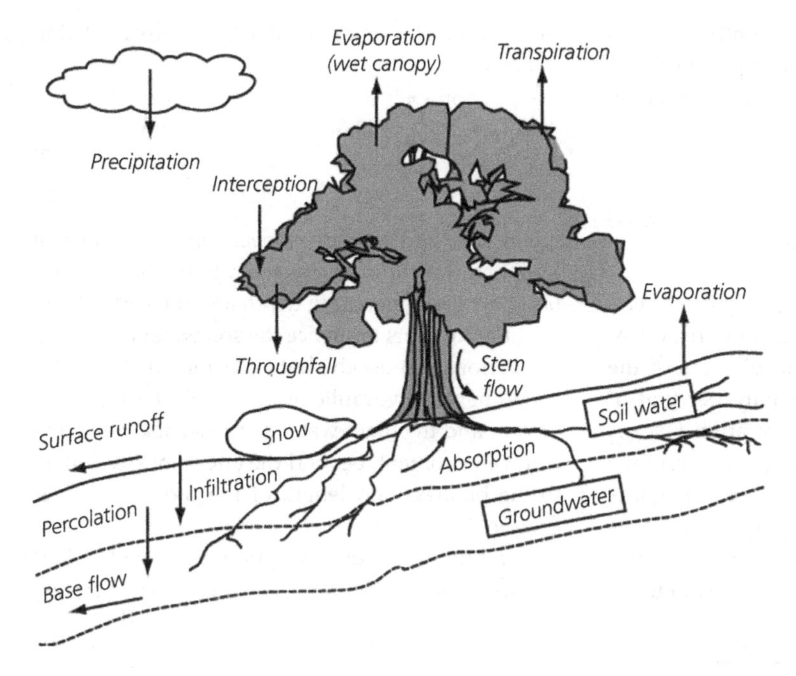

Figure 6.7 The various components of the water balance of an ecosystem as described by Waring & Running (1998). The terms of the simplified water balance presented in the text are: soil water, rainfall, evaporation, and absorption (S_t, $R_{\Delta T}$, $E_{\Delta t}$, and $W_{ab\Delta T}$, respectively). Taken from Chapin et al. (2011). Reproduced with permission from Springer.

the processes[1] involved in this balance. They will therefore not be taken into account in the following discussion.

Assuming that actual evapotranspiration during the time interval Δt is the sum of direct evaporation from the soil and the absorption of water by the vegetation, a simplified water balance can be expressed in the following manner (cf. e.g. Boulant et al., 2008):

$$SWC_{t+1} = \min[SWC_t + R_{\Delta t} - E_{\Delta t} - W_{ab\Delta T}, SWC_{FC}] \quad (6.3)$$

in which SWC_{t+1} and SWC_t correspond to the soil water contents at the times $t + 1$ and t, respectively, $R_{\Delta t}$, $E_{\Delta t}$, and $W_{ab\Delta t}$ correspond to rainfall, soil evaporation, and the absorption of water by plants during the interval of time between $t + 1$ and $t(= \Delta t)$, while SWC_{FC} represents the soil water content at field capacity (this being the quantity of water retained by a saturated soil after drainage of excess water by gravity). According to Equation 6.3, the decrease in soil water content after a rainfall event depends then on $E_{\Delta t}$ and $W_{ab\Delta t}$ (see also Eviner & Chapin, 2003). $E_{\Delta t}$ is strongly dependent on the solar radiation received per unit of soil surface area, except when rainfall intercepted by the vegetation evaporates directly (cf. Schulze et al., 2005). $E_{\Delta t}$ is inversely proportional to the leaf area index of the vegetation (Schulze et al., 2005) and the accumulation of litter (Eviner & Chapin, 2003). The effects of traits directly linked to plant growth and litter accumulation have already been extensively discussed in previous sections in this chapter, and here we will discuss principally those plant traits which are involved in the process of soil water absorption ($W_{ab\Delta t}$) by vegetation.

It has been well established that the vertical distribution of roots as well as the maximum depth to which they extend can vary greatly between coexisting species within a community (cf. Figure 3.15 in Chapter 3 for the maximum rooting depth of species in different biomes). Additionally, most soil properties also vary with their depth (Schwinning & Ehleringer, 2001, and references therein). Consequently, the amount of water absorbed over the entire soil profile is often divided into amounts of

[1] Hypotheses do exist, but these are not currently supported by quantitative data. For example, a dense and deep root mat increases the rate of soil water infiltration while the relative magnitude of soil surface leaching depends on above-ground architecture and litter production (Stokes et al., 2009).

water absorbed by strata of soil depth, and can be expressed as the sum of the absorption in each of these strata (see details of this calculation in Box 6.3):

$$W_{ab\Delta t} = \sum_{i=1}^{m} \sum_{j=1}^{n} \left[B_{root(i,j)} * SRA_{i,j} * k_{i,j} * (\Psi_{soil,i} - \Psi_j) \right] \quad (6.4)$$

In this equation, $B_{root\,(i,j)}$, $SRA_{i,j}$, $k_{i,j}$ are respectively the root biomass, the specific root area (SRA, the ratio between the surface and biomass of the roots), and the hydraulic conductivity per unit of root surface of the species j in the stratum i; $\Psi_{soil,i}$ is the water potential of the soil in the stratum i and Ψ_j is the water potential of the species j; n is the number of species in the community and m is the number of strata of soil depth being considered. If we assume that the roots have a cylindrical form, this equation becomes (see details of the calculation in Box 6.3):

$$W_{ab\Delta t} = \sum_{i=1}^{m} \sum_{j=1}^{n} \left[L_{root(i,j)} * \pi * d_{root(i,j)} * k_{i,j} * (\Psi_{soil,i} - \Psi_j) \right] \quad (6.5)$$

where L_{root} and d_{root} are respectively the mean root length and mean root diameter. Equations 6.4 and 6.5 show that the rooting depth and the vertical distribution of roots influence the soil water content in the various soil depth strata considered. They also show that the hydraulic properties of the soil-plant system and the plant water potential also affect the soil water content, beyond the effects of size linked to the biomass and length of the roots (Eviner & Chapin, 2003).

In addition, water absorption and whole plant transpiration are intimately linked (cf. Larcher,

Box 6.3 Water absorption by vegetation

Water absorption by vegetation is one of the components of the water balance of an ecosystem (see Figure 6.7). As explained in Section 6.3.4.1, this continuous process can be subdivided into the quantities of water absorbed by stratum of soil depth. The amount of water absorbed in the stratum i by the species j over the period of time dt ($W_{ab\Delta t(i,j)}$), during which all the variables involved can be considered as constant, follows Darcy's law and can be expressed in the following manner (cf. Larcher, 2003; Schwinning & Ehleringer, 2001):

$$W_{ab\Delta t(i,j)} = B_{root(i,j)} * SRA_{i,j} * k_{i,j} * (\Psi_{soil,i} - \Psi_j) \quad (6.3.1)$$

In this equation, $B_{root\,(i,j)}$, $SRA_{i,j}$, $k_{i,j}$ are respectively the root biomass, the specific root area (ratio between the surface area and biomass of the roots), and the hydraulic conductivity per unit of root surface area of species j in the stratum i; $\Psi_{soil,i}$ is the water potential of the soil in the soil stratum i and Ψ_j is the water potential of the species j (Box 6.3, Figure 1).

For reasons of simplification, the water potential Ψ_j is considered to be identical throughout the plant, which assumes a weak hydraulic resistance between the absorbing root surface and the leaves. By hypothesizing that water absorption is an additive process implicating all of the species present, the variation in the quantity of water contained in the soil due to this process can be written as:

$$W_{ab\Delta t} = \sum_{i=1}^{m} \sum_{j=1}^{n} \left[B_{root(i,j)} * SRA_{i,j} * k_{i,j} * \left(\Psi_{soil,i} - \Psi_j \right) \right] \quad (6.3.2)$$

Which is the same as Equation 6.4 presented in the main text. If we assume that the roots have a cylindrical form, their specific area can be calculated as:

$$SRA = \frac{RA}{B_{root}} = \frac{L_{root} * \pi * d_{root}}{B_{root}}$$
$$SRA = \frac{RA}{B_{root}} = \frac{L_{root} * p * d_{root}}{B_{root}} \quad (6.3.3)$$

where RA, L_{root}, and d_{root} are respectively the root surface area, root length, and the mean root diameter. Combining Equations 6.3.2 and 6.3.3 gives:

$$W_{ab\Delta t} = \sum_{i=1}^{m} \sum_{j=1}^{n} \left[L_{root(i,j)} * \pi * d_{root(i,j)} * k_{i,j} * (\Psi_{soil,i} - \Psi_j) \right] \quad (6.3.4)$$

which is the same as Equation 6.5 presented in the main text.

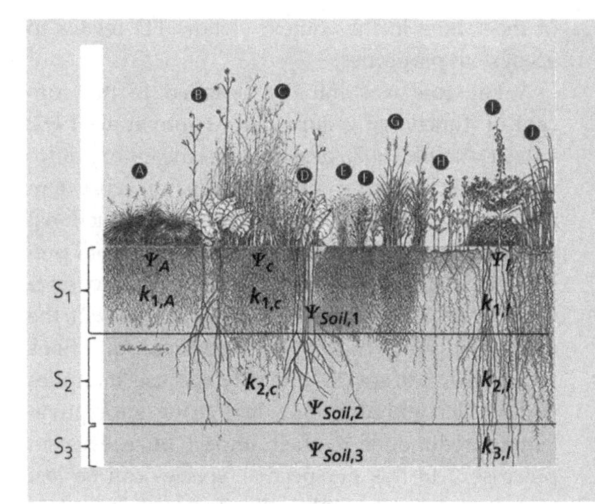

Box 6.3, Figure 1 Schematic representation of a plant community composed of 10 species (letters A–J) presenting different rooting patterns. In this figure, the soil is divided into three strata S_1, S_2, and S_3 (see the box text). The variables incorporated into the model of water absorption in Equation 6.3.1 of Box 6.3 shown on the figure are: $k_{i,j}$, the hydraulic conductivity per unit of root surface area of the species *j* in the stratum *i*; $\Psi_{soil,i}$, the water potential of the soil in the stratum i, and Ψ_j, the water potential of the species *j*. For reasons of clarity, these are shown only for three species (A, C, and I), which have differing rooting depths. The other variables involved in determining the absorption are: $B_{root(i,j)}$ and $SRA_{i,j}$, the root biomass and the specific root area of the species *j* in the stratum *i*, respectively (Equation 6.3.2 of Box 6.3), or $(RA)_{i,j}$, $(L_{root})_{i,j}$, and $(d_{root})_{i,j}$, which are respectively the root surface area, root length, and the mean root diameter of the species *j* in the stratum *i* (Equations 6.3.3 and 6.3.4 of Box 6.3).

2003; Schwinning & Ehleringer, 2001). It is thus likely that those traits controlling water fluxes from leaves to the atmosphere also have an impact on soil water content. This is for example the case for the water potential (Equations 6.4 and 6.5), but also for stomatal conductance, the sensitivity of which to atmospheric water vapour pressure deficit varies substantially among species (Oren et al., 1999).

Finally, temporal variation in water absorption should also be taken into account, which requires the integration of Equations 6.4 and 6.5 over time. It is in particular evident that temporal variation in the biomass and/or length of roots of the different species forming a community will influence the soil water content (Eviner & Chapin, 2003). Such changes depend on plant phenology and can thus occur at different times for different species, which can have important consequences for the overall functioning of the ecosystem. For example, a species which finishes its life cycle in spring will have very little influence on soil water reserves in summer, except in the case where an absence of rainfall does not replenish the soil water profile (Schwinning & Ehleringer, 2001).

6.3.4.2 Testing the dominance hypothesis

To our knowledge, the validity of Equations 6.4 and 6.5 above has not been directly tested in complex plant communities. However, some experimental studies have evaluated the impact of certain traits involved in these equations on soil water content. Gross et al. (2008) found evidence in alpine grasslands of a significant effect of the weighted mean of root length on the soil water content in the upper 15 cm depth of soil, while Mokany et al. (2008) found, in Australian grasslands, a significant negative relationship between the weighted mean of the specific root area and the soil water content calculated over the whole growth period. In an experiment carried out in an experimental garden with monocultures of 18 species originating from Mediterranean old-fields grown under low nitrogen supply, the soil water content evaluated over the whole growing period (cf. Violle et al., 2007a) was negatively correlated with rooting depth (Violle et al., 2009; cf. Figure 3.8b in Chapter 3).

Obojes et al. (2015) took a different perspective that can be used to infer some conclusions related to the same issue. These authors have assessed the relative impacts of standing biomass, abundance of certain plant functional types, and structural and functional vegetation properties on the components of water balance of high elevation grasslands in four sites in the Alps. They showed that actual evaporation increased (and thus soil water content probably decreased) with standing biomass at all sites, while the influence of other vegetation properties were site specific. For example, in three of the sites, the abundance of mat-forming graminoids tended to decrease actual evapotranspiration, which is likely to increase soil water content. The abundance of legumes was less consistent, with positive effects on evapotranspiration at one site, and negative at another one.

6.4 The complementarity hypothesis

Although an overwhelming number of studies have used species richness as a measure of biodiversity in the study of the relationships between biodiversity and ecosystem properties (Balvanera et al., 2006), theories about these relationships actually incorporate functional diversity in explanations of causation (Chapin et al., 1992; Tilman, 1997). One of the strongest hypotheses for these relationships is indeed that as biodiversity increases, so does the diversity of functional traits (Cadotte et al., 2011). By contrast with what has been discussed in Section 6.3, the niche complementarity hypothesis suggests non-additive effects among species with different trait values (Petchey & Gaston, 2006). This hypothesis proposes that it is the differences in trait values of the organisms in a community which influences ecosystem processes through mechanisms such as complementary resource use (Tilman, 2001): when there is a greater diversity in resource use strategies, it is proposed that there will be greater niche packing along a particular resource use axis, and hence the effects on resource use and biogeochemical pools and fluxes for this resource will be larger than in a less diverse community (Cadotte et al., 2011; Díaz & Cabido, 2001; Hooper, 1998). Positive relationships are therefore expected between ecosystem properties and indices capturing the various facets of community functional diversity (cf. Chapter 5, Table 5.1).

In the following, we first present how this diversity has been quantified in biodiversity-ecosystem properties studies, and second, experimental tests of this hypothesis.

6.4.1 Functional diversity and the complementarity hypothesis

Indices used to assess community functional diversity ('FD', hereafter) have been presented in Chapter 5 (see e.g. Table 5.1), but we reanalyse some of them here in the context of how FD relates to ecosystem properties.

A key role was initially assigned to the concept of 'functional groups' (cf. Chapin et al., 1992; Gitay & Noble, 1997), designed as a means of grouping species that have 'similar effects' on ecosystem properties: this actually corresponds to 'functional effect groups' in the response-effect framework presented in Chapter 2 (Gitay & Noble, 1997; Lavorel & Garnier, 2002). As stated by Chapin et al. (1992), the functional group concept raises difficult questions about the role of species diversity because 'it implies that species within a functional group are equivalent or redundant in their impact on ecosystem processes'. In this perspective, species can be lost from a community with little effects on ecosystem properties as long as each functional group is represented. Ecosystem functioning is therefore assumed to depend on: i) the number of functional groups present ('functional group richness', 'FGR', hereafter): a high FGR is expected to enhance ecosystem functioning due to greater resource use complementarity, and ii) the functional group composition, defined as which functional groups are present (e.g. C_3 vs C_4 species; grasses vs nitrogen-fixing species; early- vs late-season species, etc.).

As an expression of community functional diversity, FGR suffers from a number of limitations (e.g. Petchey et al., 2004; Walker et al., 1999). As put forward by Petchey et al. (2004), 'it relies on an arbitrary decision about the level at which interspecific differences among species are functionally significant, and it assumes that species within groups are functionally identical, that is, species within groups are entirely redundant. It also assumes that all pairs of species drawn from different functional groups are equally different'. Furthermore, interactions of effect and response traits have to be considered to assess redundancy: if two species have the same effect on ecosystems but they are active under different conditions (i.e., have different response traits), they cannot be considered as being redundant (Suding et al., 2008). A wealth of finer indices has been designed to overcome these limitations (cf. Chapter 5), and a number of them have been used to assess the impact of community functional diversity on ecosystem properties.

In Section 6.3 above, some of the mechanistic links between traits and ecosystem properties have been presented showing that several traits can be involved in the control of any single ecosystem property, while at the same time, this knowledge remains rudimentary for many properties. In addition, consensus is also lacking as to how FD should be calculated: as a consequence, what is usually referred to as FD is an approximation, based on a subset of traits, of the total FD of a community (Cadotte et al., 2011, and see Chapter 5). Based on the assumption that species that are closely related in a phylogeny tend to show more similar trait values than distantly related species (see Cavender-Bares et al., 2009; Srivastava et al., 2012, for reviews, and also Section 3.4 in Chapter 3), it has recently been postulated that phylogenetic diversity ('PD', hereafter), defined as the amount of evolutionary history represented in the species of a particular community (Mouquet et al., 2012, and references therein), is probably the most synthetic estimate of community trait space. Hence the idea that PD could be used as a proxy of unmeasured functional diversity for the purpose of assessing its connection to ecosystem functioning (Cadotte et al., 2008, 2011; Mouquet et al., 2012).

6.4.2 Testing the complementarity hypothesis

All experimental tests of the complementarity hypothesis (as a stand-alone hypothesis) that we are aware of have been conducted in manipulative experiments using herbaceous species (Table 6.3). We summarize the main results below, while further details can be found in the original studies.

The vast majority of these studies have tested complementarity effects on plant community standing biomass (AGB_{max}). Figure 6.8 synthesizes these effects for FGR, FD, PD, and species richness for 29 experiments from 11 studies (Flynn et al., 2011: study 9 in Table 6.3). It shows positive effects of all four components of diversity in decreasing order of explained variance: PD > FD > species richness > FGR. The range of explained variance is relatively narrow, however, with values comprising between 16.9 for FGR and 19.6 for PD. In addition, the effects appear to vary over time and with the composition of functional groups: Hooper and Dukes (2004) showed that the effects of FGR and functional group composition on ANPP were not significant in the short term (two years after planting). Over the longer term (seven and eight years after planting), they found that FGR and ANPP were positively related, but the effect was stronger when (i) nitrogen-fixing species were included in the analysis (also found by Tilman et al., 1997) and (ii) rainfall was high. This led the authors to conclude that a higher production with greater FGR may be restricted to particular species combinations or environmental conditions. In line with these findings, a compilation of results from 44 studies conducted over periods varying from 100 days to more than five years also show increasing effects of complementarity among species (no data on traits) through time (Cardinale et al., 2007).

A handful of studies have tested the complementarity hypothesis for processes related to nutrient cycling. Scherer-Lorenzen (2008) found significant positive effects of FGR and FD (quadratic diversity) on the rate of litter decomposition, and significant effects of functional group composition as well: mixed litters comprising legumes tended to decompose faster irrespective of the number of FGR. Hooper and Vitousek (1997, 1998) designed a very detailed experiment to assess the effects of FGR and functional group composition on patterns of soil nitrogen and phosphorus cycling, and concluded that: (i) total resource use increases with increasing FGR on a yearly timescale due to seasonal complementarity; (ii) while the presence of vegetation has a large effect on ecosystem nitrogen retention, nitrogen leaching losses do not necessarily decrease with increasing FGR, and (iii) functional group composition in general explains much more about the measured nutrient cycling processes than does FGR alone. Significant effects of FGR and functional group composition on soil NO_3^- and NH_4^+, and total plant nitrogen, were also both found to be significant by Tilman et al. (1997), but these effects were stronger for FGR.

The interpretation of these results is that ecological differentiation (captured by any of the indices used) can lead to reduced resource use overlap between species. As a consequence, these differentiated species could potentially complement each other in their resource use by differentially

Table 6.3 List of studies that have tested the complementarity hypothesis without considering dominance effects. Abbreviations: AGBmass: above-ground standing biomass; FGR: functional group richness; FGC: functional group composition (see text); FD: functional diversity index; FAD: functional attribute diversity; PD: phylogenetic diversity, *SR*: species richness; Ω and Λ denote respectively analyses of original data and reanalysis of published data. All experimental studies consist of experimental settings in which diversity levels are manipulated. See Chapter 5 for further information on indices.

Study	Some essential features of the study	Property	Metrics	Reference
1 (Ω)	Experimental plots of species from serpentine grassland in California, USA (2 years after planting)	AGBmass, inorganic nitrogen pools	FGR, FGC	Hooper & Vitousek (1997)
2 (Ω)	Experimental plots of grassland species in Cedar Creek, Minnesota, USA (2 years after planting)	AGBmass, light penetration, soil NO_3 and NH_4, plant nitrogen concentration	FGR, FGC, *SR*	Tilman et al. (1997)
3 (Ω)	Same experiment as (1)	Soil and microbial nitrogen and phosphorus pool, soil moisture	FGR, FGC	Hooper & Vitousek (1998)
4 (Ω)	Same experiment as (1), 2, 7, and 8 years after planting	AGBmass	FGR, FGC	Hooper & Dukes (2004)
5 (Λ)	6 sites of the BIODEPTH experiment (grassland species)	AGBmass	FGR, FD, FAD, *SR*	Petchey et al. (2004)
6 (Λ)	Compilation of 29 experiments from 11 studies	AGBmass	FGR, PD, *SR*	Cadotte et al. (2008)
7 (Ω)	German site of the BIODEPTH experiment (grassland species)	Litter decomposition	FGR, Q (quadratic diversity), *SR*	Scherer-Lorenzen (2008)
8 (Λ)	Reanalysis of the Cedar Creek experiment (grassland species)	AGBmass	FGR, FD (several metrics), PD, *SR*	Cadotte et al. (2009)
9 (Λ)	Irish site of the BIODEPTH experiment and Jena biodiversity experiment	AGBmass	FGR, PD, *SR*, additional 'community properties'	Connolly et al. (2011)
10 (Λ)	Compilation of 29 experiments from 11 studies (same experiments as (6))	AGBmass	FGR, PD, FD, *SR*	Flynn et al. (2011)
11 (Ω)	Grassland species from tall grass prairie in Ontario, Canada	AGBmass	PD, *SR*, imbalance across clades	Cadotte (2013)
12 (Ω)	Experimental plots of grassland species in Jena, Germany	AGBmass	FD_{Jena}, species pool composition	Ebeling et al. (2014)

capturing resources in space and/or time. Greater niche and trait differences could, in turn, translate to enhanced ecosystem functioning. It should be noted however, that the percentage of variance explained by diversity indices remains relatively low (20% at best), at least for biomass production, the property most examined in these studies. In this context, phylogenetic diversity, which is assumed to encompass most of the functional trait variation in a community, even when it is not known which traits are important, appears as an interesting proxy for FD, until a stronger rationale for trait selection is developed and FD metrics improve (Cadotte et al., 2011).

6.5 Testing simultaneously the dominance and complementarity hypotheses

The dominance and complementarity hypotheses are not mutually exclusive, and it is possible that both the traits of the dominant species and community functional diversity are important in influencing ecosystem properties (cf. Díaz et al., 2007b). The evaluation of the relative importance of the two mechanisms requires that both community means and functional diversity be assessed in the same study, which was not the case in the studies presented in the Sections 6.3 and 6.4.

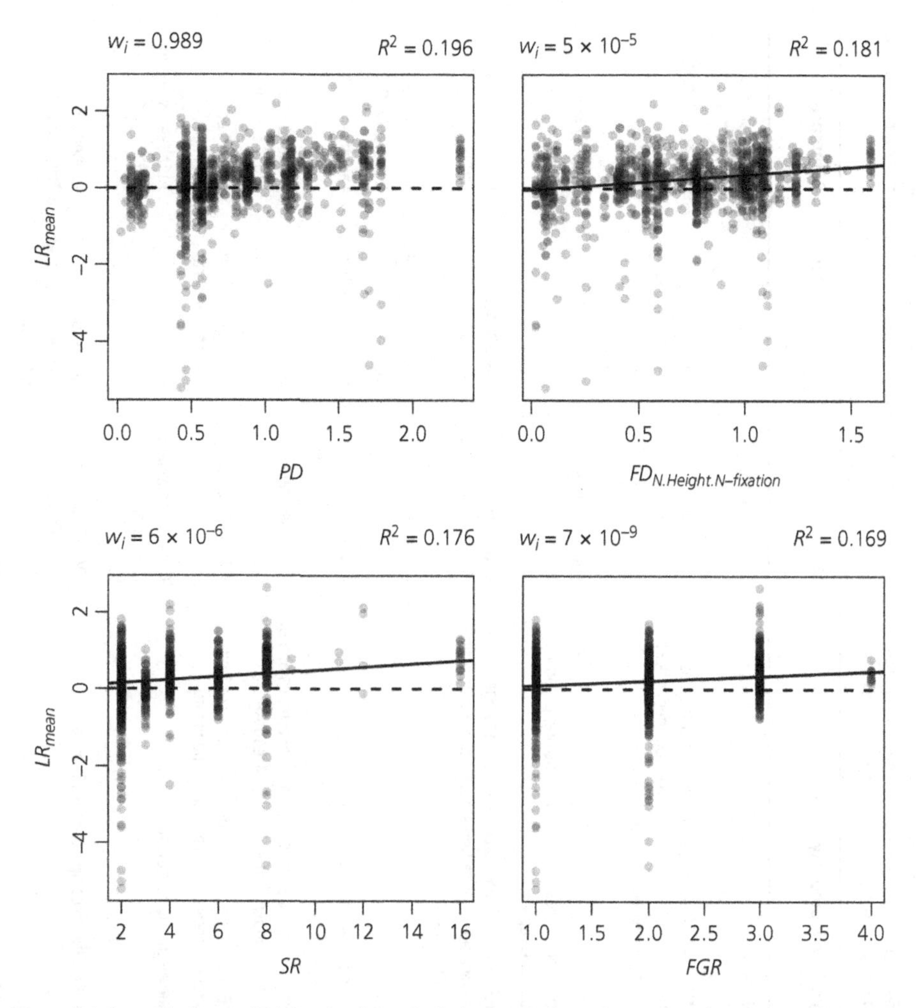

Figure 6.8 Effects of phylogenetic diversity (PD), functional diversity (FD), species richness (SR, number of species), and functional group richness (FGR, number of functional groups), across 1074 experimental units from 29 experiments. The PD shown was calculated using a molecular phylogeny, and the FD was calculated using the traits leaf nitrogen concentration, mean height, and nitrogen-fixing ability. Net biodiversity effects (LR_{mean}) are represented by the log ratio of the above-ground biomass of a polyculture to the mean biomass of the constituent species grown in monoculture. Solid lines show fits of single-variable linear mixed-effects models, with goodness-of-fit shown by Akaike weights (w_i) and model R^2. Points represent experimental units and are semi-transparent. Taken from Flynn et al. (2011). Reproduced with permission from the Ecological Society of America.

The first study we are aware of in which the dominance and complementarity effects have been formally tested in a single experiment is that by Thompson et al. (2005), in vegetation from road verges in the UK. This study showed a positive association between above-ground biomass and the community (unweighted) mean of leaf area, while relationships were *negative* with functional group richness and functional diversity. Since then, a number of studies have estimated the relative importance of the two effects in various systems, mostly *in natura* (Table 6.4). Before presenting the main conclusions that can be drawn from these studies, some methodological issues are first discussed.

6.5.1 Methodological issues

Studies aiming at assessing simultaneously dominance and complementarity effects face two major

Table 6.4 List of studies that have tested simultaneously the dominance and complementarity hypotheses. Abbreviations: ABG_{mass}: above-ground standing biomass; ANPP: above-ground net primary productivity; SANPP: specific above-ground net primary productivity; BGB_{mass}: below-ground biomass; TB_{mass}: total (above- and below-ground) biomass; NEP: net ecosystem productivity; NEE: net ecosystem CO_2 exchange; GPP: gross ecosystem productivity; AQY: canopy apparent quantum yield; NUE: nitrogen use efficiency; WUE: water use efficiency; Reco: ecosystem respiration; CArM: community arithmetic mean; CWM: community weighted mean; FGR: functional group richness; FD_Q: Rao's quadratic entropy; FD_{vg} functional divergence; FD_{var}: Mason's functional divergence; CWV: community weighted variance; *S*: species richness. When the effects of environmental factors have been taken into account in addition to those of community structure, these are mentioned in the 'Environmental factors' column ('-' denotes that no environmental factor has been considered in the analyses).

Study	Some essential features of the study	Ecosystem property	Metrics	Environmental factors	Reference
1	Vegetation from a road verge, UK (*in natura*)	ABG_{mass}	CAM, FD, FGR, *SR*	–	Thompson et al. (2005)
2	Alpine grasslands in the French Alps (*in natura*)	ANPP, SANPP, AGB_{mass}, standing litter biomass, litter decomposability, available NO_3, denitrification potential, soil water content	CWM, FD_{vg}	Nutrient limitation, disturbance, soil water holding capacity	Díaz et al. (2007b)
3	Temperate native grassland in Australia (*in natura*)	AGB_{mass}, BGB_{mass}, TB_{mass}, litter biomass, ANPP, litter decomposition rate, soil moisture, proportion light intercepted	CWM, FD, FD_Q	–	Mokany et al. (2008)
4	Upland grassland site in Central France (*in natura*)	NEP, ANPP, SANPP, total net carbon storage, belowground net carbon storage, top soil organic carbon content	CWM, FD_{vg}	Disturbance	Klumpp & Soussana (2009)
5	Semi-natural grasslands in Germany (*in natura*)	AGB_{mass}	CWM, FD_Q	Soil variables, light intensity	Schumacher & Roscher (2009)
6	*Pinus ponderosa* forest in Arizona, USA (*in natura*)	Soil nitrification potential	CWM, FD_Q	Soil texture, litter mass	Laughlin (2011)
7	German site of the BIODEPTH project (common garden)	Productivity, litter decomposition, decomposition of cotton standard, nitrogen pool size in ABG_{mass}	CWM, several FD indices	–	Mouillot et al. (2011)
8	Alpine grasslands in the French Alps (*in natura*)	ABG_{mass}, litter mass, crude protein content, soil carbon, species diversity, date of flowering	CWM, FD_{var}	Topography, land use, soil properties	Lavorel et al. (2011)
9	Natural forest and mixed-species forest plantation in Panama (*in natura* and plantation) (6 years after plantation for the latter)	Above-ground carbon stock	CWM, Functional dispersion	–	Ruiz-Jaen & Potvin (2011)
10	Experimental grasslands in New Zealand (plots in grasslands)	ANPP, litter decomposition, soil carbon sequestration	CWM, Functional dispersion	Soil resource availability, grazing intensity	Laliberté & Tylianakis (2012)
11	Experimental grassland plots of the Jena experiment in Germany (common garden)	ABG_{mass}	CWM, FD_Q	–	Roscher et al. (2012)

12	Pastures in open rangelands and *Quercus douglasii* savanna in California, USA (*in natura*)	ABG_{mass}, soil carbon	CWM, several FD indices (single and multi-traits)	Elevation, soil bulk density, slope	Butterfield & Suding (2013)
13	Forests and shrublands in the Gran Chaco, Argentina (*in natura*)	ABG standing biomass carbon, litter mass carbon, soil organic carbon, total ecosystem carbon	CWM, FD_{var}	–	Conti & Díaz (2013)
14	Grasslands from Austrian Tyrol, northern England and the French Alps (*in natura*)	AGB_{mass}, standing litter, N mineralization potential, potential leaching of soil inorganic N (NO_3 and NH_4), microbial biomass N	CWM, FD_{var}	–	Grigulis et al. (2013)
15	Experimental grassland plots from Jena experiment in Germany (common garden)	AGB_{mass}	CWM, several FD indices	Climate	Roscher et al. (2013)
16	Mediterranean rangelands in Southern France (*in natura*)	ABG_{mass} at different seasons	CWM, CWV	Temperature, soil water content	Chollet et al. (2014)
17	Experimental grassland plots from Jena experiment in Germany (common garden)	NEE, GPP, Reco, AQY, NUE, WUE	CWM, FD_Q	Soil texture and moisture, climatic conditions	Milcu et al. (2014)
18	Three rainforests in south America (*in natura*)	ABG_{mass}, biomass increment	CWM, several FD indices	–	Finegan et al. (2015)

statistical issues. The first one is that the metrics used to evaluate the two effects, namely CWM and functional diversity indices, are mathematically related (Ricotta & Moretti, 2011). The prediction is that at extreme low and high values of CWM, FD will decrease because only species with similar trait values (low or high, respectively) will be present, leading to a humped-shaped relationship between the two metrics (Dias et al., 2013). Using a simulation study, these authors showed that the actual shape of the relationship depends both on the trait(s) selected and on the metrics used to assess FD: for example, the relationship between CWM and functional divergence (calculated as Rao's quadratic entropy: cf. Table 5.1 in Chapter 5) for leaf nitrogen concentration was humped-shaped indeed, while the relationship between CWM and functional richness (cf. Table 5.1) showed discrete horizontal steps.

The second one pertains to the fact that the results of these analyses depend strongly on the manner in which the different variables are introduced into the statistical models used, and in particular to the procedures of stepwise regression. The conclusions appear very sensitive to whether the variables are treated in an independent manner or in combination, or whether the methods of variable selection are based on individual correlations (Díaz et al., 2007b), or if a combination of all of the variables are introduced into a global analysis (Schumacher & Roscher, 2009).

A striking example of such a problem is given in the study by Schumacher and Roscher (2009) in semi-natural grasslands in Germany. This study showed that, when a stepwise multiple regression procedure is used without any constraints being placed on variable selection, the maximum biomass produced by the communities is best explained by a combination of the following variables: total soil carbon, the weighted means of life cycle types, vertical distribution of the canopy, specific leaf area, and rooting depth, and the functional divergence of life cycle types, leaf surface area, and vegetative height, with all of these variables having positive regression coefficients. In contrast, when the procedure for variable selection proposed by Díaz et al. (2007b) is used, the correlation coefficients are negative for the functional divergences of the majority of the traits. These results show the strong influence of the statistical method used on the nature of the results obtained, and the necessity for further work in order to stabilize these procedures. An additional unsolved statistical question relates to whether traits should be weighted in the analyses according to their assumed relative importance to explain the specific process under study (Schumacher & Roscher, 2009).

Finally, given that the effects of community functional components on ecosystem properties have been often tested through observational studies *in natura* (cf. Table 6.4), environmental effects should be addressed as well (see Section 6.2). This has not always been the case and when this was done, the variables used were dependent on the context of the study (Table 6.4, and see discussion below).

The lack of standardization of methodological procedures (statistical and experimental) used up until now almost certainly explains the sometimes contradictory conclusions obtained in different studies. These limitations should be kept in mind during the synthesis of experimental results presented in the following section.

6.5.2 Experimental tests

A prominent feature of the studies summarized in Table 6.4 is that the ranges of experimental situations and ecosystem properties analysed is very broad, which makes a synthesis somewhat challenging. An attempt is given below with an emphasis on some selected properties.

Several studies have suggested that the effects of functional community structure on ecosystem properties may primarily be attributed to dominance rather than complementarity effects (Laughlin, 2011; Lavorel et al., 2011; Mokany et al., 2008). Studies have indeed reported positive (Mouillot et al., 2011; Schumacher & Roscher, 2009), negative (Laliberté & Tylianakis, 2012; Thompson et al., 2005) or non-significant (Díaz et al., 2007b; Finegan et al., 2015) effects of functional diversity, often in combination with both dominance and environmental effects. We illustrate below these contrasting patterns in the case of components of primary production and soil organic carbon, separating experiments conducted *in natura* and manipulative experiments conducted in common gardens (discussed in Section 6.2).

6.5.2.1 Experiments *in natura*

Concomitant positive effects of CWM and negative or non-significant effects of functional diversity on above-ground biomass have been reported in a number of studies (e.g. Chollet et al., 2014; Conti & Díaz, 2013; Finegan et al., 2015; Grigulis et al., 2013; Laliberté & Tylianakis, 2012). In two studies conducted in grazed and fertilized grasslands, positive associations were found between CWM_SLA and ABG_{mass} (Chollet et al., 2014; Laliberté & Tylianakis, 2012), while relationships between AGB_{mass} and an FD index calculated for five leaf traits (Laliberté & Tylianakis, 2012) and CWV of plant height and LDMC (Chollet et al., 2014) were negative. Negative plant height diversity effects on AGB_{mass} (also found by Conti and Díaz, 2013, in South American shrublands) suggest that dominance by tall species, rather than a set of coexisting species with diverse heights, results in greatest production. By contrast, Butterfield and Suding (2013) found that a greater range of heights among rangeland herbs was associated with higher production, while Schumacher and Roscher (2009) found significant positive effects of both components of community functional structure (taking plant height into account) on AGB_{mass}. Combining data from three rainforests in South America, Finegan et al. (2015) found that CWM of several traits (SLA, LNC, leaf resistance to fracture, plant height, and wood specific gravity) were reasonable predictors of AGB_{mass} and biomass increments, while they did not find any significant relationships with functional diversity indices.

In the case of soil organic carbon (SOC), Klumpp and Soussana (2009) have shown in semi-natural grasslands that SOC was negatively correlated with CWM specific root length (the ratio between root length and biomass: cf. Chapter 3) and root tissue density. Dense and low SRL roots generally have low nitrogen but high lignin concentrations (references in Lavorel et al., 2007) leading to the production of slowly decomposing litter (Birouste et al., 2012; Hobbie et al., 2010), resulting in the production of stable organic matter (Figure 6.6). These authors did not find any significant complementarity effects in addition to these functional effects of dominant species. Working in the Chaco forest of Argentina, Conti and Díaz (2013) found that SOC was positively related to the CWM of plant height, but that the relationship was negative with a multi-trait functional dissimilarity index. In this system, tall species may also have denser wood and greater below-ground production, and hence make a greater contribution to carbon storage.

Comparable contrasting results for other ecosystem properties have been found among the studies listed in Table 6.4, and will not be further detailed here. Environmental factors obviously play a role as well in the detection of the dominance and/or complementarity effects. The relative effects of community functional structure vs environment are likely to depend on the ranges covered by the two types of factors, and a wide array of results have been found. An extreme example is that of the study conducted by Díaz et al. (2007b) in alpine grasslands, where no significant impact of any component of community functional structure on productivity was found, while nitrogen limitation exerted a significant control on this property. More commonly, both biotic and environmental factors (when the latter are taken into account in the analyses: cf. Table 6.4) are found to explain a significant part of the variance in ecosystem properties (see individual studies for further details).

6.5.2.2 Common garden experiments

Common garden experiments in which assemblages of species with different trait values are artificially manipulated (e.g. Ebeling et al., 2014; Roscher et al., 2004; Tobner et al., 2014) have been designed precisely to detect biotic effects by minimizing environmental variations across plots with different functional structures. In three such experiments, both dominance and complementarity effects on biomass production (Mouillot et al., 2011; Roscher et al., 2013) and tree carbon storage (Ruiz-Jaen & Potvin, 2011) have been detected. In this latter study, the results obtained in the plantation did not match those obtained in a nearby natural forest: although dominance and complementarity effects were also detected in the forest, the slopes of the relationships between carbon storage and CWM of plant height and the CWM of SLA were opposite in the two situations. In addition, FD effects were respectively non-significant for plant height and significant for both SLA and proportion of nitrogen fixers in the plantation, while the exact reverse was true in the natural forest. These results led Ruiz-Jaen and Potvin (2011) to conclude that the carbon storage capacity of natural forests could not be easily predicted using data from experimental plantations.

In several of the studies mentioned, conducted either in the field or in common gardens (e.g. Conti & Díaz, 2013; Roscher et al., 2013; Ruiz-Jaen & Potvin, 2011), additional effects of particular species not accounted for by the traits measured were also found, which confirms the relevance of this step in the general framework proposed by Diaz et al. (2007b).

6.6 Dominance and complementarity effects: too early for conclusions

Studies discussed in the previous sections of this chapter provide ample evidence that the functional structure of communities has significant impacts on a variety of ecosystem properties. Dominance effects have been repeatedly reported, while functional diversity effects have been found to be either positive, negative, or non-existent, except when these have been tested individually. There is therefore currently stronger support for the dominance than for the complementarity hypothesis (see also Lavorel, 2013). The relative importance of dominance and complementarity effects therefore remains to be established however, and it might well be the case that this importance differs between systems, ecosystem properties, and experimental conditions. A synthesis is currently difficult for several, very different, reasons. These include:

1) The selection of traits used to assess the different components of community functional structure should be based on a mechanistic understanding of the relationships between traits and ecosystem properties (see Section 6.3), which is far from being the case currently.
2) A proper assessment of the relative impacts of biotic vs abiotic factors on ecosystem properties is necessary. As already stressed in Chapter 4, this requires a better quantification of relevant environmental variables to be measured across studies. Designing experiments in common gardens is another way to address this issue, but extrapolation to the field might prove problematic. In addition, scientific syntheses based on such experiments still require an adequate assessment of key environmental factors which may affect the direction and intensity of effects.
3) Whether FD indices based on traits actually describe complementarity in resource use among species present in a community is not always straightforward. A higher FD is assumed to correspond to a higher separation in niche space, but this relies on the assumption that differences in trait values actually correspond to different positions along a particular niche axis, resulting in a more complete use of resources. This is probably the case for differences in e.g. plant height with regards to light, but what do differences in leaf nitrogen concentration among species within a community tell us about niche partitioning? This issue points to the necessity of establishing more formal connections between the trait and niche concepts.
4) Although dominance effects appear to be properly captured by the CWM metrics, a number of different indices have actually been used to assess FD effects (cf. Tables 6.3 and 6.4), each one describing a particular facet of community functional diversity (cf. Chapter 5). This leads to different interpretation of results according to the index used. In addition, care should be taken in the interpretation of (i) dominance and complementarity effects, and (ii) FD vs species richness effects, as metrics used to assess these different effects are not mathematically independent (cf. Chapter 5 for the latter).

During the last two decades, research on the relationships between biodiversity and ecosystem properties has followed two almost independent paths: early trait-based approaches have favoured testing of the dominance hypothesis without considering potential effects of complementarity. By contrast, the species-based approach, which has assumed from the start that increasing species number led to an increase in the diversity of functional traits, has favoured testing of the complementarity hypothesis without considering potential dominance effects. The increasing number of studies, both observational and manipulative, which now consider both hypotheses simultaneously is central to advancing this key ecological question, as is the development of mechanistic and/or statistical models of the processes involved.

6.7 A new perspective: ecosystem allometry

The allometric approach to ecosystems constitutes a new way to represent the relationships between individuals and ecosystem functioning, which places an emphasis on size distribution within communities (Enquist et al., 2003; Kerkhoff & Enquist, 2006). It is an approach based on the use of plant traits, as size is a performance trait as defined in Chapter 2. Nevertheless, this form of analysis is carried out without distinguishing between the different species which make up a community. The theoretical basis of this approach, applied in the case of the net primary productivity of ecosystems (Kerkhoff & Enquist, 2006), is briefly presented below.

By definition the NPP is the sum of the growth of all of the individuals present in a community (cf. Box 6.2). The allometric approach stipulates that the absolute growth rate of a whole plant (AGR; see Box 3.3 in Chapter 3) can be expressed as:

$$AGR = \beta_g * M_T^{3/4} \quad (6.8)$$

in which β_g is an allometric coefficient linking the absolute growth rate of the whole plant with its total biomass (M_T). The exponent term $^3/_4$ is explained by the constraints exerted on the resource provisioning of cells (West et al., 1997). Experimental tests suggest that the term β_g is constant for a large variety of photosynthetic organisms (Niklas & Enquist, 2001), regardless of the species considered. This implies that when a wide range of sizes exists amongst a group of species, the size effect dominates and prevails over the interspecific differences in functioning between these species. On this basis, Kerkhoff and Enquist (2006) proposed to regroup individuals forming a plant community into K size classes, each with a mean biomass of $M_{T(k)}$, within which the growth rate is the same for all of the individuals. The NPP of the ecosystem can thus be written as:

$$NPP = \sum_{k=0}^{K} n_k * \beta_g * M_{T(k)}{}^{3/4} \quad (6.9)$$

in which n_k is the number of individuals in the size class k present per m^2. On the basis of this equation, it is possible to show that the NPP is theoretically proportional to the exponent $^3/_4$ of the total biomass of the community (Kerkhoff & Enquist, 2006), which is the sum of the $M_{T(k)}$ calculated for all size classes. A test of this prediction was carried out using data from highly contrasting ecosystems (a comparison *between* ecosystems), from the arctic tundra to tropical forests. As expected, this test showed evidence of a log-linear relationship between the NPP and the total biomass of the communities, but with an exponent function much lower than that predicted by theory (0.46 vs 0.75: Kerkhoff & Enquist, 2006; Figure 6.9). This deviation from the expected theoretical result is due to the incapacity of the model to take into account processes occurring within plant communities; for example asymmetries in light interception between interacting plants, which is a direct consequence of size differences. In this study, Kerkhoff and Enquist (2006) also developed an ensemble of predictions linking the total biomass of communities with their mass of mineral nutrients (nitrogen and phosphorus), which were generally very close to experimental data.

This new approach is quite radically different from that presented in the other sections of this chapter, where the difference between species plays a central role. However, it needs now to be validated in a more systematic manner, including in particular situations less contrasted than those described above: the average biomass of communities between the arctic tundra and tropical forests can vary by a factor of 600 (cf. Figure 6.9), while, in the majority of cases, the range of biomass variation between compared communities is much more restricted. For example, the maximal biomass of grasslands along an annual rainfall gradient varying between 250 mm and 1400 mm in different regions of the central USA varies only by a factor of seven over the whole of the gradient (Sala et al., 1988). Similarly, the continuous fertilization of herbaceous communities over 150 years in the context of the 'Park Grass Experiment' carried out north of London, resulted only in variation of a factor of 3 in the biomass produced between the most and least productive treatments (Crawley et al., 2005). For ranges of variation of these magnitudes, it is

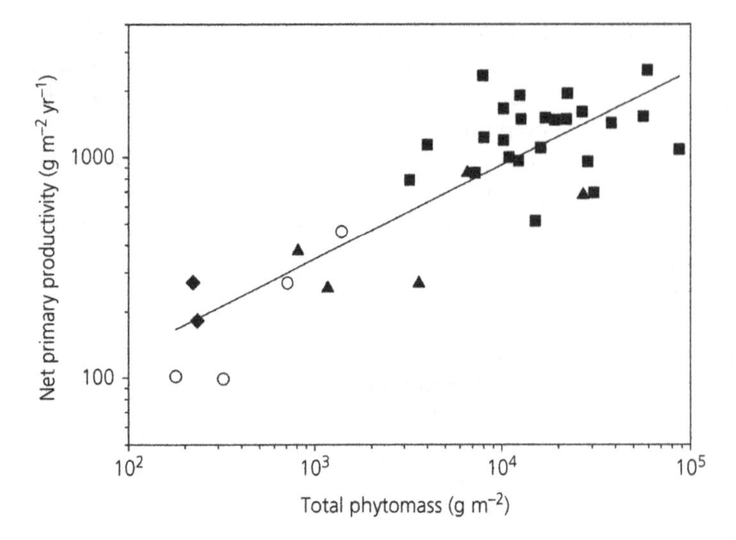

Figure 6.9 Net primary productivity varies allometrically with the total plant biomass of a community (total phytomass) in a relationship including grasslands (black diamonds), tundras (white circles), shrublands (grey triangles up), and forests (black squares). The regression equation is the following: Productivity = 14.5 × $Biomass^{0.46}$ ($r^2 = 0.60$, n = 36). Taken from Kerkhoff & Enquist (2006). Reproduced with permission from Wiley.

probable that the differing functional characteristics between species play a non-negligible role in ecosystem functioning, beyond those of simple differences in size between individuals, which are the only ones taken into account in the context of the allometric approach.

6.8 Conclusions

This chapter has focused on the important role that the plant compartment plays in the control of ecosystem properties, and brings insights into the major pending ecological questions pertaining to the understanding of the relationships between biodiversity and ecosystem functioning. The approach used in this chapter has concentrated on the analysis of the individual properties of ecosystems (biomass production, decomposition, carbon storage, etc.). This analysis will be expanded to simultaneously take into account multiple properties—'ecosystem multifunctionality'—in the following chapter.

The studies presented here suggest that the functional structure of communities has significant impacts on ecosystem properties. There is currently stronger support for the dominance than for the functional complementarity hypothesis, but the relative importance of these two mechanisms remains to be established in a wider range of situations. This allows us to understand why the intensity and direction of ecosystem processes are not correlated in a simple and unidirectional way with the community species richness (see also Grime & Pierce, 2012). The relative contributions of the controls exerted by the biotic compartment and by environmental factors also remain to be quantified. Some studies presented in this chapter suggest that the effects of the biotic compartment become weaker as the range of variation of environmental factors becomes larger (Baker et al., 2009; Díaz et al., 2007b).

Another conclusion from this chapter is that the traits involved in the functioning of ecosystems differ depending on the property considered: for example root traits have important effects on soil water content, while leaf traits have a prominent role in the control of light interception and productivity. This is true for the instantaneous functioning of ecosystems as presented in this chapter, but probably also when this functioning is considered over longer time scales, with the implication of species replacements under conditions of a changing environment. From this perspective, the effects of regeneration traits, which tend to be independent from those relating to plant physiological activity (cf. Chapters 3 and 4), and which control the dynamics of communities, are of primary importance. They are, however, only rarely taken into account, as observed by Solbrig (1993) over twenty years ago. Furthermore, the effects of complementarity between species seem to be more clearly expressed over longer time scales (e.g. Cardinale et al., 2007), which might be the consequence of species replacement as well.

From a methodological perspective, while experiments with artificial assemblages make up the core of studies that have addressed the species and functional diversity–ecosystem properties issue, the effect of other community functional components has been mostly tested through observational studies. It is now time to fully merge these two approaches and test the relative importance of these two components of community functional structure on ecosystem properties and services delivery (Dias et al., 2013). Coupling these experimental approaches with mechanistic and advanced statistical modelling should enable us to advance knowledge on this topic of paramount importance in ecology.

6.9 Key points

1. Ecosystem properties (EP), which refer to both the pools (quantities) and fluxes (flows) of materials and energy in ecological systems, depend on two categories of controlling factors: (a) independent control factors (climate, bedrock, topography, potential biological diversity, and time), and (b) interactive control factors, which act upon and are also affected by ecosystem properties (resource availability, environmental conditions, disturbances, local biological diversity, and human activities). This chapter focuses on the controls that plants, as a major component of the biotic compartment of the environment, exert on ecosystem properties.
2. Two non-exclusive hypotheses have been proposed to explain the effects of the plant compartment on EP: (a) the dominance hypothesis, which states that certain traits of locally abundant species determine the magnitude of ecosystem processes, and (b) the functional complementarity hypothesis, according to which it is primarily the coexistence of species which use environmental resources in a complementary manner which will influence EP. Numerous studies have shown that the functional structure of communities has impacts on EP, rather than the species richness of the community. There is currently stronger support for the dominance than for the functional complementarity hypothesis, but the relative importance of these two mechanisms remains to be established in a wider range of situations. In addition, the relative impacts of the biotic and abiotic components on ecosystem properties also largely remain to be quantified.
3. The traits which influence EP differ depending on the property considered: for example, certain components of primary productivity and litter decomposition depend on traits linked to the leaf economic spectrum (cf. Chapter 3), while soil water content depends rather on the vertical distribution of roots in the soil and root traits controlling water absorption. The allometric approach to ecosystems is a different approach to address the relationships between individuals and EP, which focuses on the distribution of plant size within the community, without distinguishing between the different species making up this community. This approach, partially validated for the case of large-scale comparisons between biomes, needs to be further tested in a more systematic manner and for less-contrasted ecosystems.
4. Understanding the complex relationships between functional diversity and EP requires the use of a plurality of approaches combining observations and experiments *in natura* with experiments in more controlled conditions, as well as the development of mechanistic and/or statistical models of the processes involved.

6.10 References

Aber, J. D., & Mellilo, J. M. (2001). *Terrestrial Ecosystems* (2nd ed.). Pacific Grove, CA: Brooks/Cole Publishing.

Aerts, R. (1997). Climate, leaf litter chemistry and leaf litter decomposition in terrestrial ecosystems: a triangular relationship. *Oikos, 79*, 439–449.

Aerts, R., & Chapin, F. S., III. (2000). The mineral nutrition of wild plants revisited: a re-evaluation of processes and patterns. *Advances in Ecological Research, 30*, 1–67.

Altesor, A., Oesterheld, M., Leoni, E., Lezama, F., & Rodrìguez, C. (2005). Effect of grazing exclosure on community structure and productivity of a Uruguayan grassland. *Plant Ecology, 179*, 83–91.

Andueza, D., Cruz, P., Farruggia, A., Baumont, R., Picard, F., & Michalet-Doreau, B. (2010). Nutritive value of two meadows and relationships with some vegetation traits. *Grass and Forage Science, 65*(3), 325–334. doi:10.1111/j.1365-2494.2010.00750.x

Ansquer, P., Duru, M., Theau, J. P., & Cruz, P. (2009). Functional traits as indicators of fodder provision over a short time scale in species-rich grasslands. *Annals of Botany, 103*(1), 117–126. doi:10.1093/aob/mcn215

Baker, T. R., Phillips, O. L., Laurance, W. F., Pitman, N. C. A., Almeida, S., Arroyo, L., ... Lloyd, J. (2009). Do species traits determine patterns of wood production in Amazonian forests? *Biogeosciences, 6*, 297–307.

Balvanera, P., Pfisterer, A. B., Buchmann, N., He, J.-S., Nakashizuka, T., Raffaelli, D., & Schmid, B. (2006). Quantifying the evidence for biodiversity effects on ecosystem functioning and services. *Ecology Letters, 9*, 1146–1156.

Bardgett, R. D., & Wardle, D. A. (2010). Aboveground-Belowground Linkages—Biotic Interactions, Ecosystem Processes, and Global Change. Oxford: Oxford University Press.

Berendse, F., & Aerts, R. (1987). Nitrogen-use-efficiency: a biologically meaningful definition? *Functional Ecology, 1*, 293–296.

Birouste, M., Kazakou, E., Blanchard, A., & Roumet, C. (2012). Plant traits and decomposition: are the relationships for roots comparable to those for leaves? *Annals of Botany, 109*(2), 463–472. doi:10.1093/aob/mcr297

Boulant, N., Kunstler, G., Rambal, S., & Lepart, J. (2008). Seed supply, drought, and grazing determine spatio-temporal patterns of recruitment for native and introduced invasive pines in grasslands. *Diversity and Distributions, 14*, 862–874.

Bruinenberg, M. H., Valk, H., Korevaar, H., & Struik, P. C. (2002). Factors affecting digestibility of temperate forages from seminatural grasslands: a review. *Grass and Forage Science, 57*(3), 292–301. doi:10.1046/j.1365-2494.2002.00327.x

Butterfield, B. J., & Suding, K. N. (2013). Single-trait functional indices outperform multi-trait indices in linking environmental gradients and ecosystem services in a complex landscape. *Journal of Ecology, 101*(1), 9–17. doi:10.1111/1365-2745.12013

Cadotte, M. W. (2013). Experimental evidence that evolutionarily diverse assemblages result in higher productivity. *Proc. Natl. Acad. Sci. USA, 110*(22), 8996–9000. doi:10.1073/pnas.1301685110

Cadotte, M. W., Cardinale, B. J., & Oakley, T. H. (2008). Evolutionary history and the effect of biodiversity on plant productivity. *Proc. Natl. Acad. Sci. USA, 105*(44), 17012–17017. doi:10.1073/pnas.0805962105

Cadotte, M. W., Carscadden, K., & Mirotchnick, N. (2011). Beyond species: functional diversity and the maintenance of ecological processes and services. *Journal of Applied Ecology, 48*(5), 1079–1087. doi:10.1111/j.1365-2664.2011.02048.x

Cadotte, M. W., Cavender-Bares, J., Tilman, D., & Oakley, T. H. (2009). Using phylogenetic, functional, and trait diversity to understand patterns of plant community productivity. *PLoS ONE, 4*(5), e5695. doi:10.1371/journal.pone.0005695

Cardinale, B. J., Duffy, J. E., Gonzalez, A., Hooper, D. U., Perrings, C., Venail, P., ... Naeem, S. (2012). Biodiversity loss and its impact on humanity. *Nature, 486*(7401), 59–67. doi:10.1038/nature11148

Cardinale, B. J., Srivastava, D. S., Duffy, J. E., Wright, J. P., Downing, A. L., Sankaran, M., & Jouseau, C. (2006). Effects of biodiversity on the functioning of trophic groups and ecosystems. *Nature, 443*, 989–992.

Cardinale, B. J., Wright, J. P., Cadotte, M. W., Carroll, I. T., Hector, A., Srivastava, D. S., ... Weis, J. J. (2007). Impacts of plant diversity on biomass production increase through time because of species complementarity. *Proc. Natl. Acad. Sci. USA, 104*, 18123–18128.

Cavender-Bares, J., Kozak, K. H., Fine, P. V. A., & Kembel, S. W. (2009). The merging of community ecology and phylogenetic biology. *Ecology Letters, 12*, 693–715.

Cebrián, J., & Duarte, C. M. (1995). Plant growth-rate dependence of detrital carbon storage in ecosystems. *Science, 268*, 1606–1608.

Chapin, F. S., III. (1993). Functional role of growth forms in ecosystem and global processes. In J. R. Ehleringer & C. B. Field (eds), *Scaling Physiological Processes. Leaf to Globe* (pp. 287–312). San Diego: Academic Press, Inc.

Chapin, F. S., III, Matson, P. A., & Vitousek, P. M. (2011). *Principles of Terrestrial Ecosystem Ecology* (2nd ed.). New York: Springer.

Chapin, F. S., III, Reynolds, H. L., D'Antonio, C. M., & Eckhart, V. M. (1996). The functional role of species in terrestrial ecosystems. In B. Walker & W. Steffen (eds), *Global Change and Terrestrial Ecosystems* (pp. 403–428). Cambridge: Cambridge University Press.

Chapin, F. S., III, Schulze, E. D., & Mooney, H. A. (1992). Biodiversity and ecosystem processes. *Trends in Ecology and Evolution, 7*(4), 107–108. doi:10.1016/0169-5347(92)90141-W

Chapin, F. S., III, Zavaleta, E. S., Eviner, V. T., Naylor, R. L., Vitousek, P. M., Reynolds, H. L., ... Díaz, S. (2000). Consequences of changing biodiversity. *Nature, 405*, 234–242.

Chollet, S., Rambal, S., Fayolle, A., Hubert, D., Foulquié, D., & Garnier, E. (2014). Combined effects of climate, resource availability, and plant traits on biomass produced in a Mediterranean rangeland. *Ecology, 95*(3), 737–748. doi:10.1890/13-0751.1

Choong, M. F., Lucas, P. W., Ong, J. S. Y., Pereira, B., Tan, H. T. W., & Turner, I. M. (1992). Leaf fracture toughness and sclerophylly: their correlations and ecological implications. *New Phytologist, 121*, 597–610.

Clark, D. A., Brown, S., Kicklighter, D. W., Chambers, J. Q., Thomlinson, J. R., & Ni, J. (2001). Measuring net primary production in forests: concepts and field methods. *Ecological Applications, 11*(2), 356–370. doi:10.2307/3060894

Connolly, J., Cadotte, M. W., Brophy, C., Dooley, Á., Finn, J., Kirwan, L., ... Weigelt, A. (2011). Phylogenetically diverse grasslands are associated with pairwise interspecific processes that increase biomass. *Ecology, 92*(7), 1385–1392. doi:10.1890/10-2270.1

Conti, G., & Díaz, S. (2013). Plant functional diversity and carbon storage: an empirical test in semi-arid forest ecosystems. *Journal of Ecology, 101*(1), 18–28. doi:10.1111/1365-2745.12012

Cortez, J., Garnier, E., Pérez-Harguindeguy, N., Debussche, M., & Gillon, D. (2007). Plant traits, litter quality, and decomposition in a Mediterranean old-field succession. *Plant and Soil, 296*, 19–34.

Craine, J. M., Tilman, D., Wedin, D., Reich, P., Tjoelker, M., & Knops, J. (2002). Functional traits, productivity, and effects on nitrogen cycling of 33 grassland species. *Functional Ecology, 16*, 565–574.

Crawley, M. J., Johnston, A. E., Silvertown, J., Dodd, M., de Mazencourt, C., Heard, M. S., ... Edwards, G. R. (2005). Determinants of species richness in the Park Grass Experiment. *The American Naturalist, 165*, 179–192.

de Bello, F., Lavorel, S., Díaz, S., Harrington, R., Cornelissen, J. H. C., Bardgett, R. D., ... Harrison, P. A. (2010). Towards an assessment of multiple ecosystem processes and services via functional traits. *Biodiversity and Conservation, 19*, 2873–2893.

De Deyn, G. B., Cornelissen, J. H. C., & Bardgett, R. D. (2008). Plant functional traits and soil carbon sequestration in contrasting biomes. *Ecology Letters, 11*, 516–531.

de Vries, F. T., Manning, P., Tallowin, J. R. B., Mortimer, S. R., Pilgrim, E. S., Harrison, K. A., ... Bardgett, R. D. (2012). Abiotic drivers and plant traits explain landscape-scale patterns in soil microbial communities. *Ecology Letters, 15*(11), 1230–1239. doi:10.1111/j.1461-0248.2012.01844.x

Dias, A. T. C., Berg, M. P., de Bello, F., Van Oosten, A. R., Bílá, K., & Moretti, M. (2013). An experimental framework to identify community functional components driving ecosystem processes and services delivery. *Journal of Ecology, 101*(1), 29–37. doi:10.1111/1365-2745.12024

Díaz, S., & Cabido, M. (2001). Vive la différence: plant functional diversity matters to ecosystem processes. *Trends in Ecology and Evolution, 16*, 646–655.

Díaz, S., Fargione, J., Chapin, F. S., III, & Tilman, D. (2006). Biodiversity loss threatens human well-being. *PLoS Biology* 4(8), e277. doi:10.1371/journal.pbio.0040277

Díaz, S., Lavorel, S., Chapin, F. S., III, Tecco, P. A., Gurvich, D. E., & Grigulis, K. (2007a). Functional diversity—at the crossroads between ecosystem functioning and environmental filters. In J. G. Canadell, D. E. Pataki, & L. F. Pitelka (eds), *Terrestrial Ecosystems in a Changing World* (pp. 81–91). Berlin: Springer-Verlag.

Díaz, S., Lavorel, S., de Bello, F., Quétier, F., Grigulis, K., & Robson, M. (2007b). Incorporating plant functional diversity effects in ecosystem service assessments. *Proc. Natl. Acad. Sci. USA, 104*, 20684–20689.

Duru, M., Cruz, P., Al Haj Khaled, R., Ducourtieux, C., & Theau, J.-P. (2008). Relevance of plant functional types based on leaf dry matter content for assessing digestibility of native grass species and species-rich grassland communities in spring. *Agronomy Journal, 100*, 1622–1630.

Ebeling, A., Pompe, S., Baade, J., Eisenhauer, N., Hillebrand, H., Proulx, R., ... Weisser, W. W. (2014). A trait-based experimental approach to understand the mechanisms underlying biodiversity–ecosystem functioning relationships. *Basic and Applied Ecology, 15*(3), 229–240. doi:10.1016/j.baae.2014.02.003

Enquist, B. J., Economo, E. P., Huxman, T. E., Allen, A. P., Ignace, D. D., & Gillooly, J. F. (2003). Scaling metabolism from organisms to ecosystems. *Nature, 423*(6940), 639–642. doi:10.1038/nature01671

Eviner, V. T., & Chapin, F. S., III. (2003). Functional matrix: a conceptual framework for predicting multiple plant effects on ecosystems. *Annual Review of Ecology and Systematics, 34*, 455–485.

Finegan, B., Peña-Claros, M., de Oliveira, A., Ascarrunz, N., Bret-Harte, M. S., Carreño-Rocabado, G., ... Poorter, L. (2015). Does functional trait diversity predict above-ground biomass and productivity of tropical forests? Testing three alternative hypotheses. *Journal of Ecology, 103*(1), 191–201. doi:10.1111/1365-2745.12346

Flynn, D. F. B., Mirotchnick, N., Jain, M., Palmer, M. I., & Naeem, S. (2011). Functional and phylogenetic diversity as predictors of biodiversity–ecosystem-function relationships. *Ecology, 92*(8), 1573–1581. doi:10.1890/10-1245.1

Fortunel, C., Garnier, E., Joffre, R., Kazakou, E., Quested, H., Grigulis, K., ... Zarovali, M. (2009). Leaf traits capture the effects of land use changes and climate on litter decomposability of grasslands across Europe. *Ecology, 90*, 598–611.

Gardarin, A., Garnier, E., Carrère, P., Cruz, P., Andueza, D., Bonis, A., ... Kazakou, E. (2014). Plant trait–digestibility relationships across management and climate gradients in permanent grasslands. *Journal of Applied Ecology, 51*, 1207–1217. doi:10.1111/1365-2664.12293

Garnier, E., Cortez, J., Billès, G., Navas, M.-L., Roumet, C., Debussche, M., ... Toussaint, J.-P. (2004). Plant functional markers capture ecosystem properties during secondary succession. *Ecology, 85*(9), 2630–2637.

Garnier, E., Lavorel, S., Ansquer, P., Castro, H., Cruz, P., Dolezal, J., ... Zarovali, M. (2007). Assessing the effects of land use change on plant traits, communities, and ecosystem functioning in grasslands: a standardized methodology and lessons from an application to 11 European sites. *Annals of Botany, 99*, 967–985.

Gitay, H., & Noble, I. R. (1997). What are functional types and how should we seek them? In T. M. Smith, H. H. Shugart, & F. I. Woodward (eds), *Plant Functional Types. Their Relevance to Ecosystem Properties and Global Change* (pp. 3–19). Cambridge: Cambridge University Press.

Gower, S. T. (2002). Productivity of terrestrial ecosystems. In H. A. Mooney & J. G. Canadell (eds), *The Earth system: biological and ecological dimensions of global environmental change* (pp. 516–521). New York: John Wiley & Sons.

Grace, J. B., Anderson, T. M., Smith, M. D., Seabloom, E., Andelman, S. J., Meche, G., . . . Willig, M. R. (2007). Does species diversity limit productivity in natural grassland communities? *Ecology Letters, 10*, 680–689.

Grigulis, K., Lavorel, S., Krainer, U., Legay, N., Baxendale, C., Dumont, M., . . . Clément, J.-C. (2013). Relative contributions of plant traits and soil microbial properties to mountain grassland ecosystem services. *Journal of Ecology, 101*(1), 47–57. doi:10.1111/1365-2745.12014

Grime, J. P. (1997). Biodiversity and ecosystem function: the debate deepens. *Science, 277*, 1260–1261.

Grime, J. P. (1998). Benefits of plant diversity to ecosystems: immediate, filter, and founder effects. *Journal of Ecology, 86*, 902–910.

Grime, J. P. (2001). *Plant Strategies, Vegetation Processes, and Ecosystem Properties* (2nd ed.). Chichester: John Wiley & Sons.

Grime, J. P., & Pierce, S. (2012). *The Evolutionary Strategies that Shape Ecosystems*. Oxford: Wiley-Blackwell.

Gross, N., Robson, T. M., Lavorel, S., Albert, C., Le Bagousse-Pinguet, Y., & Guillemin, R. (2008). Plant response traits mediate the effects of subalpine grasslands on soil moisture. *New Phytologist, 180*, 652–662.

Heisse, K., Roscher, C., Schuhmacher, J., & Schulze, E.-D. (2007). Establishment of grassland species in monocultures: different strategies lead to success. *Oecologia, 152*, 435–447.

Hobbie, S. E., Oleksyn, J., Eissenstat, D. M., & Reich, P. B. (2010). Fine root decomposition rates do not mirror those of leaf litter among temperate tree species. *Oecologia, 162*, 505–513.

Hooper, D. U. (1998). The role of complementarity and competition in ecosystem responses to variation in plant diversity. *Ecology, 79*(2), 704–719. doi:10.1890/0012-9658

Hooper, D. U., & Dukes, J. S. (2004). Overyielding among plant functional groups in a long-term experiment. *Ecology Letters, 7*(2), 95–105. doi:10.1046/j.1461-0248.2003.00555.x

Hooper, D. U., & Vitousek, P. M. (1997). The effects of plant composition and diversity on ecosystem processes. *Science, 277*, 1302–1305.

Hooper, D. U., & Vitousek, P. M. (1998). Effects of plant composition and diversity on nutrient cycling. *Ecological Monographs, 68*, 121–149.

Hooper, D. U., Chapin, F. S., III, Ewel, J. J., Hector, A., Inchausti, P., Lavorel, S., . . . Wardle, D. A. (2005). Effects of biodiversity on ecosystem functioning: a consensus of current knowledge. *Ecological Monographs, 75*, 3–35.

Kerkhoff, A. J., & Enquist, B. J. (2006). Ecosystem allometry: the scaling of nutrient stocks and primary productivity across plant communities. *Ecology Letters, 9*, 419–427.

Klumpp, K., & Soussana, J.-F. (2009). Using functional traits to predict grassland ecosystem change: a mathematical test of the response-and-effect trait approach. *Global Change Biology, 15*, 2921–2934.

Laliberté, E., & Tylianakis, J. M. (2012). Cascading effects of long-term land-use changes on plant traits and ecosystem functioning. *Ecology, 93*(1), 145–155. doi:10.1890/11-0338.1

Larcher, W. (2003). *Physiological Plant Ecology—Ecophysiology and Stress Physiology of Functional Groups (4th ed.)*. Berlin: Springer-Verlag.

Laughlin, D. C. (2011). Nitrification is linked to dominant leaf traits rather than functional diversity. *Journal of Ecology, 99*(5), 1091–1099. doi:10.1111/j.1365-2745.2011.01856.x

Lavorel, S. (2013). Plant functional effects on ecosystem services. *Journal of Ecology, 101*(1), 4–8. doi:10.1111/1365-2745.12031

Lavorel, S., & Garnier, E. (2002). Predicting changes in community composition and ecosystem functioning from plant traits: revisiting the Holy Grail. *Functional Ecology, 16*, 545–556.

Lavorel, S., & Grigulis, K. (2012). How fundamental plant functional trait relationships scale-up to trade-offs and synergies in ecosystem services. *Journal of Ecology, 100*(1), 128–140. doi:10.1111/j.1365-2745.2011.01914.x

Lavorel, S., Díaz, S., Cornelissen, J. H. C., Garnier, E., Harrison, S. P., McIntyre, S., . . . Urcelay, C. (2007). Plant functional types: are we getting any closer to the Holy Grail? In J. Canadell, D. Pataki, & L. Pitelka (eds), *Terrestrial Ecosystems in a Changing World* (pp. 149–164). Berlin: Springer-Verlag.

Lavorel, S., Grigulis, K., Lamarque, P., Colace, M.-P., Garden, D., Girel, J., . . . Douzet, R. (2011). Using plant functional traits to understand the landscape distribution of multiple ecosystem services. *Journal of Ecology, 99*, 135–147.

Leoni, E., Altesor, A., & Paruelo, J. M. (2009). Explaining patterns of primary production from individual level traits. *Journal of Vegetation Science, 20*, 612–619.

Long, S. P., Garcia Moya, E., Imbamba, S. K., Kamnalrut, A., Piedade, M. T. F., Scurlock, J. M. O., . . . Hall, D. O.

(1989). Primary productivity of natural grassland ecosystems of the tropics: a reappraisal. *Plant and Soil, 115*, 155–166.

Loreau, M., Naeem, S., Inchausti, P., Bengtsson, J., Grime, J. P., Hector, A., ... Wardle, D. A. (2001). Biodiversity and ecosystem functioning: current knowledge and future challenges. *Science, 294*, 804–808.

Luyssaert, S., Inglima, I., Jung, M., Richardson, A. D., Reichstein, M., Papale, D., ... Janssens, I. A. (2007). CO_2 balance of boreal, temperate, and tropical forests derived from a global database. *Global Change Biology, 13*(12), 2509–2537. doi:10.1111/j.1365-2486.2007.01439.x

McNaughton, S. J., Oesterheld, M., Frank, D. A., & Williams, K. J. (1989). Ecosystem-level patterns of primary productivity and herbivory in terrestrial habitats. *Nature, 341*, 142–144.

Milcu, A., Roscher, C., Gessler, A., Bachmann, D., Gockele, A., Guderle, M., ... Roy, J. (2014). Functional diversity of leaf nitrogen concentrations drives grassland carbon fluxes. *Ecology Letters, 17*(4), 435–444. doi:10.1111/ele.12243

Mokany, K., Ash, J., & Roxburgh, S. (2008). Functional identity is more important than diversity in influencing ecosystem processes in a temperate native grassland. *Journal of Ecology, 96*, 884–893.

Mouillot, D., Villéger, S., Scherer-Lorenzen, M., & Mason, N. W. H. (2011). Functional structure of biological communities predicts ecosystem multifunctionality. *PLoS ONE, 6*(3), e17476. doi:10.1371/journal.pone.0017476

Mouquet, N., Devictor, V., Meynard, C. N., Munoz, F., Bersier, L. F., Chave, J., ... Thuiller, W. (2012). Ecophylogenetics: advances and perspectives. *Biological Reviews, 87*(4), 769–785. doi:10.1111/j.1469-185X.2012.00224.x

Naeem, S., Bunker, D. A., Hector, A., Loreau, M., & Perrings, C. (eds) (2009). *Biodiversity, Ecosystem Functioning, and Human Wellbeing—An Ecological and Economic Perspective*. New York: Oxford University Press.

Naeem, S., Duffy, J. E., & Zavaleta, E. (2012). The functions of biological diversity in an age of extinction. *Science, 336*(6087), 1401–1406. doi:10.1126/science.1215855

Niklas, K. J., & Enquist, B. J. (2001). Invariant scaling relationships for interspecific plant biomass production rates and body size. *Proc. Natl. Acad. Sci. USA, 98*, 2922–2997.

Obojes, N., Bahn, M., Tasser, E., Walde, J., Inauen, N., Hiltbrunner, E., ... Körner, C. (2015). Vegetation effects on the water balance of mountain grasslands depend on climatic conditions. *Ecohydrology, 8*(4), 552–569. doi:10.1002/eco.1524

Oren, O., Sperry, J. S., Katul, G. G., Pataki, D. E., Ewers, B. E., Phillips, N., & Schäfer, K. V. R. (1999). Survey and synthesis of intra- and interspecific variation in stomatal sensitivity to vapour pressure deficit. *Plant, Cell and Environment, 22*, 1515–1526.

Pakeman, R. J., Eastwood, A., & Scobie, A. (2011). Leaf dry matter content as a predictor of grassland litter decomposition: a test of the 'mass ratio hypothesis'. *Plant and Soil, 342*(1–2), 49–57. doi:10.1007/s11104-010-0664-z

Petchey, O. L., & Gaston, K. J. (2006). Functional diversity: back to basics and looking forward. *Ecology Letters, 9*, 741–758.

Petchey, O. L., Hector, A., & Gaston, K. J. (2004). How do different measures of functional diversity perform? *Ecology, 85*, 847–857.

Pontes, L. D. S., Soussana, J.-F., Louault, F., Andueza, D., & Carrère, P. (2007). Leaf traits affect the above-ground productivity and quality of pasture grasses. *Functional Ecology*, 21, 844–853.

Quested, H., Eriksson, O., Fortunel, C., & Garnier, E. (2007). Plant traits relate to whole-community litter quality and decomposition following land use change. *Functional Ecology, 21*, 1016–1026.

Quétier, F., Thébault, A., & Lavorel, S. (2007). Plant traits in a state and transition framework as markers of ecosystem response to land-use change. *Ecological Monographs, 77*, 33–52.

Reich, P. B. (2012). Key canopy traits drive forest productivity. *Proceedings of the Royal Society B: Biological Sciences, 279*(1736), 2128–2134. doi:10.1098/rspb.2011.2270

Reich, P. B. (2014). The world-wide 'fast–slow' plant economics spectrum: a traits manifesto. *Journal of Ecology, 102*(2), 275–301. doi:10.1111/1365-2745.12211

Reich, P. B., Walters, M. B., & Ellsworth, D. S. (1992). Leaf life-span in relation to leaf, plant, and stand characteristics among diverse ecosystems. *Ecological Monographs, 62*, 365–392.

Reich, P. B., Walters, M. B., & Ellsworth, D. S. (1997). From tropics to tundra: global convergence in plant functioning. *Proc. Natl. Acad. Sci. USA, 94*, 13730–13734.

Ricotta, C., & Moretti, M. (2011). CWM and Rao's quadratic diversity: a unified framework for functional ecology. *Oecologia, 167*(1), 181–188. doi:10.1007/s00442-011-1965-5

Roscher, C., Schumacher, J., Baade, J., Wilcke, W., Gleixner, G., Weisser, W. W., ... Schulze, E.-D. (2004). The role of biodiversity for element cycling and trophic interactions: an experimental approach in a grassland community. *Basic and Applied Ecology, 5*(2), 107–121. doi:10.1078/1439-1791-00216

Roscher, C., Schumacher, J., Gubsch, M., Lipowsky, A., Weigelt, A., Buchmann, N., ... Schulze, E.-D. (2012). Using plant functional traits to explain diversity—productivity relationships. *PLoS ONE, 7*(5), e36760. doi:10.1371/journal.pone.0036760

Roscher, C., Schumacher, J., Lipowsky, A., Gubsch, M., Weigelt, A., Pompe, S., ... Schulze, E.-D. (2013). A functional trait-based approach to understand community assembly and diversity–productivity relationships over 7 years in experimental grasslands. *Perspectives in Plant Ecology, Evolution and Systematics, 15*(3), 139–149. doi:10.1016/j.ppees.2013.02.004

Ruiz-Jaen, M. C., & Potvin, C. (2011). Can we predict carbon stocks in tropical ecosystems from tree diversity? Comparing species and functional diversity in a plantation and a natural forest. *New Phytologist, 189*(4), 978–987. doi:10.1111/j.1469-8137.2010.03501.x

Sala, O. E., Parton, W. J., Joyce, L. A., & Lauenroth, W. K. (1988). Primary production of the central grassland region of the United States. *Ecology, 69*, 40–45.

Saugier, B., Roy, J., & Mooney, H. A. (2001). Estimations of global terrestrial productivity: converging towards a single number? In J. Roy, B. Saugier, & H. A. Mooney (eds), *Terrestrial Global Productivity* (pp. 543–557). San Diego: Academic Press.

Scherer-Lorenzen, M. (2008). Functional diversity affects decomposition processes in experimental grasslands. *Functional Ecology, 22*(3), 547–555. doi:10.1111/j.1365-2435.2008.01389.x

Schulze, E.-D., & Zwölfer, H. (eds) (1987). *Potential and Limitations in Ecosystem Analysis.* Berlin: Springer-Verlag.

Schulze, E.-D., Beck, E., & Müller-Hohenstein, K. (2005). *Plant Ecology.* Berlin: Springer-Verlag.

Schumacher, J., & Roscher, C. (2009). Differential effects of functional traits on aboveground biomass in semi-natural grasslands. *Oikos, 118*(11), 1659–1668. doi:10.1111/j.1600-0706.2009.17711.x

Schwinning, S., & Ehleringer, J. R. (2001). Water use trade-offs and optimal adaptations to pulse-driven arid ecosystems. *Journal of Ecology, 89*, 464–480.

Scurlock, J. M. O., Cramer, W., Olson, R. J., Parton, W. J., & Prince, S. D. (1999). Terrestrial NPP: toward a consistent data set for global model evaluation. *Ecological Applications, 9*, 913–919.

Scurlock, J. M. O., Johnson, K., & Olson, R. J. (2002). Estimating net primary productivity from grassland biomass dynamics measurements. *Global Change Biology, 8*, 736–753.

Shipley, B. (2010). *From Plant Traits to Vegetation Structure. Chance and Selection in the Assembly of Ecological Communities.* Cambridge: Cambridge University Press.

Smith, M. D., & Knapp, A. K. (2003). Dominant species maintain ecosystem function with non-random species loss. *Ecology Letters, 6*(6), 509–517. doi:10.1046/j.1461-0248.2003.00454.x

Solbrig, O. T. (1993). Plant traits and adaptative strategies: their role in ecosystem function. In E.-D.Schulze & H. A. Mooney (eds), *Biodiversity and Ecosystem Function* (pp. 97–126). Berlin, New York: Springer-Verlag.

Srivastava, D. S., Cadotte, M. W., MacDonald, A. A. M., Marushia, R. G., & Mirotchnick, N. (2012). Phylogenetic diversity and the functioning of ecosystems. *Ecology Letters, 15*(7), 637–648. doi:10.1111/j.1461-0248.2012.01795.x

Stokes, A., Atger, C., Bengough, A., Fourcaud, T., & Sidle, R. (2009). Desirable plant root traits for protecting natural and engineered slopes against landslides. *Plant and Soil, 324*(1–2), 1–30. doi:10.1007/s11104-009-0159-y

Storkey, J., Brooks, D., Haughton, A. J., Hawes, C., Smith, B. M., & Holland, J. M. (2013). Using functional traits to quantify the value of plant communities to invertebrate ecosystem service providers in arable landscapes. *Journal of Ecology, 101*(1), 38–46. doi:10.1111/1365-2745.12020

Suding, K. N., Lavorel, S., Chapin, F. S., III, Cornelissen, J. H. C., Díaz, S., Garnier, E., . . . Navas, M.-L. (2008). Scaling environmental change through the community-level: a trait-based response-and-effect framework for plants. *Global Change Biology, 14*, 1125–1140.

Tardif, A., & Shipley, B. (2013). Using the biomass-ratio and idiosyncratic hypotheses to predict mixed-species litter decomposition. *Annals of Botany, 111*(1), 135–141. doi:10.1093/aob/mcs241

Tardif, A., Shipley, B., Bloor, J. M. G., & Soussana, J.-F. (2014). Can the biomass-ratio hypothesis predict mixed-species litter decomposition along a climatic gradient? *Annals of Botany, 113*(5), 843–850. doi:10.1093/aob/mct304

Thompson, K., Askew, A. P., Grime, J. P., Dunnett, N. P., & Willis, A. J. (2005). Biodiversity, ecosystem function and plant traits in mature and immature plant communities. *Functional Ecology, 19*, 355–358.

Tilman, D. (1997). Distinguishing between the effects of species diversity and species composition. *Oikos, 80*, 185.

Tilman, D. (2001). Functional diversity. In S. A. Levin (ed.), *Encyclopedia of Biodiversity*, Vol. 3 (pp. 109–120). San Diego: Academic Press, Inc.

Tilman, D., Knops, J., Wedin, D., Reich, P., Ritchie, M., & Siemann, E. (1997). The influence of functional diversity and composition on ecosystem processes. *Science, 277*(5330), 1300–1302. doi:10.1126/science.277.5330.1300

Tobner, C. M., Paquette, A., Reich, P. B., Gravel, D., & Messier, C. (2014). Advancing biodiversity-ecosystem functioning science using high-density tree-based experiments over functional diversity gradients. *Oecologia, 174*(3), 609–621. doi:10.1007/s00442-013-2815-4

Vile, D., Shipley, B., & Garnier, E. (2006). Ecosystem productivity can be predicted from potential relative growth rate and species abundance. *Ecology Letters, 9*, 1061–1067.

Violle, C., Garnier, E., Lecœur, J., Roumet, C., Podeur, C., Blanchard, A., & Navas, M.-L. (2009). Competition, traits, and resource depletion in plant communities. *Oecologia, 160*, 747–755.

Violle, C., Lecoeur, J., & Navas, M.-L. (2007a). How relevant are instantaneous measurements for assessing resource depletion under plant cover? A test on light and soil water availability in 18 herbaceous communities. *Functional Ecology, 21*, 185–190.

Violle, C., Navas, M.-L., Vile, D., Kazakou, E., Fortunel, C., Hummel, I., & Garnier, E. (2007b). Let the concept of trait be functional! *Oikos, 116*, 882–892.

Vitousek, P. M., & Hooper, D. U. (1993). Biological diversity and terrestrial ecosystem biogeochemistry. In E.-D.Schulze & H. A. Mooney (eds), *Biodiversity and Ecosystem Function* (pp. 3–14). Berlin: Springer-Verlag.

Walker, B., Kinzig, A., & Langridge, J. (1999). Plant attribute diversity, resilience, and ecosystem function: the nature and significance of dominant and minor species. *Ecosystems, 2*, 95–113.

Wardle, D. A. (2002). *Communities and Ecosystems. Linking the Aboveground and Belowground Components*. Princeton, USA: Princeton University Press.

Wardle, D. A., Bardgett, R. D., Klironomos, J. N., Setala, H., van der Putten, W. H., & Wall, D. H. (2004). Ecological linkages between aboveground and belowground biota. *Science, 304*(5677), 1629–1633. doi:10.1126/science.1094875

Waring, R. H., & Running, S. W. (1998). *Forest Ecosystems: Analysis at Multiple Scales*. New York: Academic Press.

West, G. B., Brown, J. H., & Enquist, B. J. (1997). A general model for the origin of allometric scaling laws in biology. *Science, 276*, 122–126.

Whittaker, R. H. (1975). *Communities and Ecosystems* (2nd ed.). New York, USA: Macmillan Publishing Co., Inc.

CHAPTER 7

Functional diversity and ecosystem services

7.1 Introduction

The concept of ecosystem services was initially proposed to promote the protection of biodiversity by demonstrating the ways in which biodiversity was useful to human society, as biodiversity was increasingly threatened by human activities (Ehrlich & Mooney, 1983; Westman, 1977). In the year 2000, this idea received major political recognition through establishment of the Millennium Ecosystem Assessment (MEA, 2005; http://www.maweb.org/en/Index.aspx) by the United Nations. The first action of this international assessment was to define ecosystem services as 'benefits that human society derives from ecosystems'. The MEA was also mandated to summarize the direct and indirect benefits of biodiversity, to formalize methods for the analysis and evaluation of services, and to establish an initial consensus regarding the role of biodiversity in the supply of services.

The simplicity of the concept and its capacity to crystallize the essence of ecosystem goods and services, as well as their interdependence with human activities, ensured its rapid success (Vihervaara et al., 2010). From the beginning, the concept has been resolutely multidisciplinary. Coupling ecological, economic, and sociological approaches allowed evaluation of services supplied under highly contrasted situations in terms of political and social regulations, and of population access to resources (MEA, 2005). The MEA results were alarming and confirmed earlier studies. While changes in ecosystem use over the last fifty years contributed to considerable increases in the provision of food and energy, there has been an explosion in the cost of exploiting ecosystems. As a consequence, the Earth's environment is considerably degraded, placing access to services in peril for future generations. 'The Economics of Ecosystems and Biodiversity' (TEEB: http://www.teebweb.org/) was established in 2007 by Germany and the European Commission, whose objective was to evaluate the economic costs of biodiversity losses, as well as the costs and benefits of actions to reduce these losses, recognizing that biodiversity does not respond to classical economic logic. More ambitiously, the WAVES initiative of the World Bank 'Wealth Accounting and the Valuation of Ecosystem Services' (http://www.wavespartnership.org/) has since 2010 proposed tools for the evaluation of services, based on indicators of economic growth, that include the resulting environmental consequences. Another international initiative, the EU Biodiversity Strategy to 2020, calls member states to map and assess the state of ecosystems and their services in their national territory. To that aim, categories of ecosystem services are defined in the Common International Classification of Ecosystem Services (CICES: http://biodiversity.europa.eu/maes/common-international-classification-of-ecosystem-services-cices-classification-version-4.3).

The concept is also important in promoting sustainable development and for the establishment of management policies, or biodiversity offsetting measures, founded on realistic evaluations of the effects of biodiversity on the supply of services and the consequences to be expected from biodiversity losses. The aim of the present chapter is not to encompass all of this complexity, but to explain the biological and ecological basis of ecosystem services. We also demonstrate how the use of a functional approach to diversity is relevant to identify and quantify services, especially in a changing environment. These issues are of major research importance in order to allow economic evaluations of services (for France: Chevassus-au-Louis et al., 2009; for the European Union: Maes et al., 2012) at a time when the assessment of ecosystem services by stakeholders is becoming a major policy tool (Lamarque, 2012). After presenting the major types of currently recognized ecosystem services, we investigate the linkages among traits, processes, and services, before discussing how the plant trait approach can be relevant for evaluating and managing services. We concentrate primarily on the role of plant traits, with some discussion of other trophic groups, recognizing that many services rely on these groups.

7.2 Types of ecosystem services

The MEA identified different categories of services, which, though criticized, remain generally recognized (Seppelt et al., 2011; Vihervaara et al., 2010): 'provisioning services', which are the products obtained from ecosystems (foods, materials, and fibres, clean water, bioenergy, etc.); 'regulating services', which have the capacity to modulate in a manner favourable for humanity phenomena such as the climate, the occurrence and virulence of diseases, or differing aspects of the water cycle (floods, flows, physico-chemical quality, etc.); and 'cultural services', corresponding to the use of ecosystems for recreation, aesthetic, or spiritual values. A debate currently exists regarding the designation of a fourth category, 'supporting services', not directly used by humans but which are necessary for ecosystem functioning (nutrient cycling, biomass production, etc.). As those services correspond to processes necessary for the provision of the three other types of services, they cannot be evaluated in their own right, and therefore were not incorporated into TEEB, or in IPBES, etc.

A review of studies evaluating ecosystem services carried out before 2010 demonstrates the wide variety of approaches used to define and evaluate services (Crossman et al., 2013b; Seppelt et al., 2011; Vihervaara et al., 2010). The review shows that the concept of ecosystem service is often used in a vague, simplistic, and sometimes erroneous manner, and there is a need for standardization of evaluation methods. There are two categories of studies addressing different questions. The first category aims at identifying services within particular ecosystems. Figure 7.1, for example, shows the redefinition of services for agro-ecosystems (Le Roux et al., 2009; Zhang et al., 2007), explicitly distinguishing those services linked with agricultural income and those with other values, extending beyond the context of cultivated areas (see also Figures 8.1 and 8.6 for similar studies in the cases of grasslands and crop weeds). A second category evaluates the impact of management or land-use scenarios on services in particular geographic or socio-economic contexts. This approach has been largely used by geographers or developers to project the future dynamics of a region under different management scenarios, and comprise more than two thirds of the studies referenced by Seppelt et al. (2011). The method most commonly used in these studies is to map ecosystem services by attributing to each type of land use an estimated value for the supplied services, often based on empirical relationships developed in other systems (Eigenbrod et al., 2010). These forms of estimation have been heavily criticized for two principal reasons (Lamarque et al., 2011). The first is related to error propagation, resulting from errors in land-use or land-cover mapping and/or from the transfer of estimations developed in different systems. The second is related to the overestimation of the impact of human activities on ecosystems via the identification of land-use categories as the major classifier, while ignoring the biological mechanisms underlying ecosystem properties, as well as their spatial and temporal dynamics (Lavorel et al., 2011).

The poor biophysical realism of current services assessment, due to the lack of understanding of the role of diversity in the provision of services, has been noted by numerous researchers (Cardinale et al., 2012; Crossman et al., 2013a; Lavorel, 2013; Perrings et al., 2010; Seppelt et al., 2011; Vihervaara et al., 2010). Determining the relationships between diversity and processes linked to ecosystem services, particularly in the context of

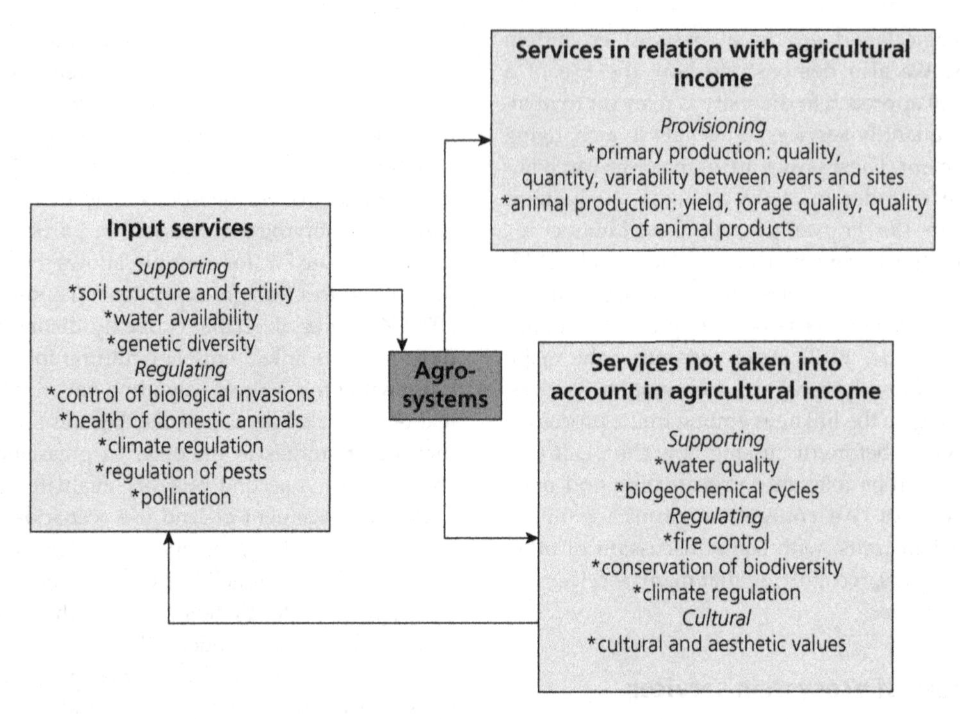

Figure 7.1 Ecosystem services defined in an agricultural context. Three categories are proposed, based on agricultural activities. These categories regroup those services initially represented in the four types proposed by the MEA (indicated here in italics). Based on Zhang et al. (2007) and Le Roux et al. (2009).

changing land uses and environmental conditions, is therefore a research topic in which progress is essential to advance the evaluation of ecosystem services (Crossman et al., 2013a; Lavorel, 2013). Additionally, since the current understanding of diversity impact on services is partial, it is often based on a uniquely taxonomic approach to diversity. It is surprising to realize that the experts asked to evaluate the importance of the role of biodiversity in the provision of services 'forget' or underestimate the role of functional diversity in comparison to the role of the number of species, or that of a few emblematic species (Quijas et al., 2012). The aim of the remainder of this chapter is to outline the contributions that functional ecology can bring to the quantification and management of ecosystem services.

7.3 A functional approach to services

As described above, the ecological determinants of the provision of ecosystem services are currently a very active research field (Cardinale et al., 2012; Lavorel, 2013; Perrings et al., 2010; Vihervaara et al., 2010). This field continues the line of work which has established that the functional composition of a community plays a major role in explaining relationships between biodiversity and ecosystem functioning (Díaz et al., 2006). Before reviewing this area of research, the actual importance that biodiversity plays in determining the definition of services and their value remains to be established.

7.3.1 Cascading effects and the determination of services and their value

The role of diversity in the provision of ecosystem services can be represented as a cascade of effects (Haines-Young & Potschin, 2010; Lamarque et al., 2011; and Figure 7.2). The services produced by an ecosystem are linked to the existence of functions, which in turn depend on the types of organisms constituting the ecosystem, as well as the processes controlling their interactions with the environment. The identification of these functions is fundamental,

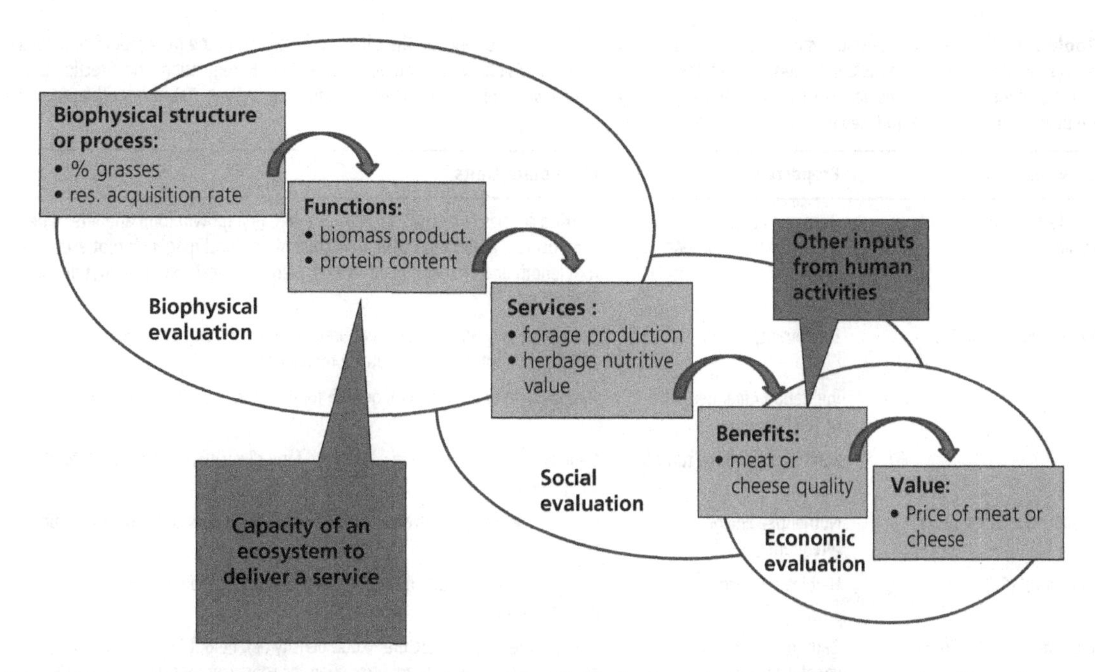

Figure 7.2 A cascade of effects leading to the production of ecosystem services, for the particular example of the production of forage in mountain grasslands. The phases of this cascade taken into account by the different types of evaluation (biophysical, social and economic) are regrouped within the ellipses. See text for further details. Res. acquisition rate: resource acquisition rate. Adapted from Haines-Young & Potschin (2010) and Lamarque et al. (2011).

as it allows for the quantification of the capacity of an ecosystem to provide a given service. This first chain of relationships among structure/process–functions–services is characterized by means of a biophysical evaluation, which detects the impact of biodiversity on a given service. The benefit which humanity derives from the produced services is then modified by social, constructed, or technological human activity, in order to fulfil the needs of beneficiaries. For example, the quality of a cheese is certainly dependent on the composition and the quality of the milk, but is also due to the technological processes involved in its fabrication, as well as the alimentary preferences of consumers. Furthermore, the availability of a good in a locality where it is not produced depends on available infrastructure and transport. Consequently, the link between services and benefits does not necessarily have a strictly ecological component, as it also depends on social evaluations. Finally, the relationships between the benefits drawn from a service and their value are directly assessed by an economic evaluation. This cascade of effects illustrates the key role which biological diversity plays in a given system, by modulating processes involved in the final production of services. It also illustrates why the notion of biodiversity is often associated with that of ecosystem services, even when the relationships between the two are not always explicitly taken into account.

7.3.2 Relationships between functional structure, processes, and services

The previous chapter identified the role of functional diversity in determining ecosystem properties linked with fluxes of carbon, nutrients, and water. Recent reviews have well described the important role of functional diversity in these cycles, but also in other processes more linked with ecosystem regulation (Balvanera et al., 2006; Cardinale et al., 2006; Hooper et al., 2005; Kremen et al., 2007). The traits by which organisms consume or transform resources, modify habitat structure or environmental chemistry, or interact with other organisms are *a priori* good candidates to be used for the prediction of ecosystem properties.

Table 7.1 Significant relationships among plant traits, ecosystem properties, and services. The category, according to the MEA typology, to which each ecosystem service belongs is indicated between the parentheses: P: provisioning; C: cultural; S: supporting; R: regulation. The direction of the effect of the trait on the considered process(es) can be positive (+), negative (-) or variable when no indication is given. Taken from the review by de Bello et al. (2010). Reproduced with permission from Springer.

Services	Properties	Main plant traits
Soil fertility and nutrient cycling (R, S)	Decomposition, mineralization, nutrient mobilization	N fixing species (+); LDMC (-); SLA (+); LNC (+); growth form and litter type composition; association with arbuscular mycorrhizal fungi (+); root exudates; root length and biomass; plant size (-); plant chemical composition; time of flowering (-)
Water regulation (R, S)	Evapotranspiration	Canopy density and size (+); leaf area (+); growth form composition (C_3 / C_4); phenology; root depth; stomatal conductance
	Infiltration / maintenance of soil humidity	Canopy density and size (+); growth form composition; litter amount (-)
	Surface water flow/ run-off	Canopy complexity and size (-); growth form diversity (-); and growth form composition
Water purification (R, S)	Nutrient/sediment retention	Decomposability (-); growth form diversity (+) and growth form composition; leaf area (+)
Biocontrol (R, S)	Herbivory control	Tissue chemistry including LNC (-); SLA (-); LDMC (+); leaf toughness (+); phenology; ruderality
Climate regulation (R, S)	Carbon sequestration in vegetation and soil	Canopy size and architecture; wood density (+); height (+); flowering phenology (-); growth form; litter quantity; root size; root density (+); RGR (+); LNC (+)
	Heat exchange	Canopy complexity; growth form composition; photosynthetic rate
Soil stability (R, S)	Erosion prevention	Canopy size and architecture (+); growth form; root depth/density
Soil formation (R, S)	Soil formation	Growth form composition; litter quality and quantity; root size and complexity
Pollination (R, S)	Pollinator provision	Flower traits : accessibility (+); abundance (+); attractiveness (+); density (+); variety of nectar type (+); size (+)
Invasion resistance (R)	Persistence and resistance of habitat and processes	Diversity and composition of growth forms
Natural hazard prevention (R)	Fire risk prevention	Flammability (-); growth form (woody, resprouters); resin, terpene, and oils content (-); ramification (-); twig/leaf dry matter content; and leaf area (-)
	Avalanche prevention	Leaf toughness (-) and litter amount (+)
	Hurricanes/wind damage resistance	Canopy complexity and size; root depth
Disease control (R)	Allergy prevention	Pollen type; tissue chemical composition (secondary metabolites)
Forage production (P)	Accumulation of standing biomass	Canopy size (+); SLA (+); leaf size (+), LDMC (-); LNC (+); leaf toughness (-); diversity (+) and composition of growth form; root density (+)
	Consumption and health of livestock	Diversity in chemical compounds and nutritional value (+); LNC (+)
Fibre production (P)	Accumulation of standing biomass	Canopy architecture; size (+); wood density (+); diversity (+) and composition of growth form
Aesthetic values, cultural and sense of place (C)	Accumulation of standing biomass and species coexistence	LNC (+); leaf toughness (+)

De Bello et al. (2010) and others (e.g. Harrison et al., 2014) proposed syntheses of the contribution of traits to the provision of ecosystem services (Table 7.1). If the relationships between structural traits and decomposition, mineralization of organic matter, biomass production, evapotranspiration, herbivory or pollination are beginning to be well characterized, this is not the case for many other services, for example resistance to fires or storms, or the determination of cultural or aesthetic values. Despite this reservation, this initial synthesis shows that those plant traits having a major contribution to the provision of numerous services are those presented in the previous chapters (Figure 7.3): the size and architecture of the above-ground plant parts are involved in the regulation of local climate and the water cycle, as well as soil stability; the size and architecture of the root system are involved in the water cycle and soil stability; traits identified in the leaf economic spectrum (leaf structure and chemical composition) influence the production of biomass and soil fertility. Combinations of certain of these traits can exercise multiple controls on a given service: for example, forage production depends not only on leaf structural and chemical traits, but also on phenological traits.

Beyond this plant-centred approach, it should not be forgotten that the production of numerous services requires contributions from many types of organisms. For example, the decomposition rate of organic matter varies as a function of the size and feeding preferences of detritivorous animals present in the soil animal community, as well as litter characteristics (Hättenschwiler & Gasser, 2005), but also as a function of the relative proportion of bacteria and fungi in the soil microbial community (Grigulis et al., 2013). More generally, soil properties linked to carbon and nitrogen cycles depend not only on the functional characteristics of plants but also on those of below-ground trophic networks (de Vries et al., 2013). Another example is that of above-ground biomass accumulation, which depends on the type and activity of the herbivores present, both invertebrate and vertebrate, as well as on the traits of living leaves (Lavorel & Grigulis, 2012).

Recent studies of the impact of plant–pollinator interactions on the service of pollination were based on a new approach to the trophic system (Lavorel et al., 2013). This approach is based on the identification of multi-trophic controls on the provision of services: traits at each trophic level respond to the environment and modulate ecosystem services,

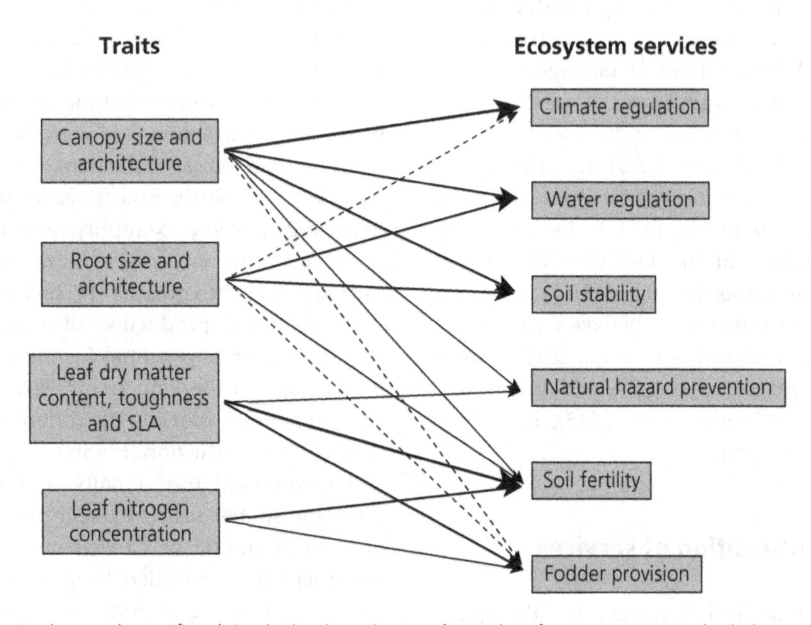

Figure 7.3 Plant traits shown to be significantly involved in the realization of a number of ecosystem services. The thickness of the arrows reflects the significance of the relationships. Taken from de Bello et al. (2010). Reproduced with permission from Springer.

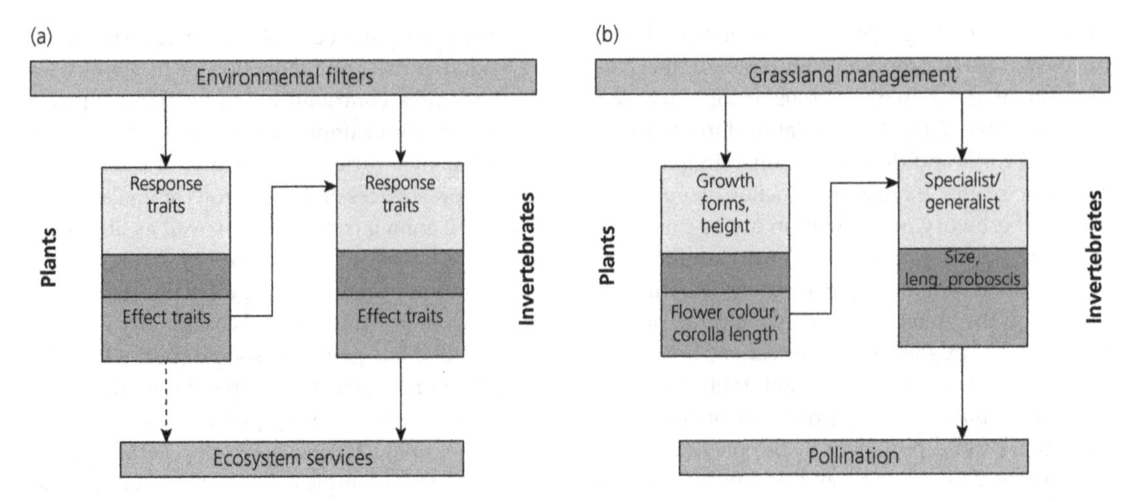

Figure 7.4 Multi-trophic control of the provision of ecosystem services. (a) General framework. The traits of each of the trophic levels considered (in this case plants represent the first level and invertebrates the second level) respond to filters represented by environmental factors and certain traits (which can be different to those of the first type) determine the ecosystem services. However, the traits of the two trophic levels are not independent as the response traits of the invertebrates are also under the influence of the plant effect traits. The dashed arrow indicates that there is not necessarily a direct effect of plants on the particular service considered (b) The example of pollination. In these two figures, darkers areas in the middle of the boxes correspond to traits involved both in response and effect. Taken from Lavorel et al. (2013). Reproduced with permission from Wiley.

but relationships also exist among trophic levels which are expressed by the traits of the second level being modulated by those of the first level (Figure 7.4). In the case of pollination, intensification of grassland management changes vegetation height and the growth forms present, inducing parallel changes in the colour and size of flower corollas present. Intensification of management also modifies the relative abundances of specialist and generalist insect pollinators, inducing changes in the body size of pollinators and also the size of their proboscis, the organ used for the suction and thus acquisition of nectar, both traits which are strongly correlated with the floral characteristics of the vegetation (Campbell et al., 2012). Frameworks of these types of interactions between weeds and their dependent fauna (Brooks et al., 2012; Storkey et al., 2013), and between grassland plants and microorganisms (Grigulis et al., 2013), have been successfully developed.

7.4 The quantification of services

Responding to societal demands for the quantification of ecosystem services raises important research questions. A first priority is to identify the stakeholders concerned and their expectations in terms of ecosystem services. This question, while essential, will not be discussed here as it relates more to the social sciences. A second aspect to take into account is that the quantification of services needs to account for multifunctionality. Multifunctionality is here defined as the occurrence of multiple services, and in particular as the trade-offs between certain services in time and space as well as between the expectations of various regional stakeholders (Bennett & Balvanera, 2007; Raudsepp-Hearne et al., 2010; Rodríguez et al., 2006). The following questions exemplify these types of trade-offs: Can we reconcile agricultural production and carbon storage in a grassland? How can we increase the agricultural production of a landscape without decreasing its potential for the conservation of biodiversity in the long term? How can we manage a region to reconcile the different objectives of agricultural production, biodiversity conservation, and recreational use? Finally, a third point concerns the spatial scale at which the quantification of services should be carried out. While services are essentially controlled by processes defined at the level of the ecosystem, their quantification is often required at the level of the landscape or the region, scales at which the majority of management or development actions and policy prescriptions occur. 'Off-site' effects also exist when there is a spatial disconnect between the decision-making site and the site where impacts on ecosystem services are experienced (Hein et al., 2006); for example, in the situation of the agricultural abandonment of an alpine grassland leading to changes in flood and water quality regulation felt downstream (Inauen et al., 2013).

These questions are explicitly taken into account by the functional approach to evaluation of ecosystem services. A first stage consists of establishing the processes linked to the different target services in a given system and the traits involved in modulating these processes. As seen earlier, a service can depend on multiple traits, sometimes linked to organisms from different trophic levels, with these traits differing in the amplitude and direction of their impact on the processes depending on local environmental conditions. In the following sections, we present and discuss the current methodologies used for the selection of traits and to take into account trade-offs among services at the landscape level.

7.4.1 Selection of traits acting on ecosystem processes and services

Díaz et al. (2007) proposed a methodology to identify the most parsimonious combination of abiotic factors and components of diversity acting on ecosystem properties, and thus on services in a given area (Figure 7.5). As indicated in Chapter 6, the latter authors proposed identifying the effect on ecosystem processes of, first, abiotic environmental factors, and then that of the functional structure of communities estimated by the community weighted means and functional divergence of trait values, to take into account that certain species can have particular impacts on a system independently of their abundance. The four grassland studies that have applied this method showed that certain processes linked with biomass production (such as NPP and above-ground SNPP) can be completely explained by abiotic variables (in this case, the nitrogen status of grasslands estimated by their Nitrogen Nutrition Index). Other properties such as litter decomposition or soil nitrification depend solely on the community weighted means

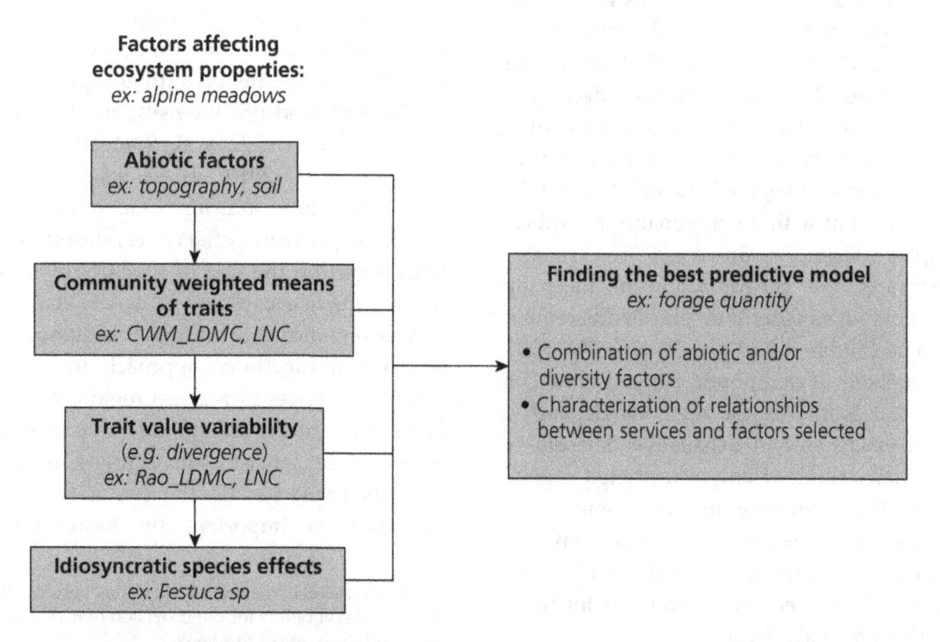

Figure 7.5 Diagram showing the selection of factors, and in particular of traits, acting on the provision of ecosystem services. The diagram shows the example of the provision of forage by mountain grasslands. The two principal components of the community functional structure are CWM (the community weighted mean) and the functional divergence estimated by the Rao index. LDMC: leaf dry matter content, LNC: leaf nitrogen concentration. After Díaz et al. (2007).

of certain traits, in particular of LNC, and not on their functional divergence. A third group of properties, for example soil water reserves, depend on a combination of the two types of variables. The simple models obtained via these analyses can be further used to evaluate the production of certain properties and services under different scenarios of environmental change. This methodology, which has been extensively used since its proposal, has confirmed the major role of functional structure in the provision of ecosystem services, in particular through the establishment of relationships between the functional structure of grassland communities and primary production or soil carbon storage (Lavorel et al., 2011).

An alternative approach is that of structural equation modelling, which allows the comparison of the influence of the functional structure of communities, estimated by the community weighted mean, and functional divergence of different traits, on targeted ecosystem processes (Lienin & Kleyer, 2012; Mokany et al., 2008; Mouillot et al., 2011). These analyses are based on hypothetical models of the network of relationships between the different studied variables established on the basis of previous studies, which are then quantified. For example, Mokany et al. (2008) established that all of their tested properties (biomass production, decomposition, light interception, and soil water quantity) in temperate grasslands were perfectly explained by the community weighted mean of a number of traits, consistent with the dominance hypothesis (see Chapter 6; Grime, 1998).

Nevertheless, the generalization of these approaches to systems other than grasslands remains largely to be validated (but see Conti & Díaz, 2013 along a gradient of xerophytic forests). One constraint is that both of these methods depend on the use of extensive and exhaustive datasets, allowing the modelling of only a restricted number of services. Their extension to more complex systems, including species for which traits have not been measured, and/or for services poorly linked to processes, clearly remains a challenge for future research (Naeem et al., 2012).

7.4.2 Multifunctionality, trade-offs between services, and changes in scale

The question of joint identification of multiple services, linked to the multifunctionality of ecosystems and landscapes, is more difficult to resolve, as it depends on the characterization of positive (synergies) or negative (trade-offs) associations among services which can exist over space and time (Cardinale et al., 2012). In reality, ecosystem services are not independent of each other (MEA, 2005) and are interconnected in 'bundles of services', defined as groups of services which appear in a repeated fashion in time, in space, or in the representations of stakeholders[1] (Raudsepp-Hearne et al., 2010).

The first method to address this question consists of defining if there exist 'hot spots' of ecosystem services, which are geographic zones providing a large number of services (Egoh et al., 2008), and respectively 'cold spots', where few services are provided. Such studies have shown that, in general, there are not hot spots of services (Perrings et al., 2010). There is contradictory evidence showing both positive and negative (and sometimes no) relationship between biodiversity and ecosystem services. No one at this point really knows where, when, and why. Some reasons for discrepancies across studies include scale, the breadth and position along a gradient of land-use intensity, the biodiversity of the biota, etc. (Dickie et al., 2011; Egoh et al., 2008; O'Farrell et al., 2010). These results have undermined the idea of using local biodiversity as a proxy for provision of services, despite extrapolations based on the role of specific richness on the functioning of ecosystems (Zavaleta et al., 2010).

A second means to address this question consists of using a functional approach to evaluate the potential services of a given region in an additive fashion. In a study carried out in alpine grasslands, a first stage consisted of defining those groups of traits implicated in the provision of services, identified as important by local stakeholders

[1] A discussion of the trade-offs existing between the expectations of stakeholders for the provision of ecosystem services is beyond the scope of this book.

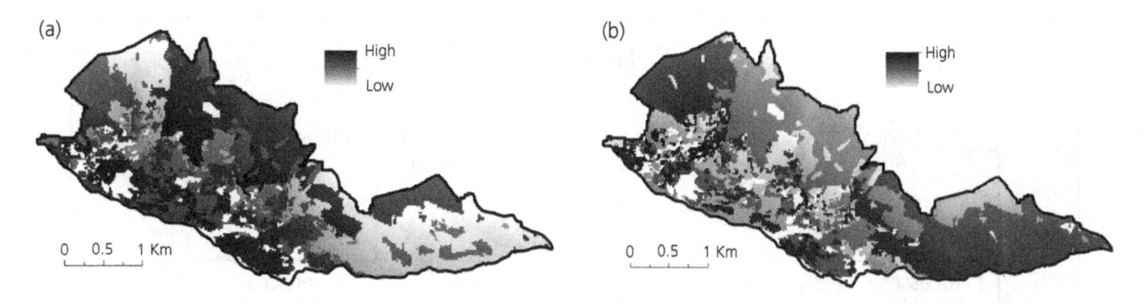

Figure 7.6 An example of the mapping of the provision of ecosystem services at the scale of the landscape. In the case of this study carried out in the local government area of Villar d'Arène (French Alps), an initial mapping exercise allowed the identification of the spatial repartition of major ecosystem characteristics such as green biomass production, its protein content and flowering phenology and then that of the associated ecosystem services. The maps show the spatial distribution of ecosystem properties essentially associated with (a) cultural services (b) the production of forage. The value of the services ranges from low (white) to high (black). Taken from Lavorel et al. (2011). Reproduced with permission from Wiley.

(Lavorel et al., 2011; Quétier et al., 2007): the agronomic value was associated with the quantity of green biomass produced, its nutrient concentration, and its flowering phenology, while cultural value was positively linked with floristic diversity and negatively with litter mass. In this case, the traits linked with the quality and quantity of forage were also those implicated in defining cultural value, linked to the presence of a 'traditional' alpine landscape. The joint evaluation of multiple services in an area is then carried out using additive models that include components of the functional structure of each ecosystem. It is then possible to create landscape-level maps showing areas of high or low provision of ecosystem services as a function of the distribution of major ecosystem characteristics (Figure 7.6).

Nevertheless, the use of simple rules for the addition of services in an area, even weighted by stakeholder assessments (Fagerholm et al., 2012; Gos & Lavorel, 2012), can only be an initial approximation for estimating trade-offs among services. This is why these same authors proposed investigating this question further via the analysis of functional trade-offs existing among traits. The hypothesis is simple: functional trade-offs, for example those associated with the continuum between the acquisition and conservation of resources identified in the leaf economics spectrum, or which structure the three axes of the LHS strategies (see Chapter 3), translate into trade-offs at the level of processes, and thus in services at the ecosystem level (Lavorel & Grigulis, 2012). For the example of alpine grasslands (Figure 7.7), these authors hypothesized that differences in functional structure among communities subject to differing management regimes, corresponding either to strong conservation or a rapid acquisition of resources, translate into differences in ecosystem processes. They opposed on the one hand grasslands marked by high amounts of litter accumulation and slow biogeochemical cycles, and on the other hand grasslands with high productivity and high palatability. These differences among grasslands further translate respectively into a contrast in the provision of services oriented either towards carbon storage or towards high forage production. An experimental test of this model in the alpine grasslands of Lautaret suggests that two traits which vary independently of one another in this particular situation, plant height and leaf nutrient concentration, respectively underlie conservation capacity and resource acquisition, while the community weighted means of these traits, calculated at the level of the landscape, allow the prediction of trade-offs existing locally among services. The same approach was applied successfully in the case of services dependent on trophic networks: the functional structure of grasslands, and of their associated communities of soil microorganisms, explained the trade-offs existing among productivity, carbon storage, and soil nitrogen retention, evaluated at the ecosystem level (Grigulis et al., 2013).

The transferability of this method for trade-off evaluation to other systems and services remains to be established; it requires a detailed understanding of the biophysical and ecological processes

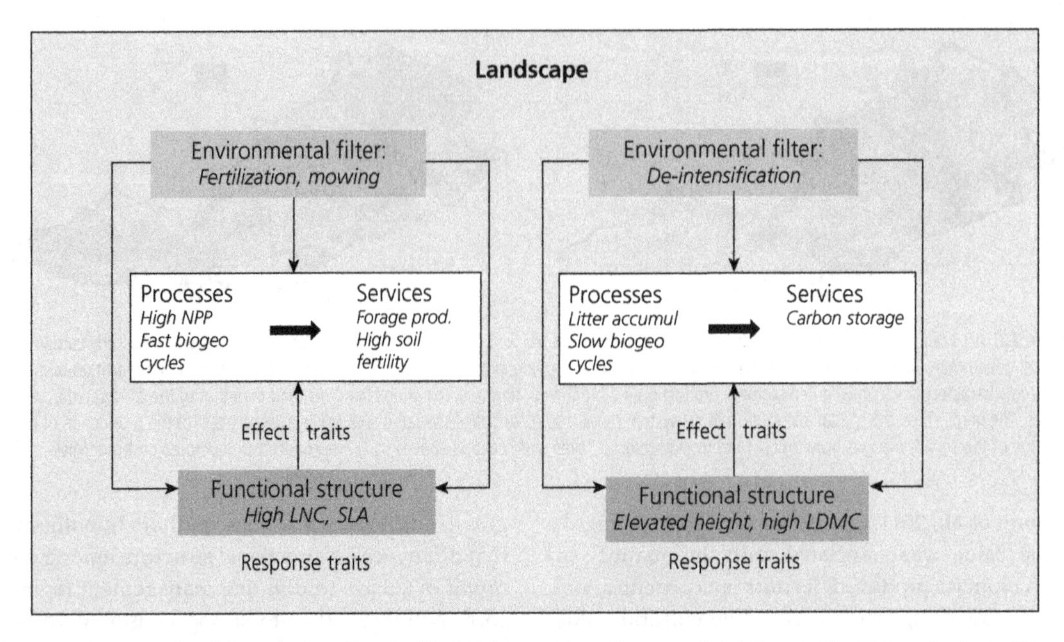

Figure 7.7 Representation of the method used for quantification of ecosystem services as applied at the scale of an alpine valley (Lavorel & Garnier, 2002; Lavorel et al., 2011; Lavorel & Grigulis, 2012). The different communities identifiable at the level of the landscape are subject to differing environmental conditions, for example as a function of topography, soil type or management regime. Two types of grasslands are represented here; these differ in their management regime, which corresponds to a gradient of management intensity: to the left, management practices are intensive and include fertilization and mowing, to the right, there is de-intensification with abandonment of the practices of fertilization and mowing. These two grassland types have different functional structures, with high values for traits associated with the acquisition of resources for those on the left, and high values for traits associated with the conservation of resources on the right. These traits directly influence ecosystem processes and the resulting ecosystem services, with, on the left, a high net primary productivity and rapid biogeochemical cycles, leading to high forage production and high soil fertility, and to the right, increases in litter production and slow biogeochemical cycles leading to increases in the soil storage of carbon. Key: NPP: Net Primary Productivity; LNC: Leaf Nitrogen Concentration; SLA: Specific Leaf Area; LDMC: Leaf Dry Matter Content

operating in the studied systems and the constitution of extensive databases. Our understanding of functional trade-offs remains limited, even though it is more and more apparent that universal relationships such as that of the leaf economic spectrum exist also at the level of roots and shoots (Lavorel, 2013). Taking into account root traits would allow improved predictions of the trade-offs between above- and below-ground processes, such as biomass production and soil carbon storage (Butterfield & Suding, 2013); however, this approach requires a better knowledge of the links among these traits and processes (see Chapters 3 and 6). In another example, taking into account the existing trade-offs between the functioning of leaves and tree trunks (Baraloto et al., 2010) could be used to estimate the production of complementary services in tropical forests. However, the direct transfer of trait trade-offs to trade-offs in services is only possible for services evaluated at the same spatial scale; in the case of the above example, at the level of the ecosystem. In other cases, it appears more problematic, for example when the aim is to conjointly evaluate services linked to agricultural production at the level of the field, and regulation services at the level of the landscape. From another perspective, it is probable that those processes explaining trade-offs between services change with the spatial scale studied. In particular, it is probable that trade-offs established at the level of the region or continent are based on differing types of traits than those which are relevant at the level of the ecosystem, for example those linked with differences in the regeneration between different vegetation types and/or types of communities.

Whichever method is used to estimate the service bundles provided by each ecosystem, it is subsequently possible to obtain maps of the distribution of services at the landscape level which are relatively simple to interpret (Figure 7.6). These maps reflect the fact that the higher the structural diversity of ecosystems at the landscape level, the higher the diversity of services provided by that landscape. For example, a mosaic of cultivated and semi-natural areas at the level of a region allows for the optimization of agricultural production and of pollination, the regulation of pest species, and the maintenance of cultural services associated with these types of landscapes (Tscharntke et al., 2012). The landscape is the scale at which it is possible to manage the provision of services incompatible at the level of the individual ecosystem, with spatial distribution influencing the diversity of ecosystems present. This is the reasoning behind proposals by ecologists to plan the spatial distribution of crops and semi-natural elements in agricultural landscapes so as to increase their compatibility with the demands for multiple ecosystem services associated with these areas (Fahrig et al., 2011).

7.5 Conclusions

The overview presented in this chapter demonstrates that the functional approach is a powerful tool for identifying the role of biological diversity in the provision of ecosystem services and for predicting their future in a changing world (Cardinale et al., 2012; Naeem et al., 2012; Perrings et al., 2010). In particular, the identification of particular trait–process–service relationships, and/or of trade-offs among services, is an essential step in establishing mechanistic models for the prediction of services at the scales of the ecosystem and the landscape. Identifying these relationships is necessary to establish the biological bases of the provision of services and their prediction under the variable conditions, linked with current global changes. It is also needed to inform policies for environmental protection and biodiversity offsetting. In the following chapter, the notion of services is illustrated in a more operational manner, in the context of the actual management of agricultural systems.

7.6 Key points

1. Ecosystem services, which are the benefits that humanity receives from ecosystems, are classified into four categories: (a) supporting services, which are necessary for the functioning of ecosystems, (b) provisioning services, which lead to the production of economic goods, (c) regulation services, which modulate natural processes in a manner favourable for humanity, and finally (d) cultural services. Their identification and quantification have become major challenges in the context of informing the actions of managers and decision makers.
2. The evaluation of services raises numerous research questions at the interface of a number of disciplines, as it is based on jointly taking into account biophysical, economic, and social determinants. In this chapter, we have examined to what extent the functional approach can be used to improve the understanding of biophysical determinants. On the basis of conclusions from Chapters 5 and 6, we conclude that the use of traits allows for a better quantification of services than methods based on the use of empirical values established for broad land-use categories, which was initially the most commonly used approach.
3. The services provided by an ecosystem result from functions which depend on the types of organisms constituting the ecosystem, as well as the processes controlling their interactions with the environment. Traits which are linked to the manner in which organisms use resources and interact with each other, including between trophic levels, are good candidates for predicting a large number of processes fundamental for the production of ecosystem services. However, ecosystems can deliver multiple services, and each of these can depend on one or a number of processes, which can in turn depend on varying combinations of traits.
4. From a more operational point of view, the determination of the biophysical bases of services requires the identification of the main abiotic factors and components of diversity affecting in a major way those processes underlying the provision of services in a given ecosystem. As taxonomic diversity on its own does not allow the local quantification of the level of multiple services or the trade-offs among these services, a new approach consists of analysing how

functional trade-offs existing at the level of traits translate to trade-offs at the level of services.

7.7 References

Balvanera, P., Pfisterer, A. B., Buchmann, N., He, J.-S., Nakashizuka, T., Raffaelli, D., & Schmid, B. (2006). Quantifying the evidence for biodiversity effects on ecosystem functioning and services. *Ecology Letters, 9*, 1146–1156.

Baraloto, C., Paine, C. E. T., Poorter, L., Beauchene, J., Bonal, D., Domenach, A. M., ... Chave, J. (2010). Decoupled leaf and stem economics in rain forest trees. *Ecology Letters, 13*(11), 1338–1347. doi:10.1111/j.1461-0248.2010.01517.x

Bennett, E. M., & Balvanera, P. (2007). The future of production systems in a globalized world. *Frontiers in Ecology and the Environment, 5*(4), 191–198. doi:10.1890/1540-9295

Brooks, D. R., Storkey, J., Clark, S. J., Firbank, L. G., Petit, S., & Woiwod, I. P. (2012). Trophic links between functional groups of arable plants and beetles are stable at a national scale. *Journal of Animal Ecology, 81*(1), 4–13. doi:10.1111/j.1365-2656.2011.01897.x

Butterfield, B. J., & Suding, K. N. (2013). Single-trait functional indices outperform multi-trait indices in linking environmental gradients and ecosystem services in a complex landscape. *Journal of Ecology, 101*(1), 9–17. doi:10.1111/1365-2745.12013

Campbell, A. J., Biesmeijer, J. C., Varma, V., & Wackers, F. L. (2012). Realising multiple ecosystem services based on the response of three beneficial insect groups to floral traits and trait diversity. *Basic and Applied Ecology, 13*(4), 363–370. doi:10.1016/j.baae.2012.04.003

Cardinale, B. J., Duffy, J. E., Gonzalez, A., Hooper, D. U., Perrings, C., Venail, P., ... Naeem, S. (2012). Biodiversity loss and its impact on humanity. *Nature, 486*(7401), 59–67. doi:10.1038/nature11148

Cardinale, B. J., Srivastava, D. S., Duffy, J. E., Wright, J. P., Downing, A. L., Sankaran, M., & Jouseau, C. (2006). Effects of biodiversity on the functioning of trophic groups and ecosystems. *Nature, 443*, 989–992.

Chevassus-au-Louis, B., Salles, J.-M., & Pujol, J. -L. (2009). *Approche économique de la biodiversité et des services liés aux écosystèmes. Contribution à la décision publique*. Paris, France: Centre d'Analyse Stratégique.

Conti, G., & Díaz, S. (2013). Plant functional diversity and carbon storage: an empirical test in semi-arid forest ecosystems. *Journal of Ecology, 101*(1), 18–28. doi:10.1111/1365-2745.12012

Crossman, N. D., Bryan, B. A., de Groot, R. S., Lin, Y.-P., & Minang, P. A. (2013a). Land science contributions to ecosystem services. *Current Opinion in Environmental Sustainability, 5*(5), 509–514. doi: 10.1016/j.cosust.2013.06.003

Crossman, N. D., Burkhard, B., Nedkov, S., Willemen, L., Petz, K., Palomo, I., ... Maes, J. (2013b). A blueprint for mapping and modelling ecosystem services. *Ecosystem Services, 4*(0), 4–14. doi: http://dx.doi.org/10.1016/j.ecoser.2013.02.001

de Bello, F., Lavorel, S., Díaz, S., Harrington, R., Cornelissen, J. H. C., Bardgett, R. D., ... Harrison, P. A. (2010). Towards an assessment of multiple ecosystem processes and services via functional traits. *Biodiversity and Conservation, 19*, 2873–2893.

de Vries, F. T., Thébault, E., Liiri, M., Birkhofer, K., Tsiafouli, M. A., Bjørnlund, L., ... Bardgett, R. D. (2013). Soil food web properties explain ecosystem services across European land use systems. *Proc. Natl. Acad. Sci. USA, 110*(35), 14296–14301. doi:10.1073/pnas.1305198110

Díaz, S., Fargione, J., Chapin, F. S., III, & Tilman, D. (2006). Biodiversity loss threatens human well-being. *PLoS Biology 4(8)*, e277. doi:10.1371/journal.pbio.0040277

Díaz, S., Lavorel, S., de Bello, F., Quétier, F., Grigulis, K., & Robson, M. (2007). Incorporating plant functional diversity effects in ecosystem service assessments. *Proc. Natl. Acad. Sci. USA, 104*, 20684–20689.

Dickie, I. A., Yeates, G. W., St John, M. G., Stevenson, B. A., Scott, J. T., Rillig, M. C., ... Aislabie, J. (2011). Ecosystem service and biodiversity trade-offs in two woody successions. *Journal of Applied Ecology, 48*(4), 926–934. doi:10.1111/j.1365-2664.2011.01980.x

Egoh, B., Reyers, B., Rouget, M., Richardson, D. M., Le Maitre, D. C., & van Jaarsveld, A. S. (2008). Mapping ecosystem services for planning and management. *Agriculture Ecosystems & Environment, 127*(1–2), 135–140. doi:10.1016/j.agee.2008.03.013

Ehrlich, P. R., & Mooney, H. A. (1983). Extinction, substitution, and ecosystem services. *Bioscience, 33*(4), 248–254.

Eigenbrod, F., Armsworth, P. R., Anderson, B. J., Heinemeyer, A., Gillings, S., Roy, D. B., ... Gaston, K. J. (2010). The impact of proxy-based methods on mapping the distribution of ecosystem services. *Journal of Applied Ecology, 47*(2), 377–385. doi:10.1111/j.1365-2664.2010.01777.x

Fagerholm, N., Kayhko, N., Ndumbaro, F., & Khamis, M. (2012). Community stakeholders' knowledge in landscape assessments: mapping indicators for landscape services. *Ecological Indicators, 18*, 421–433. doi:10.1016/j.ecolind.2011.12.004

Fahrig, L., Baudry, J., Brotons, L., Burel, F. G., Crist, T. O., Fuller, R. J., ... Martin, J.-L. (2011). Functional landscape heterogeneity and animal biodiversity in agricultural landscapes. *Ecology Letters, 14*(2), 101–112. doi:10.1111/j.1461-0248.2010.01559.x

Gos, P., & Lavorel, S. (2012). Stakeholders' expectations on ecosystem services affect the assessment of ecosystem services hotspots and their congruence with biodiversity. *International Journal of Biodiversity Science, Ecosystem Services & Management*, *8*(1–2), 93–106. doi:10.1080/21513732.2011.646303

Grigulis, K., Lavorel, S., Krainer, U., Legay, N., Baxendale, C., Dumont, M.,... Clément, J.-C. (2013). Relative contributions of plant traits and soil microbial properties to mountain grassland ecosystem services. *Journal of Ecology*, *101*(1), 47–57. doi:10.1111/1365-2745.12014

Grime, J. P. (1998). Benefits of plant diversity to ecosystems: immediate, filter, and founder effects. *Journal of Ecology*, *86*, 902–910.

Haines-Young, R., & Potschin, M. (2010). The links between biodiversity, ecosystem services, and human well-being. In D. Raffaelli & C. Frid (eds), *Ecosystem Ecology: A New Synthesis* (pp. 110–139). Cambridge: Cambridge University Press.

Harrison, P. A., Berry, P. M., Simpson, G., Haslett, J. R., Blicharska, M., Bucur, M., ... Turkelboom, F. (2014). Linkages between biodiversity attributes and ecosystem services: a systematic review. *Ecosystem Services*, *9*(0), 191–203. doi:10.1016/j.ecoser.2014.05.006

Hättenschwiler, S., & Gasser, P. (2005). Soil animals alter plant litter diversity effects on decomposition. *Proc. Natl. Acad. Sci. USA*, *102*(5), 1519–1524.

Hein, L., van Koppen, K., de Groot, R. S., & van Ierland, E. C. (2006). Spatial scales, stakeholders and the valuation of ecosystem services. *Ecological Economics*, *57*(2), 209–228. doi: http://dx.doi.org/10.1016/j.ecolecon.2005.04.005

Hooper, D. U., Chapin, F. S., III, Ewel, J. J., Hector, A., Inchausti, P., Lavorel, S., ... Wardle, D. A. (2005). Effects of biodiversity on ecosystem functioning: a consensus of current knowledge. *Ecological Monographs*, *75*, 3–35.

Inauen, N., Körner, C., & Hiltbrunner, E. (2013). Hydrological consequences of declining land use and elevated CO_2 in alpine grassland. *Journal of Ecology*, *101*(1), 86–96. doi:10.1111/1365-2745.12029

Kremen, C., Williams, N. M., Aizen, M. A., Gemmill-Herren, B., LeBuhn, G., Minckley, R., ... Ricketts, T. H. (2007). Pollination and other ecosystem services produced by mobile organisms: a conceptual framework for the effects of land-use change. *Ecology Letters*, *10*(4), 299–314.

Lamarque, P. (2012). *Une approche socio-écologique des services écosystémiques. Cas d'étude des prairies subalpines du Lautaret*. Thèse de Doctorat, Université Joseph Fourier, Grenoble.

Lamarque, P., Quetier, F., & Lavorel, S. (2011). The diversity of the ecosystem services concept and its implications for their assessment and management. *Comptes Rendus Biologies*, *334* (5–6),441–449. doi:10.1016/j.crvi.2010.11.007

Lavorel, S. (2013). Plant functional effects on ecosystem services. *Journal of Ecology*, *101*(1), 4–8. doi:10.1111/1365-2745.12031

Lavorel, S., & Garnier, E. (2002). Predicting changes in community composition and ecosystem functioning from plant traits: revisiting the Holy Grail. *Functional Ecology*, *16*, 545–556.

Lavorel, S., & Grigulis, K. (2012). How fundamental plant functional trait relationships scale-up to trade-offs and synergies in ecosystem services. *Journal of Ecology*, *100*(1), 128–140. doi:10.1111/j.1365-2745.2011.01914.x

Lavorel, S., Grigulis, K., Lamarque, P., Colace, M.-P., Garden, D., Girel, J.,... Douzet, R. (2011). Using plant functional traits to understand the landscape distribution of multiple ecosystem services. *Journal of Ecology*, *99*, 135–147.

Lavorel, S., Storkey, J., Bardgett, R. D., de Bello, F., Berg, M. P., Le Roux, X., ... Harrington, R. (2013). A novel framework for linking functional diversity of plants with other trophic levels for the quantification of ecosystem services. *Journal of Vegetation Science*, *24*(5), 942–948. doi:10.1111/jvs.12083

Le Roux, X., Barbault, R., Baudry, J., Burel, F., Doussan, I., Garnier, E., ... Trommetter, M. (eds). (2009). *Agriculture et biodiversité. Valoriser les synergies*. Versailles: Quae.

Lienin, P., & Kleyer, M. (2012). Plant trait responses to the environment and effects on ecosystem properties. *Basic and Applied Ecology*, *13*(4), 301–311. doi:10.1016/j.baae.2012.05.002

Maes, J., Hauck, J., Paracchini, M.-L., Ratamäki, O., Termansen, M., Perez-Soba, M., ... Bidoglio, G. (2012). A spatial assessment of ecosystem services in Europe: methods, case studies, and policy analysis—phase 2. Synthesis report. Ispra: Partnership for European Environmental Research.

Millennium Ecosystem Assessment. (2005). *Ecosystems and Human Well-being: Synthesis*. Washington DC: Island Press.

Mokany, K., Ash, J., & Roxburgh, S. (2008). Functional identity is more important than diversity in influencing ecosystem processes in a temperate native grassland. *Journal of Ecology*, *96*, 884–893.

Mouillot, D., Villéger, S., Scherer-Lorenzen, M., & Mason, N. W. H. (2011). Functional structure of biological communities predicts ecosystem multifunctionality. *PLoS ONE*, *6*(3), e17476. doi:10.1371/journal.pone.0017476

Naeem, S., Duffy, J. E., & Zavaleta, E. (2012). The functions of biological diversity in an age of extinction. *Science*, *336*(6087), 1401–1406. doi:10.1126/science.1215855

O'Farrell, P. J., Reyers, B., Le Maitre, D. C., Milton, S. J., Egoh, B., Maherry, A., ... Cowling, R. M. (2010). Multi-functional landscapes in semi-arid environments: implications for biodiversity and ecosystem services. *Landscape Ecology, 25*(8), 1231–1246. doi:10.1007/s10980-010-9495-9

Perrings, C., Naeem, S., Ahrestani, F., Bunker, D. E., Burkill, P., Canziani, G., ... Weisser, W. (2010). Ecosystem Services for 2020. *Science, 330*(6002), 323–324. doi:10.1126/science.1196431

Quétier, F., Lavorel, S., Thuiller, W., & Davies, I. (2007). Plant-trait-based modeling assessment of ecosystem-service sensitivity to land-use change. *Ecological Applications, 17*(8), 2377–2386.

Quijas, S., Jackson, L. E., Maass, M., Schmid, B., Raffaelli, D., & Balvanera, P. (2012). Plant diversity and generation of ecosystem services at the landscape scale: expert knowledge assessment. *Journal of Applied Ecology, 49*(4), 929–940. doi:10.1111/j.1365-2664.2012.02153.x

Raudsepp-Hearne, C., Peterson, G. D., & Bennett, E. M. (2010). Ecosystem service bundles for analyzing tradeoffs in diverse landscapes. *Proc. Natl. Acad. Sci. USA, 107*(11), 5242–5247. doi:10.1073/pnas.0907284107

Rodríguez, J. P., Beard, J. T. D., Bennett, E. M., Cumming, G. S., Cork, S. J., Agard, J., ... Peterson, G. D. (2006). Trade-offs across space, time, and ecosystem services. *Ecology and Society, 11*(1), 28 [online].

Seppelt, R., Dormann, C. F., Eppink, F. V., Lautenbach, S., & Schmidt, S. (2011). A quantitative review of ecosystem service studies: approaches, shortcomings, and the road ahead. *Journal of Applied Ecology, 48*(3), 630–636. doi:10.1111/j.1365-2664.2010.01952.x

Storkey, J., Brooks, D., Haughton, A., Hawes, C., Smith, B. M., & Holland, J. M. (2013). Using functional traits to quantify the value of plant communities to invertebrate ecosystem service providers in arable landscapes. *Journal of Ecology, 101*(1), 38–46. doi:10.1111/1365-2745.12020

Tscharntke, T., Tylianakis, J. M., Rand, T. A., Didham, R. K., Fahrig, L., Batáry, P., ... Westphal, C. (2012). Landscape moderation of biodiversity patterns and processes: eight hypotheses. *Biological Reviews, 87*(3), 661–685. doi:10.1111/j.1469-185X.2011.00216.x

Vihervaara, P., Ronka, M., & Walls, M. (2010). Trends in ecosystem service research: early steps and current drivers. *Ambio, 39*(4), 314–324. doi:10.1007/s13280-010-0048-x

Westman, W. E. (1977). How much are nature's services worth? *Science, 197*(4307), 960–964. doi:10.1126/science.197.4307.960

Zavaleta, E. S., Pasari, J. R., Hulvey, K. B., & Tilman, G. D. (2010). Sustaining multiple ecosystem functions in grassland communities requires higher biodiversity. *Proc. Natl. Acad. Sci. USA, 107*(4), 1443–1446. doi:10.1073/pnas.0906829107

Zhang, W., Ricketts, T. H., Kremen, C., Carney, K., & Swinton, S. M. (2007). Ecosystem services and dis-services to agriculture. *Ecological Economics, 64*(2), 253–260. doi: http://dx.doi.org/10.1016/j.ecolecon.2007.02.024

CHAPTER 8

Functional diversity in agriculture: grasslands and crop weeds as case studies

8.1 Introduction

The preceding chapters have presented the functional approach and outlined how the identification of pertinent traits allows for the understanding of community structure, explains ecosystem functioning, and can identify the biological determinants of ecosystem services provided to human societies. The aim of this chapter is to show how traits can be used to address more applied objectives. The functional approach has already been used to address a large number of these types of challenges: determining the responses of communities to particular environmental conditions (Ceulemans et al., 2011; Lizée et al., 2011; Tall et al., 2011), restoring degraded areas (Gomez-Aparicio et al., 2009; Herault et al., 2005; Kardol & Wardle, 2010; Pywell et al., 2003), and understanding the provision of ecosystem services (Brym et al., 2011; Sinclair & Hoffmann, 2003; Walker et al., 2009).

However, applications linked to agriculture are undeniably the most numerous. While needing to increase agricultural production to feed a growing human population, it is also necessary to reduce the negative impacts of agriculture on the environment by implementing strategies to reduce the amounts of inputs, and using inputs with lower impacts on the environment and on local biodiversity (Doré et al., 2011; Le Roux et al., 2009; Robertson & Swinton, 2005). One possible approach is to increase the diversity in currently low-diversity agricultural systems. This would be achieved by increasing biological diversity so as to restore largely missing ecological functions, which would allow for the improved provision of multiple agricultural and environmental services (Andueza et al., 2010; Gaba et al., 2015; Koohafkan et al., 2012). The objective would be to restore, within agricultural systems, certain functional components of natural systems. An example of this approach is in agro-forests with high specific diversity which mimic the functioning of naturally establishing forests (Malézieux et al., 2009). Another example is in cropping systems where regulatory interactions are favoured because of complex biotic relationships between pest species and their predators or parasites (Altieri, 1999). Another possible approach would be the development of agricultural systems conforming to new agricultural, economic, societal, and environmental objectives, based on a detailed understanding of the links existing between the structural components of agricultural systems, their performance, and their required environmental conditions (Doré et al., 2011; Gaba et al., 2015; Meynard et al., 2012). Development of these approaches is particularly advanced in the case of managed permanent grasslands and in the management of weeds in major crops. Consequently, we have chosen to present these two cases as they illustrate many of the concepts discussed in the previous chapters, as well as extending the functional approach to new systems and questions.

In the following sections, we will successively treat, for each example: i) the identification of environmental gradients, resulting from climatic and soil conditions or to management practices, and the response traits linked with these gradients, whether these are generic or specific to the particular environmental conditions found in agricultural environments; ii) the functional structure of these managed communities whether these are planned (grassland) or not (weeds); iii) the relationships between effect traits and processes or services of agricultural interest; and iv) the potential for the use of these traits, or of proxys, for modifying or managing the system being considered. Finally, we will discuss the implications of these aspects in the larger context of the overall management of agricultural systems.

8.2 Managed permanent grasslands

The large majority of studies that have characterized the functional structure of communities along environmental gradients, and most of those cited in this book, have been carried out in grasslands. This explains why the functional approach appeared very early on as an interesting option, potentially leading to new management methods in these agricultural systems. Grasslands are also of considerable importance in the context of current land use changes in agricultural systems. They have a role in limiting the negative effects of losses of diversity and increasing greenhouse gas emissions, which explains their importance in the 'greening measures' of the common agricultural policies of the European Union (Pe'er et al., 2014).

Figure 8.1 represents how the use of response and effect traits can be applied in the case of permanent grasslands (Garnier & Navas, 2011). In these systems, in addition to local climatic and soil conditions, the environmental factors which filter species according to the values of their response traits relate to management intensity characterized by the defoliation regime due to grazing or mowing, the level of mineral and organic fertilization, and the possible establishment of sown seeds. The functional structure resulting from the actions of management filters strongly affects certain ecosystem processes, such as the production of biomass, water quality, and nutrient cycling, which have direct impacts on important ecosystem services such as forage production, soil carbon storage, and the conservation of keystone or emblematic grassland species and their associated species. This diagram will be further developed in the remainder of this section, using results drawn from the literature. Note that since the emphasis is placed on permanent grasslands, management practices used in temporary grasslands

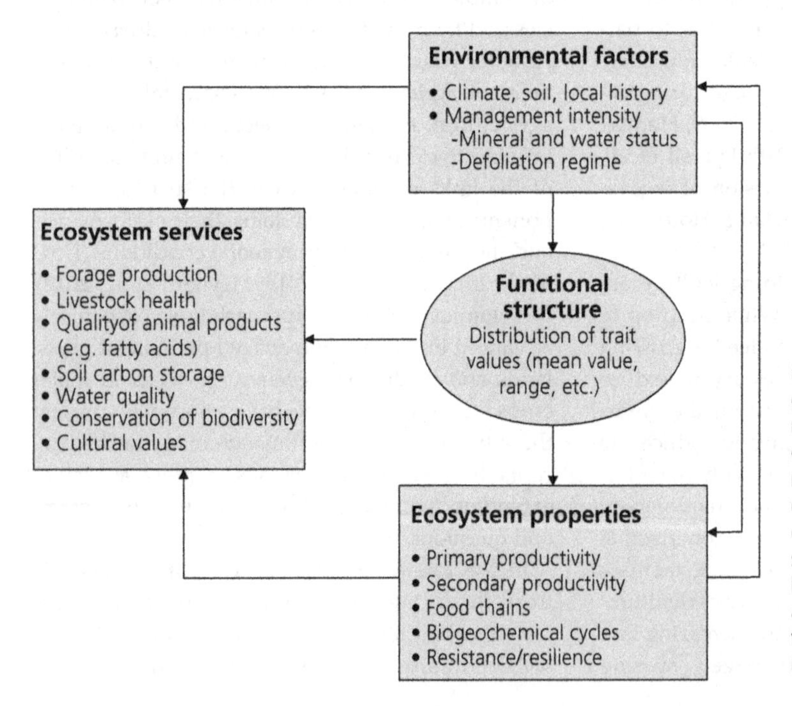

Figure 8.1 The functional structure of a grassland results from the differential responses of individuals to environmental conditions, whether these are linked to management or not. These responses depend on their trait values, which affect ecosystem processes and thus ecosystem services. This general schema is illustrated here using examples from grassland systems and is developed further in the text. Adapted from Díaz et al. (2007a) and Garnier & Navas (2011).

(such as the initial choice of seeds to be sown) will not be further considered.

8.2.1 Characterizing environmental gradients

The environmental gradients considered in this section are essentially those affected by management practices, via the defoliation or fertilization of grasslands.

8.2.1.1 Defoliation

The characterization of disturbance due to defoliation in experimental work is generally rudimentary, and comparisons are often made between grasslands subject to highly contrasting situations, for example between grazed and abandoned grasslands, or between grasslands differing greatly in grazing pressure (Evju et al., 2009; Louault et al., 2005). However, a finer characterization is necessary for comparative studies including different regions and/or production systems.

Quantifying defoliation is complex as it is based on the estimation of multiple variables so as to fully take into account the magnitude of the disturbance experienced. Based on the disturbance descriptors proposed by White and Pickett (1985; see also Section 4.2 in Chapter 4), Duru et al. (1998), and Kleyer (1999), protocols were established in the context of a European research project involving 11 different sites, in order to obtain a better characterization of the impact of defoliation on grassland ecosystems (Garnier et al., 2007). Five parameters were used to describe the disturbance regime: (1) the previous management regime, which corresponds to the usage of the field prior to its current usage; (2) the type of defoliation (e.g. grazing, type of herbivore, mowing); (3) the date of the disturbance (in Julian days); (4) the intensity of the disturbance (expressed as the quantity of biomass removed as a proportion of the total biomass); and (5) the periodicity of the disturbance (in years) at a given site. While it was possible to correctly measure the first three parameters at the 11 sites, evaluating the intensity was more difficult in the majority of cases. In reality, this requires the regular evaluation of standing biomass so as to take into account seasonal variation in growth and the impact of the date of defoliation on the magnitude of regrowth (McNaughton et al., 1996), necessitating the use of grazing exclosures in the case of grazed pastures. As for characterizing the periodicity of the disturbance, this proved to be particularly difficult. For example, a relatively short period of grazing, corresponding to, e.g., 10 days spread over one year, results in a more rapid calculated return interval than for the case of five mowing events carried out each year. This result is not satisfactory, especially given that the differing spatial scales at which these disturbance regimes apply is not taken into account. It is thus necessary to further refine these protocols and to develop new measures and indices to allow for the correct evaluation of the impacts of defoliation.

8.2.1.2 Mineral nutrition

Characterizing the status of the mineral nutrition of grasslands turns out to be also quite difficult. A first way in which this is done is to measure soil parameters such as the pools of nitrogen or phosphorus. However, these are not necessarily pertinent estimates of the quantity of resources available to plants as the relationships between these two types of variables depends on other soil characteristics such as the texture, the pH, or the amount of organic matter. Another way to proceed, initially developed for sown pastures, consists of comparing the concentration of nitrogen and phosphorus of the vegetation cover to that of a monoculture growing under non-limiting nutrient conditions. The nitrogen nutrition index (NNI, Lemaire & Gastal, 1997) gives an indication of the nitrogen limitation of a vegetation sward as a function of the quantity of biomass produced per unit of area. This index has been successfully used to measure the nitrogen status of mixed crops (Cruz & Soussana, 1997) and permanent grasslands (Duru et al., 1997). However, it cannot be used in the presence of legumes without corrections, and when the biomass of the vegetation is lower than one tonne per hectare, which makes it barely usable in low productivity grasslands, such as arid Mediterranean grasslands. The phosphorus nutrition index (PNI, Duru & Ducrocq, 1997), based on the same principle, is, within a given site, positively correlated with soil phosphorus as measured using the Olsen method, which corresponds to the soluble fraction of this element available for plants (Garnier et al., 2007; Jouany et al., 2002). However, this relationship does not seem to be conserved between sites because of differences in soil capacity to fix phosphorus. Further work is therefore also necessary to develop new protocols and indices to allow for the correct evaluation of phosphorus, and also of potassium, limitation in grasslands.

Amongst the principal environmental factors which affect the various elements presented in Figure 8.1, the effects of temperature and water availability should not be forgotten, as they strongly limit possible management practices. This constraint will become more important in the context of climate change where the predicted extremes of temperature and rainfall will considerably modify the composition and functioning of grasslands. More specifically, local climatic conditions can be described using a large range of different variables, amongst which many are redundant due to their high inter-correlation, such as, for example, temperature and solar radiation (Pakeman et al., 2009). Synthetic indices based on combinations of variables can also be used, such as aridity or rainfall for the Martonne or Thornthwaite indices (Thornthwaite, 1948), which were used in the previously cited European programme for the identification of defoliation gradients (Garnier et al., 2007). Other authors have also proposed the use of multivariate analyses incorporating all relevant climatic variables (Pakeman et al., 2009).

8.2.2 Response traits to management

The relative importance of management factors as compared to other environmental factors is far from being negligible: in a research programme carried out across 11 European sites, they were found to be responsible for 15% of the variation in recorded trait values, while climate and edaphic factors explained respectively 40% and 21% of this variation (Pakeman et al., 2009). Amongst these practices are found defoliation either by animals or by mowing, and fertilization, but also water management, the importance of which will increase due to current climatic changes. However, the importance of stochastic factors, such as the dispersal of species, in the assemblage of these grasslands should not be forgotten, as these can have a major effect on the distribution of traits and the functioning of grasslands (von Gillhaussen et al., 2014).

8.2.2.1 Defoliation

Defoliation by large herbivores, both domestic and wild, influences a large number of plant traits (cf. Table 4.1 in Chapter 4): a strong grazing pressure favours annual species as compared to perennial ones, small plants as compared to large ones, prostrate life forms as opposed to erect ones, rosette or stoloniferous species as opposed to tussocks, while phenology in general and leaf senescence in particular is strongly affected by the date of defoliation (Díaz et al., 2007b; Duru et al., 1998; Hodgson et al., 2005; Louault et al., 2005; Pakeman, 2011) (Figure 8.2).

The type of herbivore also influences the functional structure of communities. Comparisons of the grazing behaviour of domestic animals have shown that all species preferentially graze legumes and dicotyledons compared to grasses, and small plants as compared to large ones (Rook et al., 2004). However, these different animal species differ in their grazing preferences as a function of their size and their dental and digestive anatomical particularities. Horses, being monogastric, can satisfactorily graze on species of low digestibility and ingest a higher proportion of tall and fibrous grasses than can cows. In contrast, sheep and goats are more selective, due to the form of their jaws and their teeth and shorter digestive system, and graze preferentially on smaller herbaceous species that are not grazed by horses or cows (Ginane et al., 2008; Rook et al., 2004). Non-domesticated ruminants have slightly different characteristics; for example deer choose species with a higher SLA than do cows (Lloyd et al., 2010). Additionally, invertebrate herbivores also have impacts on vegetation structure, as a function of their habitat localization and also their type. For example, underground herbivores preferentially graze root grasses as opposed to those of dicotyledons (Stein et al., 2010) while the opposite is true for grasshoppers (Scherber et al., 2010).

Mowing tends to favour species with a stoloniferous or rosette life form (Pakeman et al., 2009), dicotyledons as opposed to graminoids and small plants with dense tissues (Quétier et al., 2007b). Surprisingly, a large increase in mowing intensity leads to enhanced representation of species with high LDMC and low LNC, but without any discernible effect on SLA (Pontes et al., 2007).

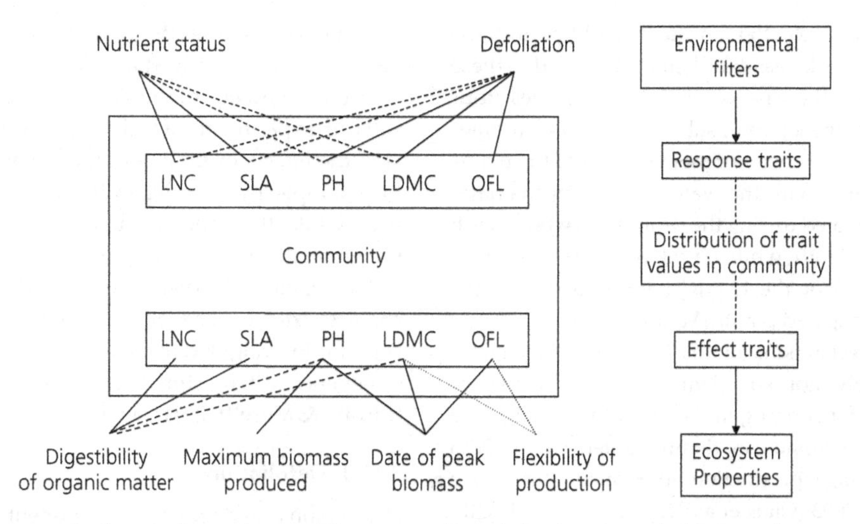

Figure 8.2 Response traits of grasslands to environmental filters generated by agricultural practices and effect traits influencing ecosystem properties. The upper part of the figure shows the positive (solid lines) and negative responses (dashed lines) of traits to the nutrient status and defoliation regime of grasslands. The lower part of the figure shows the positive (solid lines) and negative effects (dashed lines) of the community weighted means of traits on grassland ecosystem properties. The dotted lines indicate a positive relationship between divergence in the date of flowering or the leaf dry matter content and production flexibility (the period of time around the peak of biomass production during which forage can be exploited). Key: LNC: leaf nitrogen content; SLA: specific leaf area; PH: plant height; LDMC: leaf dry matter content; OFL: onset of flowering. Taken from Garnier & Navas (2011).

When defoliation ceases, such as in abandoned grasslands, species of large stature and whose leaves have a high LDMC and low nitrogen concentrations become dominant (Kahmen & Poschlod, 2004). However it is apparently seed traits (mass, length, and longevity) which are the most different in plants from abandoned or lightly grazed grasslands (Pakeman & Marriott, 2010). It is also important to note that certain traits such as the LDMC (Pakeman & Marriott, 2010), seed mass, growth form, or the date of seed dispersal (Saatkamp et al., 2010) show nonlinear responses to modifications in defoliation, which makes comparisons of variation in trait values between sites even more complex.

8.2.2.2 Nutrients

As expected according to the leaf economics spectrum, increases in nutrient availability, especially that of nitrogen, lead to increases in individuals with high SLA and LNC (Lavorel et al., 2007; Lavorel & Garnier, 2002; Ordoñez et al., 2009) and low LDMC (Al Haj Khaled et al., 2005; McIntyre, 2008) (Figure 8.2, and see Table 4.1).

8.2.2.3 Water constraints and drought

The mechanisms which underlie the responses of species to water stress, whether these be the conditions under which stomata close, rooting depth, plant resistance to embolism, or the capacity of plants to regrow after drought, remain poorly characterized for the majority of grassland species (Soussana & Duru, 2007). Four strategies of increasing tolerance have nevertheless been established for cultivated forage species, depending on the mechanisms used by the plants to cope with water stress (Volaire et al., 2009): i) increasing water acquisition via a deep root system; ii) decreasing water losses via timing phenology, in particular by an early cessation of leaf growth and the beginning of senescence, prior to the appearance of drought; iii) a tolerance of the meristems to dehydration, via a maintenance of hydration by maintaining water potential through the accumulation of hydrosoluble carbohydrates; iv) a summer dormancy marked by the total cessation of growth in summer with leaf senescence and the dehydration of the basal tissues even when plants are irrigated, followed by a resumption of growth after the drought period. Experimental work has identified traits linked to these responses: in the case of five forage species (four grasses and one legume) subject to intense summer water stress, the post-summer survival of the plants was correlated with the water content of the basal tissues estimated during the drought as well as with the rooting depth, while no relationship was found with the SLA or the LDMC measured during the spring on irrigated plants (Volaire, 2008).

These mechanisms of tolerance to water stress are probably not sufficient to maintain current grasslands functioning in the face of the recurrence of climatic extremes, as demonstrated by the 30% drop in forage production in France during the drought of 2003 (Ciais et al., 2005). It is thus highly probable that the flora will be profoundly modified, with water and temperature stresses amplifying the negative effects of defoliation and competition (Kéfi et al., 2007). These future modifications follow on from changes in grassland flora that have already occurred during the last century, characterized by increases in the representation of species with rapid growth and a short life span (Van Calster et al., 2008; Walker et al., 2009). One of the research questions to be addressed is thus to predict changes in functioning and/or composition under differing climatic scenarios, given that it is difficult to achieve this using short-term experiments (Sandel et al., 2010), due to the strong interactions between environmental factors.

8.2.3 Functional structure of grasslands

As indicated in Chapter 5, in order to explain patterns of community functional structure, it is important to be able to disentangle the effects of abiotic and biotic environmental factors. In the case of grasslands, this separation is difficult as the communities defoliated by animals are also fertilized by these same animals, and the most productive grasslands are also those which are mown the most frequently. The current functional structure of a grassland is also a result of its land-use history: a study carried out in Sweden showed that it depends not only on the type and intensity of grazing but also on the distance from historical human settlements, as well as the age of the roads and paths between fields (Reitalu et al., 2010).

In the following discussion, we will use the results of experiments which have explicitly evaluated the effects of different management practices, for example by varying only one of these practices while the others remain constant. We will preferentially use experiments in which variation in the functional structure of grasslands has been characterized by changes in relevant traits using community weighted means (CWM) and functional divergence, often estimated by the Rao coefficient (Ricotta & Moretti, 2011), as discussed in Chapter 5.

8.2.3.1 Defoliation

Defoliation acts on the two components of the functional structure of grasslands. In productive environments, a decrease in the intensity of defoliation leads to major changes in community weighted mean of traits, with decreases in SLA values and leaf digestibility, and increases in the density of plant tissues, plant height, and seed mass (Klumpp & Soussana, 2009; Louault et al., 2005). Similar results have also been found in less productive environments, when overall productivity is restrained by water stress (Cingolani et al., 2007).

Functional divergence also varies greatly with grazing, in particular with the type of domestic herbivore, depending on the different impacts they have on inter-plot spatial heterogeneity (Ginane et al., 2008). Herbivores, when they have a choice between patches of vegetation of differing height and quality, favour dietary quality and preferentially graze dicotyledon species or re-sprouting vegetation in recently grazed areas, which leads over the course of time to an increasing accentuation of intra-plot spatial heterogeneity due to defoliation (Rossignol et al., 2011). This phenomenon is more marked in areas grazed by horses as compared to those grazed by cows, as well as in grasslands of low diversity. It is also observed under arid conditions, but in this case animals select the highest plants, which are often those with the leaves of the lowest dietary quality (Mládek et al., 2013). In contrast, this increasing heterogeneity of the vegetation disappears when the grazing pressure is intense and animals remove the majority of the biomass present (Dumont et al., 2007). In this case, functional divergence decreases drastically, as has been shown in the heavily grazed grasslands of Mongolia (Sasaki et al., 2009). A cessation of grazing can then lead to an increase in functional diversity, as a result of a higher rate of recruitment of certain species (Mason et al., 2010).

8.2.3.2 Nutrients and water

The effects of these resources on the functional structure of grasslands can be illustrated by two experiments carried out at the INRA experimental station at La Fage, in which different grazing regimes have been constantly maintained since 1979, both in terms of the timing and intensity of grazing, but also of herbivore type (Box 8.1; Molénat et al., 2005). These experiments show a strong effect of soil resources on the distribution of traits related to the acquisition and conservation of resources (see Chapter 5).

The continued fertilization since 1979 of the rangelands subject to intense spring grazing allows for the identification over the long term of the effects of nitrogen and phosphorus addition on the production and functional structure (Figure 8.3) of the communities. The mean spring biomass production increased over time, rising from one to six tonnes per hectare (Chollet et al., 2014; Molénat et al., 2005). This increase is accompanied by a major change in floristic composition, characterized by the replacement of perennial grasses and small chamaetophytes with low SLA and a delayed reproductive phenology, typical of Mediterranean

Box 8.1 The INRA experimental station at La Fage

Compiled from Fayolle (2008), Molénat et al. (2005), and Bernard-Verdier et al. (2012).

The La Fage experimental station is located in the southern part of France on the western edge of the Causse du Larzac, between 760 m and 830 m altitude, some 100 km to the north-west of Montpellier, near the town of Roquefort sur Soulzon. It experiences a sub-humid montane climate with Mediterranean influence, and has a pronounced summer drought. The station has some 278 ha of rangelands on which an experimental herd of 300 meat-producing sheep (of the Romane breed) grazes all year round (Box 8.1, Figure 1).

Research on these rangelands began in 1972, with the aim of investigating the conditions required for the establishment of an economically viable sheep enterprise in this low-productivity area. The vegetation of these rangelands grows on dolomitic limestone and is dominated by the grass *Bromus erectus* with interspersed shrubs such as *Buxus sempervirens* and *Juniperus communis,* which remain ungrazed. Two types of environmental gradients have been identified at the station. The first gradient consists of three types of grazing regimes that have been maintained since 1978. The first regime is applied to an area of approximately 13 ha which has been fertilized since 1978 (65 kg per hectare per year of mineral nitrogen applied in two events and 40 kg of phosphorus oxide applied once every three years) so as to increase biomass production and advance the growth season of the vegetation. This area is subject to intense grazing during spring. The second regime corresponds to the establishment in 1987 of a non-grazed control paddock of approximately 4 ha. The third regime is applied over the remainder of the station (approximately 260 ha): these remaining paddocks (average area of 15 ha) are grazed with a grazing pressure that has been constant since 1972, with an average grazing period varying between five and 10 days per paddock depending on the season. Since the establishment of the experiment, floristic composition has been measured using permanent transects, with a survey completed every year in the fertilized areas, and once every five years in the unfertilized areas. These measurements are also coupled with estimations of biomass production.

The second environmental gradient being studied is that of a soil depth gradient occurring along a topographic gradient: the soils in the upper parts of the station and on slopes are derived from dolomite and have a depth varying between 20 cm and 60 cm, while the colluvial soils at the lower parts of the station (dolines) have a depth varying between 80 cm and 100 cm. These differences in soil depth are associated with variations in resource availability as the soil water capacity varies from 20 mm to 120 mm, the nitrogen nutrition index varies from 20 to 49, and the range of biomass production ranges from 0.5 t/ha to 2.5 t/ha. Along this second gradient, assessments of floristic composition were carried out using transects during the spring and autumn.

continued

Box 8.1 *Continued*

(a)

(b)

Box 8.1, Figure 1 (a) The southern cliffs of the Larzac plateau near the La Fage experimental station (photograph by E. Garnier); (b) a herd of 'Romane' breed sheep (photograph by A. Fayolle).

rangelands, by a more common flora composed largely of annual species, with a high SLA (Chollet et al., 2014; Fayolle, 2008; Figure 8.3). Thus, over the course of time there is a progressive replacement of species showing an efficient resource conservation strategy by species exhibiting rapid resource acquisition. This results in an increase in the community weighted mean of SLA in these communities. The temporal dynamics of the within community variance in SLA is more complex: after an increase during the first years of fertilization, this variance reduces to stabilize at a low and constant level over time, an effect largely explained by the progressive exclusion of less competitive species as the vegetation cover becomes progressively closed and dense (Chollet et al., 2014).

Another experiment was carried out in the non-fertilized rangelands along a soil depth gradient, generating large variations in the availability of water and nutrients (Box 8.1; Bernard-Verdier et al., 2012; Pérez-Ramos et al., 2012). As soil depth increases, the vegetation community is characterized

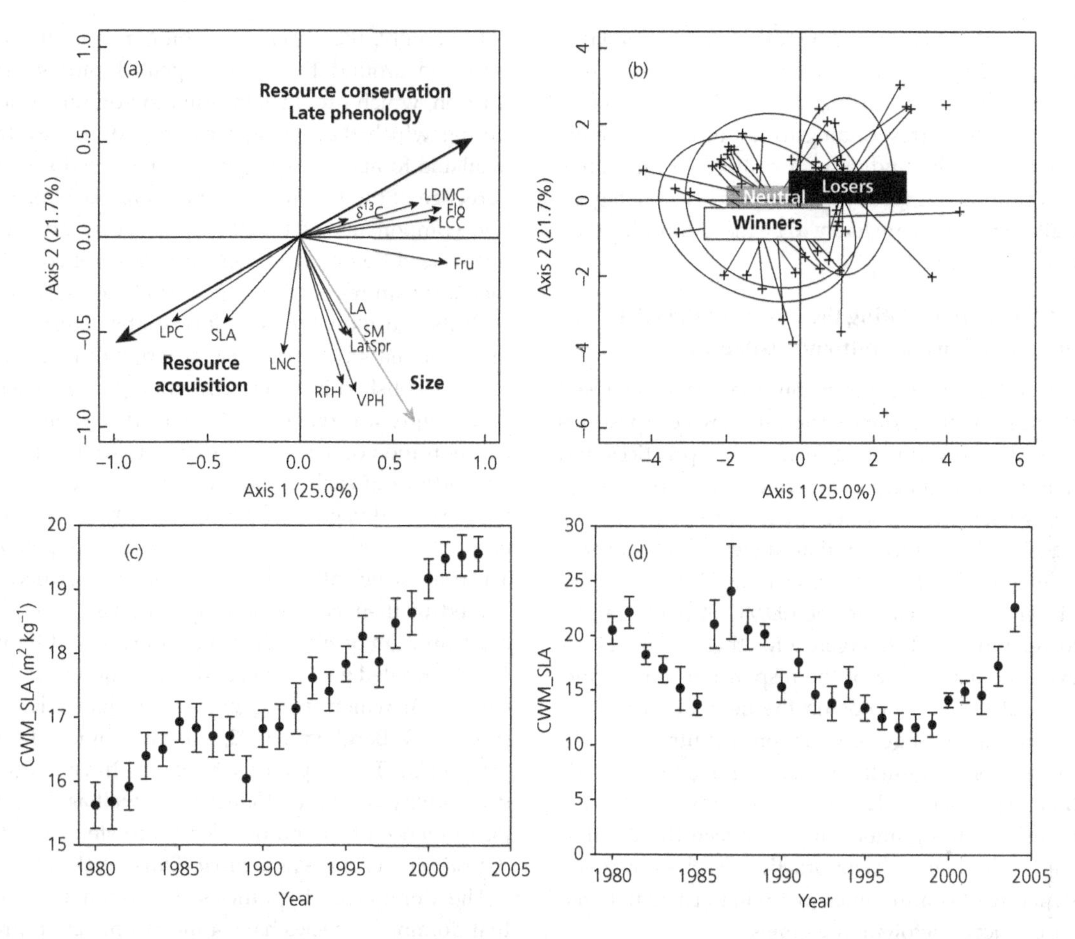

Figure 8.3 The functional structure of rangelands located on the Causse du Larzac (INRA La Fage experimental station: cf. Box 8.1), which have been intensively grazed, and consistently fertilized since 1979. The two upper panels show the traits which characterise those species which have increased in abundance (winners), decreased in abundance (losers) or remained stable (neutral) over time (taken from Fayolle, 2008). The two lower panels show the changes in the functional structure of the rangelands over time as evidenced by: (c) the community weighted mean (CWM), and (d) the community weighted variance (CWV) of SLA. (a) Relationships between traits as shown by a principal components analysis carried out using the species–traits matrix; the percentage of variation explained by each axis is shown in parentheses; (b) position of the 78 species grouped according to their status in multi-dimensional space. The classification is based on the positive (winners), negative (losers), or null (neutral species) values of the Spearman correlation coefficient for the relationship between the mean annual abundance of the species and time. The thick black and grey arrows indicate respectively the position of traits linked to the conservation or acquisition of resources or to size. Key: $\delta^{13}C$: leaf ^{13}C isotopic composition; OFL: flowering date; Fru: fruiting date; LA: leaf area; LatSpr: lateral spread; LDMC: leaf dry matter content; LCC: leaf carbon concentration; LNC: leaf nitrogen concentration; LPC: leaf phosphorus concentration; RPH: reproductive plant height; SLA: specific leaf area; SM: seed mass; VPH: vegetative plant height. Data taken from Chollet et al. (2014).

by increases in the weighted means of the SLA and LNC and decreases in the LDMC. This last trait shows a large degree of divergence in resource-limited environments, where a number of differing strategies in response to low water availability apparently coexist. In contrast, the other traits show maximal divergence in intermediate zones, which is more in conformity with theoretical expectations (Navas & Violle, 2009). As explained in Chapter 5, the functional structure of a community is expected to directly reflect the equilibrium between abiotic filters, imposed by environmental constraints, and biotic filters, whose importance is expected to increase in productive environments. Therefore, the functional structure was hypothesized to display a limited level of variation over a gradient of biomass production varying from 0.5 to 2.5 tonnes per hectare, corresponding to rather unproductive conditions. The finding of a relatively large range of variation suggests that the processes acting on trait distribution are likely to be more complex than initially assumed.

8.2.3.3 Differentiating the effects of the type of defoliation and nutrient availability

One of the rare experiments that have analysed changes in functional structure between pastures subjected to contrasting fertilization practices and defoliation regimes was carried out in low mountain grasslands (Ansquer, 2006; Duru et al., 2014a). This study confirmed that increases in nutrient availability led to an increase in the SLA and height of plants, a decrease in the LDMC of plants, and a lower functional divergence for these traits. However, the magnitude of the response of these traits to fertilization depends on the defoliation regime: for the same range of variation in nutrient levels, this response is much greater in grazed grasslands than in grasslands that are mown for hay. There are, therefore, strong interactions between these different axes of management practices on the functional structure of communities, with, in particular, a major impact of defoliation regimes.

8.2.4 Effect traits and processes linked to provisioning ecosystem services

Traits can be used to evaluate certain important processes which characterize forage provision, as the majority of the relationships existing between estimators of primary productivity and traits (cf. Chapter 6) have been established for grasslands. Consequently, the different characteristics of forage production such as the amount of biomass produced, the date of peak production, and the digestibility of the forage produced can be estimated on the basis of the community weighted means of a combination of leaf traits, notably the LDMC, plant height, and floral phenology (Al Haj Khaled et al., 2006; Ansquer et al., 2009b; Pontes et al., 2007) (Figure 8.2). In particular, low-stature vegetation produces its peak biomass and flowers earlier than tall vegetation (Ansquer et al., 2009b). Additionally, the variation in biomass production observed around the date of peak biomass production, which provides information on the period during which the manager can use optimally the available forage, is explained by the functional divergence of the LDMC (Ansquer et al., 2009b). The two components of functional structure are thus linked to different processes, indicative of complementary features of the forage production service.

Traits also allow the estimation of the digestibility of the biomass produced. For a range of monocultures established in an experimental garden, tissue digestibility was related to the density and nitrogen concentration of leaves and stems, as well as to the proportions of each of these two types of organs in the harvested vegetation (Figure 8.4; Al Haj Khaled et al., 2006; Duru et al., 2008; Pontes et al., 2007). The traits which affect the digestibility of organs are related to their chemical composition such as the concentration of fibre, (hemi) celluloses, and lignin (Al Haj Khaled et al., 2006; and see Section 3.2.1.2 in Chapter 3), which are higher in stems than in leaves (Poorter & Bergkotte, 1992; see also Figure 3.4 in Chapter 3). This explains why stems have a lower digestibility compared to leaves, as well as the major influence of the ratio of leaves to stems on the digestibility of the whole plant (Duru et al., 2008).

The dominance hypothesis, based on the idea that dominant species have a major impact on processes comparatively to other species, appears to apply in explaining the digestibility of the biomass of multi-species communities (cf. Chapter 6). Some experiments have shown that overall digestibility of a community varies negatively with the community weighted mean of LDMC (Ansquer et al., 2009b; Gardarin et al., 2014) and plant height (Ansquer et al., 2009b) (Figure 8.4).

8.2.5 Effect traits linked to other services

In addition to forage production, grasslands can provide a number of other services, largely related to soil functioning and biodiversity value (Huyghe, 2010). Apart from forage production and quality (Figure 7.3; Table 7.1), there exist very few examples of the evaluation of services provided by grasslands using the functional approach. Soil carbon storage is dependant almost entirely on the production of root

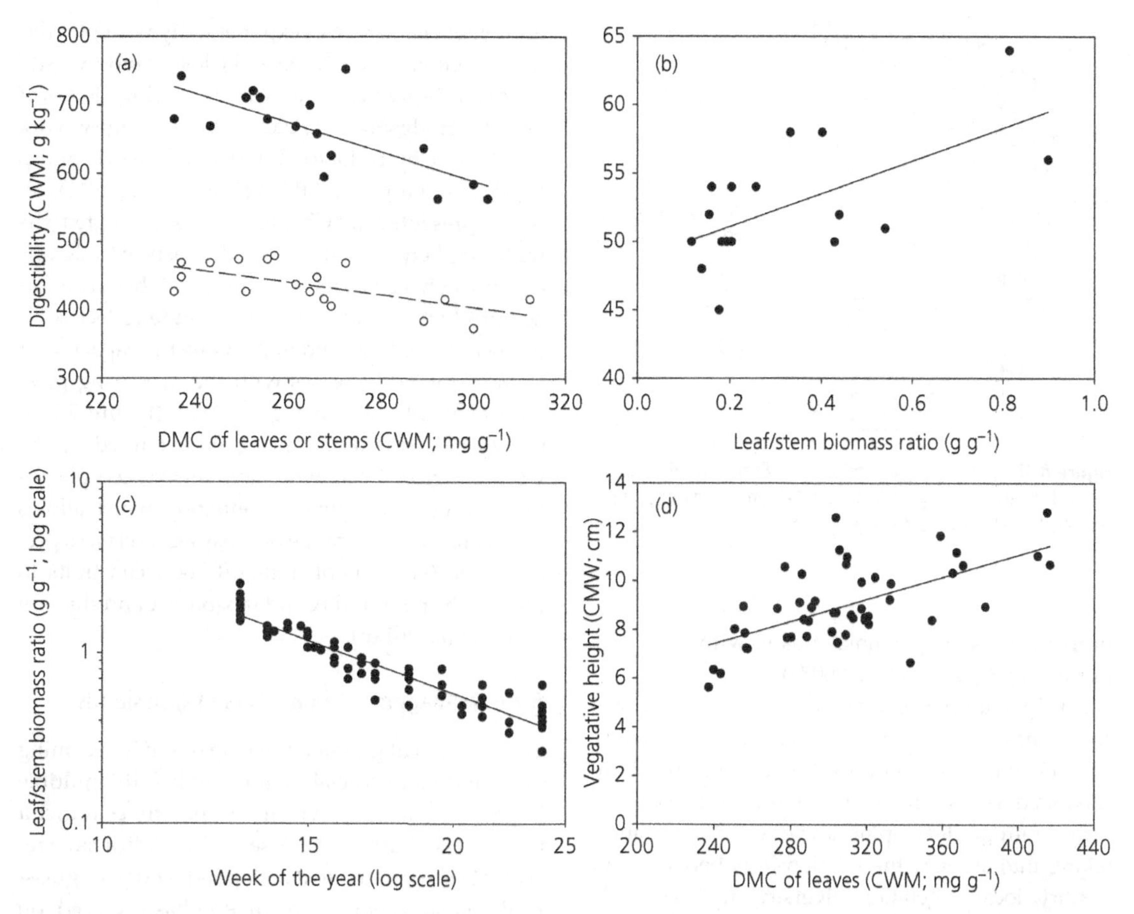

Figure 8.4A–D Decomposition of the suite of traits explaining the negative relationship between digestibility and plant size: (a) the negative relationship found between the digestibility of a community and the dry matter content of its tissues (leaves: solid symbols, $r^2 = 0.59$; stems: unfilled symbols, $r^2 = 0.45$); (b) the digestibility of a vegetation increases with the ratio of the proportion of biomass between leaves and stems ($r^2 = 0.38$), as leaves are more digestible than stems; (c) the decrease in the ratio in the proportion of biomass between leaves and stems during plant growth ($r^2 = 0.86$) and the radial expansion of plants explains the drop in digestibility over time; (d) the positive relationship between the vegetative plant height and leaf dry matter content of tissues ($r^2 = 0.67$) explains the drop in digestibility of plants with increasing size. Origin of the data: (a)–(c) grasses from grassland communities in the French Pyrénées (Ansquer, 2006; Duru et al., 2010); (d) rangelands on the Causse du Larzac (from Fayolle, 2008). See Figure 8.4E for a synthetic representation of these relationships.

biomass rather than litter (Orwin et al., 2010), which explains the contradictory results observed in the literature depending on the productivity of the environment. In fact, carbon storage can be limited in highly productive environments as a result of the rapid decomposition of roots, while it can be high in less productive environments as a result of the production of less abundant but slowly decomposing litter (Díaz et al., 2009; Duru et al., 2014b). Soil erosion also varies with local productivity: it is lower when grasslands are composed of perennial species with slow rates of growth and biomass production (McIntyre, 2008). Another documented service is pollination. The role of grasslands in the maintenance of pollination is often linked to the presence of certain legumes such as white clover and sainfoin, which greatly increase honey production potential, while it is decreased by the presence of certain grasses (Maillet-Mezeray & Gril, 2010). Finally, the aesthetic value of mown grasslands is linked to the diversity of flowering dates and to floral structures, while avalanche risk increases with increasing

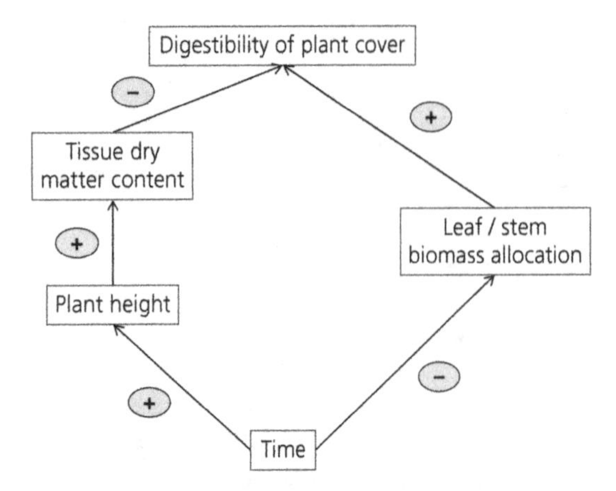

Figure 8.4E Synthetic diagram showing the different relationships involved in the digestibility of plant cover; '+' and '−' indicate positive and negative relationships, respectively.

proportions of tall perennial grasses with high litter production (Quétier et al., 2007a).

Taking multiple services into account simultaneously has rarely been done, even though such issues are at the core of the questions posed by many stakeholders. An initial study has been carried out across Europe developing a method of evaluation taking into account the relationships between soil fertility, local vegetation diversity, and economic viability (Hodgson et al., 2005b). In this method, fertility is estimated using a combination of leaf traits linked to the leaf economics spectrum and weighted by the local presence of different types of plants, while economic viability is evaluated by the quantity and quality of the animal products obtained as a function of the range of fertility found in the studied grasslands. This work established that the relationship between the fertility and economic viability of a given plot, estimated via its land use, varies between regions: in northern Europe, the rate of conversion of grasslands to crops increases with their fertility, while in the Mediterranean zone, the contrary is true, as it is only the fertile grasslands which are maintained as grasslands and not abandoned to forest conversion (Hodgson et al., 2005a). The trade-offs inherent in the production of different services, notably those between the production of animal products and the conservation of biodiversity, vary with the fertility of the environment, as economic viability rises exponentially with plot fertility, while an acceleration in the loss of biodiversity has been observed at intermediate and high levels of fertility (Hodgson et al., 2005b). More recently, work carried out by S. Lavorel and her research group (Lavorel & Grigulis, 2012; Lavorel et al., 2011), already presented in Chapter 7, has estimated the trade-offs between services on the basis of trade-offs existing between traits: the strong link between forage production and soil fertility is the reflection of high values of LNC and SLA, while the capacity for carbon storage increases with the increasing presence of plants with a high LDMC (Figure 7.7 in Chapter 7). This model has been extended by the work of agronomists who have introduced phenological aspects of biomass production, which allows for extension of the range of ecosystem services predicted on the basis of trade-offs between traits to include the temporality and flexibility of production (Duru et al., 2014b).

8.2.6 Management indicators in grasslands

The question of grassland management is becoming more and more crucial as grasslands fulfil multiple important functions, which are sometimes difficult to reconcile (Figure 8.1). A series of studies have recently been initiated to develop indicators for grassland management, with an emphasis placed on production. This research has identified the LDMC and the date of flowering as easily measured traits whose distributions are predictive of the quality and quantity of biomass produced (Ansquer et al., 2009b) (Figure 8.2). However, directly transferring this approach to non-academic grassland managers is difficult—even unlikely—as the characterization of the functional structure of a grassland requires a great deal of time and the meaning of a trait such as LDMC remains abstract to non-scientists[1] (Duru et al., 2014b).

[1] This is not the case for vegetation height, which is linked with the ability of a plant to acquire resources, and thus whose link with the quantity of biomass produced is quite intuitive. This trait is quite easy to estimate using a comparison to the height of, for example, a boot (grassland management advice given by the Institut de l'Elevage http://www.inst-elevage.asso.fr/). The fact that this method is not widely used is related to the fact that plant height is less well related to agricultural services than are traits linked with leaf structure and reproductive phenology.

Box 8.2 Tools for grassland management based on the functional approach

Taken from Duru et al. (2010a, 2010b, 2011).

Two tools for the management of grasslands have been developed by the AGIR[2] research team in Toulouse, France. They are based on the use of a classification which divides grasses into five groups depending on their growth rate, phenology, and the values of six traits: SLA, LDMC, leaf life span, mechanical leaf resistance, flowering date, and maximum plant height (Box 8.2, Figure 1; Duru et al., 2008).

Herb'type© is used to define the agricultural value of a grassland. This estimation requires the quantification of the relative abundances of five groups of grasses and all of the dicotyledon species present in the grassland being studied. These estimations are made visually, with 10 observations being made in a square quadrat of 40 × 40 cm, with a score given for each of the dominant species for each grass group[3], for the legume group, and for the other remaining species. Coupling these abundance values with the functional characteristics of these different groups of plants, as defined by the classification, allows the automatic determination of the functional structure of the sampled grasslands. The use of robust relationships between the community weighted mean and functional divergence of the traits and the components of agricultural value such as production, temporality of production, flexibility of grassland use, and digestibility allows for the estimation of these latter values for the studied grasslands. These relationships have been established from field-collected data for a number of regions in the South-West of France.

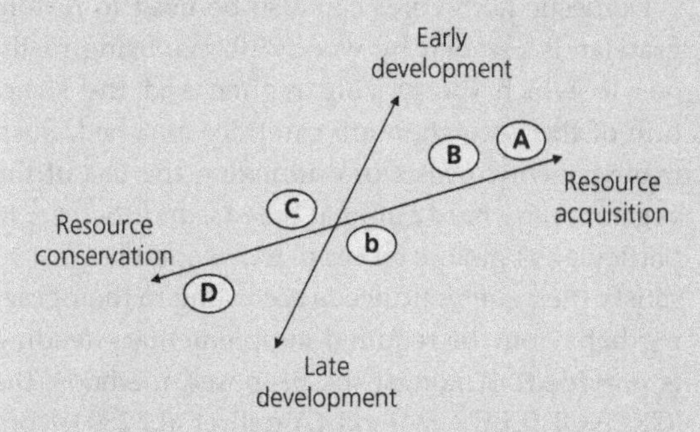

Box 8.2, Figure 1 The five groups of grasses, identified by letters, correspond to different functional strategies, depending on their position along the axes of resource acquisition/conservation on the one hand, and timing of plant development on the other hand.

[2] AGIR: Agroecology, Innovation, Territory (Agrosystèmes and Agricultures, Gestion des Ressources, Innovations, and Ruralités). A joint laboratory of INRA and INP-Toulouse.

[3] The dominant grasses are identified as representative of the five groups on the basis of their trait values.

The tool has been tested in a diverse number of environments and grazing systems. It enables the differentiation of grasslands established under contrasting environmental situations. In reality, species having a strategy of resource capture (Types A and B) are dominant in mown fertile grasslands, while those with a strategy of resource conservation (Type D) are favoured in low fertility and grazed environments. Testing of the tool in numerous situations has shown that on its own, the quantification of the proportion of Type A and B grasses is a good indicator of biomass production, and thus the potential numbers of grazing animals able to be supported by the grassland.

Herb'type© also takes into account the production potential of grasslands in terms relevant to farmers. In particular, it is possible to define two types of grazing farms: a first type in which all of the pastures have the same functional structure characterized by a large functional divergence, which provides considerable flexibility of usage; and a second type with considerable functional homogeneity but with major differences from one pasture to another, which requires management based on understanding and exploiting the complementarities between them.

This tool is also used to implement models representing the impact of different management practices on the provision of targeted ecosystem services (Duru et al., 2014b).

Herb'opti© is a graphical tool which presents the management practices of a farmer in reference to both the forage resources available, and the amount of forage he seeks to produce. The tool is based on a graph with the dates of mowing along the x-axis and the percentage of late developing grasses on the y-axis, with curves showing the optimal dates for obtaining good quality hay for grasslands of differing grass compositions. It clearly shows any possible mismatches between the simulated development of the grassland, as a function of the management practices of the farmer, and the forage needs of the grazing animals, and is a useful tool to facilitate discussions between farmers and advisory officers.

This realization has led to the proposal of a number of methods for simplifying the quantification of the functional structure of grasslands. The first consists of using a functional classification of species based on traits characterizing their growth rate and phenology, the values of which have been measured under controlled, non-nutrient limited conditions (Al Haj Khaled et al., 2006; Ansquer et al., 2004). This classification consists of five groups of grasses distinguished on the basis of their strategies of resource acquisition or conservation (see Box 8.2, and in particular Figure 1 in that box); it has been successfully used for predicting forage production in European (Duru et al., 2008) and South American (Cruz et al., 2010)[4] grasslands. A second method of simplification is based on the evaluation of ecosystem services by distinguishing the effect of grasses from that of dicotyledons. This procedure has been proposed as grasses are often dominant in grasslands and thus play a major role in the determination of ecosystem services, as expected from the dominance hypothesis (Grime, 1998). In addition, while the traits of grasses and dicotyledons respond in similar manners to gradients of fertilization and defoliation, their mean values differ: grasses generally show delayed phenological development and have leaves with higher LDMC and lower digestibility than those of dicotyledons (Ansquer et al., 2009a; Duru et al., 2010b) (Figure 8.5). Added to the five previously described groups of grasses is a sixth group comprising of all of the dicotyledons. The whole methodology, including the two phases of simplification, then leads to two tools (Box 8.2) allowing the characterization of the services provided by a grassland, and in particular its productivity, developmental timing, and nutritive value (Duru et al., 2010a).

Management indicators can also be developed to evaluate the success of the restoration of grasslands subject to agricultural abandonment. The success of restoration has been often assessed by comparing the spectrum of CSR strategies (Grime, 1974) or the traits associated to the strategies proposed by Westoby (1998) (see Chapter 3) present in the restored areas with that of reference areas. These indicators have shown that there is little similarity between restored grasslands and reference grasslands from the same areas, even after 60 years of restoration. These differences are explained by the long-term persistence of the effects of cropping, in particular that of phosphorus fertilization (Fagan et al., 2008), and by considerable temporal variation of the vegetation induced by the restoration techniques used (Moog et al., 2005; Römermann et al., 2009). A study comparing the impact of different restoration techniques applied over 25 years further illustrates these results. While restoration treatments based on the use of mulch or vegetation mowing resulted in an equivalent floristic richness to grazed grasslands, there remained major functional differences between the vegetation types. Grazed grasslands contained more rosette species and stoloniferous species, and fewer species with large seeds, than did grasslands restored using mulching or mowing (Kahmen et al., 2002).

Domestic herbivores can also be used to restore grasslands invaded by woody plants, using methods in which the grazing regime and the structure of the vegetation are carefully matched. Such management consists of optimizing the use of the vegetation by hardy animal breeds, and the use, in particular, of mobile fences to move animals so as to satisfy their energetic needs according to their grazing behaviour (if required supplementary feeding is provided). Amongst the proposed methods, the 'GRENOUILLE'[5] method (Agreil et al., 2004) consists of modifying the use of the environment by animals as a function of the availability of four categories of plants: legumes, which are preferentially grazed and which attract animals and promote their exploration of the environment; tussock-forming plants and tender-leaved shrubs; followed by large-leaved grasses and spreading shrubs which are grazed successively; and finally non-grazed species

[4] The success of this classification, which is the basis for the grassland management tools presented in Box 8.2, has led to the developers of these tools to consider taking legumes into account, which should lead to an even more precise characterization of the agricultural value of grasslands.

[5] The name of the method is linked to the graphical representation of the tool which regroups four classes of vegetation in a form resembling the body of a frog.

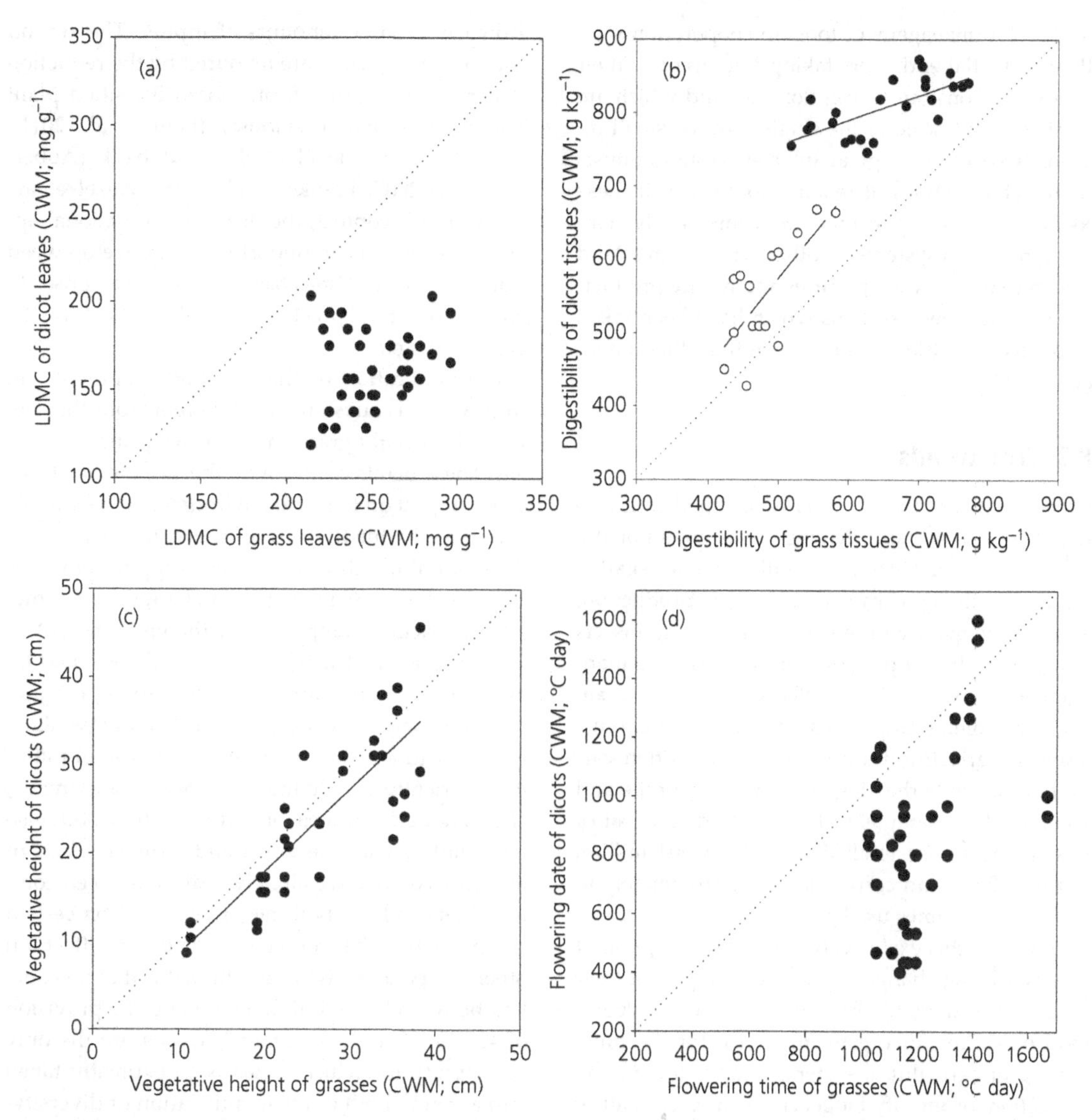

Figure 8.5 The community weighted means of the traits of grasses and dicotyledons respond in a similar manner to environmental gradients, but differ in the ranges of their values. In comparison to grasses, grassland dicotyledons have (a) lower leaf dry matter contents and (b) higher leaf (black symbols, $r^2 = 0.47$) and stem (white symbols, $r^2 = 0.53$) digestibility. Additionally, the two types of plants have similar vegetative heights (panel c: $r^2 = 0.67$), but grasses generally flower later (as shown here on a degree days basis: panel d). Only the r^2 of the linear relationships are shown. Each point corresponds to a comparison of the weighted means of a given trait obtained for the grasses and dicotyledons from the same community. The different communities correspond to permanent grasslands in the Pyrénées representing a management gradient generated by the application of two levels of fertilization and three intensities of defoliation. Data taken from Ansquer et al. (2009a).

with secondary metabolites or overly tough leaves. Despite the use of a fairly crude vegetation classification, these methods are generally highly successful, mostly because they are easy to implement, which explains their use in a large variety of grazed ecosystems, including those which are the least environmentally limiting (Guérin et al., 2009).

8.2.7 Conclusions

As shown by the wealth of studies discussed above, permanent grasslands are a good model system in which to develop methods for the transfer of the functional approach to human-managed ecosystems. One important approach is to develop management tools in cooperation with their potential end-users taking into account their needs and constraints (e.g. Box 8.2), and which are applied at the scale of the single farm or small region, two scales of great interest to agronomists (Duru et al., 2014b). It remains to be seen if these tools are transferable to other more intensively managed grassland systems, or other types of grassland systems such as sown pastures and highly productive hay meadows, systems which have been taken into account little in current studies (Plantureux et al., 2009).

8.3 Crop weeds

Crop weeds are those plants established in a cultivated field without having been intentionally planted or sown[6]. They generally have a negative impact on the crop by reducing crop production, altering the quality of the crop, harbouring insects that prey on the crop, or by complicating crop management (Box 8.3). These effects on crops explain why controlling crop weeds is a major economic issue in agricultural areas. This is still often carried out using herbicides, especially in Europe and America. For example, in France the direct cost of this herbicide use totalled more than 904 million Euros in 2008, and corresponded approximately to 10% of the volumes used globally[7].

However, this exclusively negative perception of weeds is being changed by the rediscovery of their positive functions (Altieri, 1999): they can serve as food or shelter for organisms useful for the functioning of agricultural systems, play a role in soil protection, or modify biogeochemical cycles, all of which has led to new strategies for their management (Petit et al., 2011). These new methods of management are focused on the two major types of weed communities that have been created by the development of modern agricultural practices. A first category of weeds mimic cultivated plants and establish within intensively managed crops grown with large amounts of inputs. The second kind of weed species are favoured by the reduction of herbicide use, and form more diversified plant communities than previously (Petit et al., 2011), including at the level of the seed bank (Aubertot et al., 2007; Franke et al., 2009; Wezel et al., 2009). In this context, the use of the functional approach represents an opportunity to develop weed management strategies based on their responses to given agricultural practices or on their effects on the agro-ecosystem.

The application of the trait-based approach to crop weeds is substantially different from that described above for grasslands, for two principal reasons. These plants coexist with the crop, which has a major impact on the local environmental conditions both directly due to the resources consumed by the crop, but also indirectly via the cropping practices associated with its management. Crops are a somewhat particular component of the environment as their functional structure is largely determined by the decisions of the farmer, and the crop also represents the large majority of standing biomass. Analysing the functional structure of weed communities in an agricultural field thus requires considering the effects not only of local abiotic conditions, but also those linked with the direct and indirect effects of the crop. As we have already indicated, weed communities can have both negative effects on certain components of the ecosystem and positive effects on other components. While it is probable that this dual link between biological diversity and the provision of services is quite common[8], weeds are the only community for which it has been explicitly taken into account, both in the identification of diversity-processes-services cascades (Zhang et al., 2007) and in new management practices (Petit et al., 2011).

Figure 8.6 represents how the use of the response and effect framework can be applied in the case of crop weeds. As well as local soil and climatic conditions, management practices linked to the type of crop being grown influence local resources and

[6] http://mots-agronomie.inra.fr/mots-agronomie.fr/index.php/Adventice

[7] http://www.uipp.org/Chiffres-cles/Reperes-monde-et-Europe

[8] For example, the presence of forests in the upper parts of a watershed increases carbon storage and limits downstream flooding, but at the same time significantly decreases the quantity of downstream water available for crops and human populations.

Box 8.3 Different types of weeds

Taken from Navas (1991).

A weed is a plant species which is not native to the indigenous flora of an area into which it has been accidently introduced and in which it can establish and reproduce. More specifically, in agronomy, a crop weed is a plant established in a cultivated field, without having been intentionally sown there by the farmer. These definitions, being based on their interactions with human objectives, are difficult to use to identify weeds with respect to their ecological roles. It is nevertheless possible to define a list of habitats in which weeds are found, and the types of ways in which they interfere with human activities (Box 8.3, Table 1).

Initially strictly defined on the basis of negative effects, weeds can also have positive effects on ecosystems. They play an important role in structuring trophic networks, in soil protection, and in biogeochemical cycles, and can also have a strong cultural value. The recognition of these effects has led to the proposal to base weed management on a joint assessment of their negative and positive impacts.

Box 8.3, Table 1 Habitats (with the level of anthropization for each indicated by the number of stars) in which weeds can be found and the types of damage which they can cause.

Habitat		Potential effects		
Type	**Anthropization**	**Public health and cultural**	**Ecological**	**Economic**
Rivers	*	X	X	X
Pastures	**	X	X	X
Crops	***	X	X	X
Urban zones and parks	***	X	X	X
Road edges, disturbed areas	**	X	X	X
Canals	***	X	X	X
Roads, railways, tracks	***	X		X
Natural habitats	*	X	X	
Protected natural habitats	(*)	X	X	

disturbance regimes, and filter weeds depending on the values of their response traits. The functional structure resulting from the action of these filters strongly affects certain ecosystem processes linked to services: the biomass production of the weed community can produce a competitive effect on the crop and decrease its biomass production, while the regulation of biotic interactions and biogeochemical cycles can directly impact on various services such as food provision, trophic regulation, soil cover, and the conservation of beneficial weed species or other species with a high cultural value.

8.3.1 Characterization of environmental filters

The different environmental factors linked to either local conditions or management practices do not have the same effect on weed communities. Climatic conditions, edaphic conditions, or land-use history are independent variables which constrain and modulate agricultural practices. For example, the climatic conditions of an area limit the growth period for all organisms, independent of the agricultural practices used. This is why it is more pertinent to present these two groups of factors in a separate and hierarchical manner (Navas, 2012).

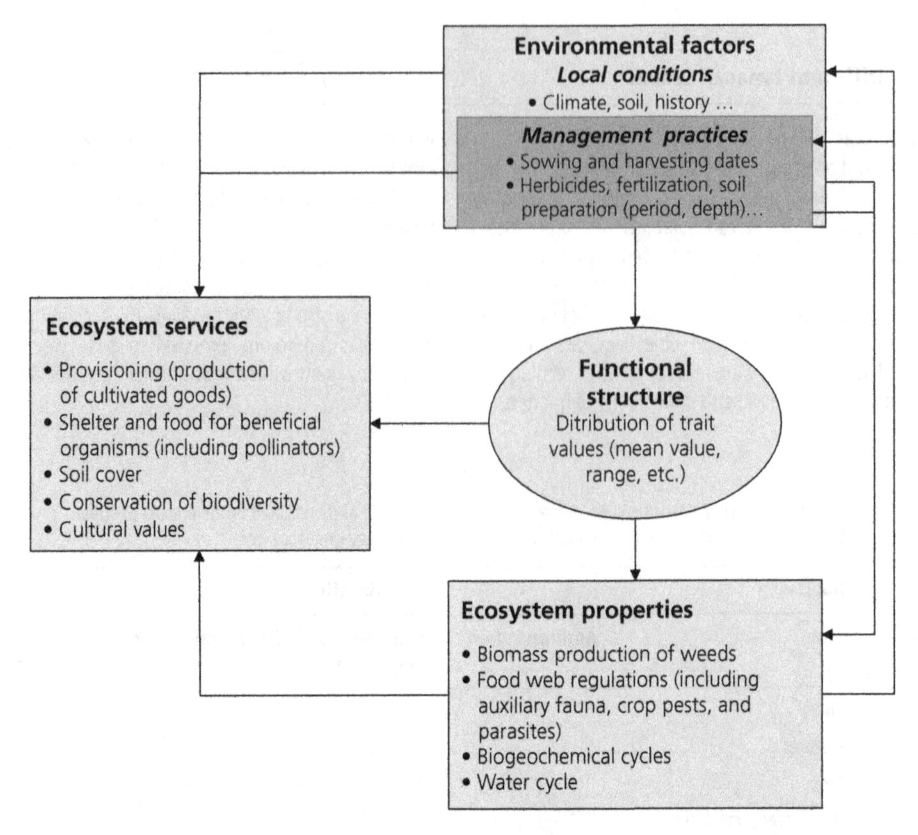

Figure 8.6 The functional structure of a crop weed community is the result of the differential responses of species to environmental conditions, either local or linked to the crop, depending on their trait values, and has onward effects on ecosystem properties and thus on ecosystem services. These properties and services are linked not only to the cultivated component (for example the competition imposed by weeds limiting the service of food provision) but also the non-cultivated component (for example sheltering insects). The impacts of management practices on the environment depend on local conditions, which is why these two types of environmental factors are represented in a nested manner. After Diaz et al. (2007a) and Navas (2012).

Amongst local conditions, it is essentially those of the soil which determine the distribution of local species: the pH and soil texture play major roles compared to that of climatic variables such as precipitation (Fried et al., 2008; Lososova et al., 2004). Historical factors linked to past land use can also have an influence due to their profound modifications of soil conditions. For example, the location of ancient Gallo-Roman farms can still be perceived from the presence of particular types of weed communities (Sciama et al., 2009).

However, it is the type of crop, the agricultural history, the level of soil drainage, and the type of soil ploughing (and also its depth) which explain the major differences among weed communities (Butler et al., 2009; Fried et al., 2008; Hawes et al., 2009; Smith, 2006). The expression 'crop type' refers to the effects of the date of sowing, the herbicides used, and the fertilization regime applied: by affecting local environmental conditions, these factors directly affect the phenology and other components of the performance of weeds (Gunton et al., 2011; Lososova et al., 2008). The importance of this latter group of factors in structuring weed communities explains why weed communities are often named after the type of crop they are associated with (see Navas, 2012, for recent references). The spatial and temporal factors structuring weed communities have also been taken into account, but in fewer studies. The spatial dimension refers to the organization of the landscape (existence of hedges, field edges, other semi-natural areas, etc.), which due to its heterogeneity influences the diversity and patterns of species abundances (Petit et al., 2011). The temporal dimension also has effects via the impacts of crops included in rotational agricultural systems which can modify the later presence of weed species (Smith & Gross, 2007), such as the example of forage plants which limit the subsequent expansion of large, highly competitive annual species (Meiss et al., 2010).

However, a remaining major question is that of quantifying the environmental gradients created by agricultural practices. Despite the large number of studies characterizing the relationships between weeds and their environment, it is not currently possible to directly compare the environmental factors explaining the structure of weed communities between crops. In reality, the majority of published studies describe the relationships between weed communities and the effects of a given management technique via its action on one or a number of environmental factors, or compare the local impact of different management techniques, such as different methods of soil ploughing. To be able to carry out larger-scale comparisons including management interventions creating different types of disturbances (for example ploughing and herbicide application), it is necessary to develop a standardized method for quantifying the gradients created by the different management practices. In an approach comparable to that presented in Chapter 4, and applied to grasslands in Section 8.2.1, it is possible to analyse the impact of these interventions on the intensity and frequency of mechanical and chemical disturbances both above ground and below ground, as well as on the levels of resource availability for plants (Gaba et al., 2014b; Navas, 2012). This research is only in its beginnings (Fried et al., 2008; Gunton et al., 2011), but represents a major challenge for better understanding and predicting the dynamics of the assembly of weed communities.

8.3.2 Response traits of weeds to management

The first characterization of weeds using traits was formalized some 50 years ago: Baker (1965) emphasized the opportunist character of these species, and the generalist nature of their genotype ('general purpose genotype'). He then hypothesized a portrait of an 'ideal weed', bringing together the ensemble of characteristics found in these species (Baker, 1974): seeds with a long life span and able to germinate in a large number of environments; a high growth rate with the rapid production of numerous seeds which can also be produced by self-pollination; non-specialized pollination; long duration of seed production, which is possible in a large number of environments; seed dispersal over both short and long distances; high competitiveness; and vigorous vegetative reproduction for perennial species. This manner of categorizing weeds has been somewhat forgotten over the course of time (but see Navas, 1991, for other references), especially as the question of weed resistance to herbicides became a major preoccupation. It was brought back to attention with the renewed emphasis on integrated weed control, due to the need to understand the factors controlling the dynamics of weed communities (Booth & Swanton, 2002). This is when the first functional characterization was carried out for 21 common weed species in the United Kingdom based on their life cycle, seed size, maximum height, and flowering date (Figure 8.7). These traits can be used to regroup these species on the basis of their ability to grow in managed environments, their competitiveness in relation to the crop, their response to nitrogen fertilization, and their capacity to invade agricultural fields from the boundaries (Storkey, 2006). A more recent analysis of the French weed flora confirms and adds precision to this information by showing that the most problematic species in crops have a high potential growth rate, an annual life cycle, and an opportunist strategy allowing them to escape from herbicide applications, in particular due to a phenology which responds to the dates of management interventions (Gunton et al., 2011).

While these general characteristics are shared by the majority of weeds, there however exist differences between communities depending on the level of intensification of the crop in which they are found. The flora currently found in the most intensively managed systems is highly specialized in terms of resource use, phenology, and herbicide

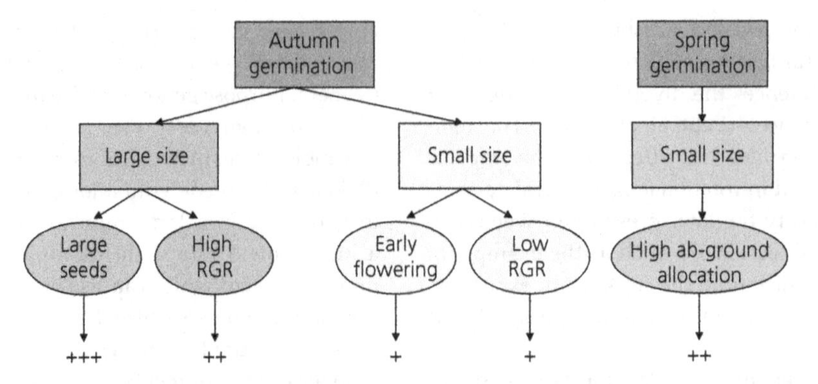

Figure 8.7 Five groups of weeds, defined on the basis of 20 traits of 21 common annual weeds of wheat crops in the United Kingdom. The number of '+' signs increases with the competitiveness of the species forming each group; competitiveness is defined as the number of plants per m^2 required to cause a loss of 5% of production in the infested wheat crop. Ab-ground: above-ground. Adapted from Storkey (2006).

tolerance. For example, the weeds which have increased in abundance in sunflower crops since the 1970s have a type of ecological functioning closely mimicking that of sunflowers: they are nitrophilic and sunlight demanding, very insensitive to herbicides used in sunflower crops, and have a summer life cycle regardless of their phylogenetic history (Fried et al., 2009). In contrast, weeds found in extensively managed systems, characterized by disturbances of irregular intensity and of variable spatial extent, are generally of a small size, have large seeds, and have later flowering in comparison to species from intensively managed systems (Lososova et al., 2006). These species are restricted to these environments as their small size does not allow them to be competitive under highly productive conditions and their low fecundity leads to low tolerance to herbicides due to the small number of individuals produced in a plot, which limits the probability of them escaping from herbicide application. Logically, these are also the attributes which characterize those weed species which have disappeared from cereal and soya crops in the United Kingdom as a result of their non-adaptation to the intensification of agricultural practices (Storkey et al., 2010).

The overall responses of weeds to cropping practices are based on the responses of a large number of traits. For example, herbicide tolerance is partially explained by leaf architecture and anatomy, the thickness of the cuticle, and phenology, but is also dependent on physiological characteristics which are not easily measurable (Gaba et al., 2014b). It can also be related to changes in the life cycle: for example weeds present in a wheat crop have delayed germination, allowing them to escape the action of herbicides (Fried et al., 2012). The response of plants to the date of sowing or to ploughing depends on their germination strategy (Smith, 2006), which can be difficult to characterize simply, in particular due to the large amounts of variation between populations. Recent studies, however, have shown that the seed bank dynamics of crop weeds can be predicted by a suite of traits: the mortality of seeds in the soil is negatively related to the thickness of the seed coat (Gardarin et al., 2010b); seed germination varies with the mass and the surface area of the seed as well as with its lipid content (Gardarin et al., 2011); and finally seed dormancy depends not only on seed mass but also on the diameter of the hypocotyl or the coleoptile, and differs between monocotyledons and dicotyledons (Gardarin et al., 2010a).

8.3.3 Functional structure of weed communities

Analysing the functional structure of weed communities requires resolving questions about its links with the crop. The crop is a particular plant species within the agro-ecosystem because, as already indicated above, it constitutes the major part of the biomass of the local vegetation, has a highly homogeneous functional structure[9], and filters the local weed community by its direct and indirect effects on the local environment. For example, the light available for weed growth depends on that intercepted by the crop throughout its growth period; the date of weed emergence depends on the timing and the nature of the disturbances linked to the preparation for and sowing of the crop. 'Mimicking' weed species are an extreme example of this relationship as they resemble the associated cultivated species in all respects, except for the attributes of at least one trait which allows their persistence in the system: for example wild rice has the same morphology and the same phenology as cultivated rice except that seed dispersal occurs slightly earlier (Ling-Hwa & Morishima, 1997).

The generality of this filtering effect of the crop on weeds has been well established in a study comparing the relationships between the traits of weeds and the traits of cultivated species for a large number of crops in France (Gunton et al., 2011). This study showed that weeds found in crops established late in the season are large in size, have delayed germination and flowering, and have a shorter flowering period than those found in other crops established earlier in the season. In particular, the maximum height of the weeds depends on that of the crop (Figure 8.8 a–c); the mean date of weed emergence increases with the date of the sowing of the crop, while the divergence of the date of weed emergence decreases, which leads to weed emergence being grouped around the date of crop establishment for late planted crops (Figure 8.8 d). Finally, more competitive crops contain weed species with contrasting strategies of response to competition, which results in a high degree of functional divergence of certain traits. For example, the divergence of biological types of weeds is larger in tall crops than in short crops: in the former, the majority of species are late germinating annuals, but the proportion of hemicryptophytes and of geophytes is far from being negligible.

[9] However, even a homogenous crop can present phenotypic differences linked to, for example, the effects of the positioning of the individual (edge of the field, soil heterogeneity, or the positioning of the seed, etc.).

8.3.4 Effect traits of weeds and the provision of agricultural services

The main impact of weeds on the ecosystem is to depress crop production, both in the short term via their competitive effects on cultivated plants, and over the longer term via their demographic dynamics (impacts on the seed bank, seed dispersal, etc.) (Figure 8.6). Predicting crop losses due to weeds has major economic importance, and has resulted in the production of a considerable number of modelling studies (Doyle, 1997). When the flora is dominated by one or a number of particularly problematic species, demographic models have been based on mechanistic approaches predicting current and future weed densities under different management options (Petit et al., 2011). The most recent of these include traits (Colbach et al., 2014; Gardarin et al., 2011; Storkey & Cussans, 2007), and can be used to predict the weed composition of an environment and its impact on the crops in which it develops. However, such models require considerable effort to develop, as they use a large number of parameters, with data available only for a few major species. Other approaches directly model the competitive effect of weeds on crops by taking into account the differences in the weighted means of the traits existing between the two communities. For example, the difference in leaf area between weeds and cultivated plants, measured at an early stage of crop development, is the major parameter influencing the final yield of the crop in the currently most frequently used models (Booth & Swanton, 2002; Kropff et al., 1992). A recent study showed that the differences in height between weeds and the crop measured at the beginning of the season is a very good indicator of the intensity of their interactions and of the consequences on both the yield of the crop and the seed production of the weeds (McDonald et al., 2010). This result is important as it raises the possibility that quite simple models, developed to predict crop-yield losses for the current year, could ultimately be used to predict the dynamics of the involved weeds.

Such models are, however, less predictive in less intensively managed systems, within which the variability of the functional structure of weed communities is higher. In these systems, there exist

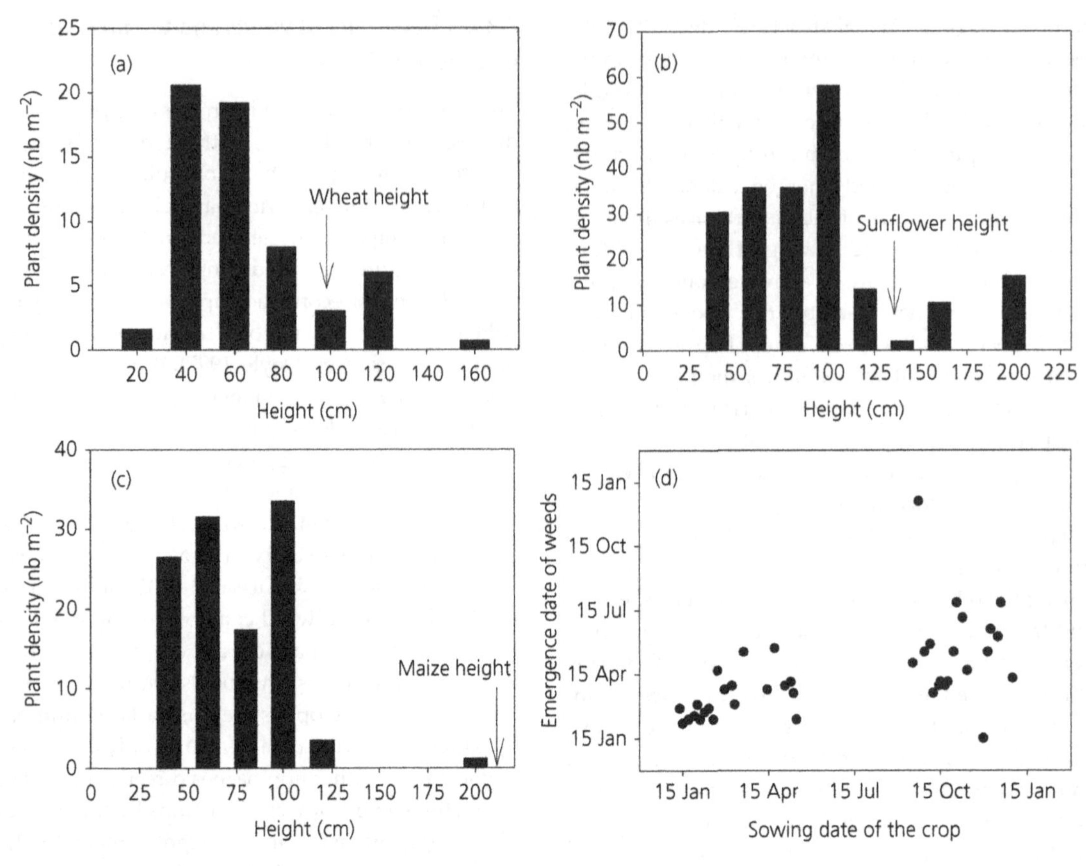

Figure 8.8 The traits of crops constrain the functional structure of weed communities, as illustrated here for two traits: maximum height (a–c) and the date of crop sowing (d). The maximum heights of the large majority of weeds present in a field are lower than that of the cultivated plants (indicated by an arrow in each graph) for the cases of (a) wheat, (b) sunflowers, and (c) maize (data taken from Fried, 2007); (d) the date of weed emergence in a cultivated field depends on the date of crop establishment; note that for sowing dates later than 15 July, the emergence date of weeds occurs during the year following that of crop sowing (data taken from Gunton et al., 2011).

weeds which have little effect on the crop as they use different pools of edaphic resources to those used by the cultivated plants due to differences in the depth and architecture of their root systems (Deen et al., 2003; Smith et al., 2009). Favouring these types of weeds to reduce the numbers of others with a functioning closer to that of the cultivated plants would limit the intensity of competition between weeds and cultivated plants and thus limit yield losses (Smith et al., 2009). This observation suggests that differences in trait values linked to resource use between weeds and cultivated plants could be of major importance in predicting crop yields: distinct ranges of trait values corresponding to a complementarity in resource use between weeds and the crop over time or in space could result in lower impacts of weeds on the production of the cultivated plants. In other terms, the hypothesis of limiting functional similarity which explains the coexistence of species in 'natural' environments could also explain a neutral coexistence between weeds and cultivated plants. However, this hypothesis remains to be tested in a more systematic manner in agricultural environments.

8.3.5 Traits of weeds and other services

Asking the question of what ecosystem services are provided by weeds can be surprising given that we generally only consider their negative impacts on crops. These species can have non-negligible positive effects, in particular on other components of local biodiversity (Altieri, 1999). For example, weeds can serve as food, shelter, and overwintering or reproductive sites for numerous species such as pollinators, granivorous arthropods, herbivores and omnivores, birds, and mammals (Bàrberi et al., 2010). Weed functional types, characterized by categorical traits, can be used to infer the traits of invertebrate species (Hawes et al., 2009). A recent study has also established relationships between weed traits and beetles found around the edges of fields in samples collected from a large number of regions and crops in Great Britain (Brooks et al., 2012). This study showed that weeds with small seeds germinating in autumn coexist with small, granivorous, springtime beetles, while weed species with large seeds and germinating in springtime were associated with larger autumnal animals with a more diversified dietary regime. Additionally, weed communities with high values of SLA and early flowering generally sheltered a larger diversity of invertebrates than those composed of more competitive species (Storkey et al., 2013; and see the review by de Bello et al., 2010 for other identified relationships). It is therefore likely that the diversity of herbivores feeding on weeds, and thus able to regulate their populations, depends on the chemical composition of the weeds' tissues, the morphology of their leaves, and their seed production. Also, the service of pollination, still rarely studied for weeds, should vary as a function of traits defining the accessibility and attractiveness of flowers within a cultivated vegetation as well as the production of nectar.

Some of the species sheltered by weeds are the natural enemies of other organisms which have negative impacts on crops: increasing their diversity by maintaining the diversity of weeds can lead to better control of such pest species (Hawes et al., 2005). In addition, some weed species are highly attractive to crop pests and can serve as traps when they are maintained in areas adjoining cultivated zones, with the management of weed species thus being implemented at the level of landscape and not only at that of the field (Bàrberi et al., 2010). It should, however, not be forgotten that this role of a reservoir can also serve to shelter other crop pests such as bacteria, viruses, or fungi which are not desirable (Franke et al., 2009), and which would require the management of certain weed species. In this last case, there exists little clear information on the traits explaining this reservoir effect, apart from the observation that when weed phenology is shifted as compared to that of the crop, this allows the maintenance of a primary inoculate over the entire year (Navas et al., 1998).

Finally, weeds are sometimes valued for their effects of soil protection, which depend largely on their size, architecture, and growth form (de Bello et al., 2010). They can even be encouraged and used in some cropping systems as cover plants (Damour et al., 2014).

8.3.6 Towards more efficient management methods for different types of weeds

This issue of weed management can be considered within the framework of defining the composition and functioning of the flora that is considered desirable for a particular area. This composition results from trade-offs which change as a function of location and time. The objective of 'zero weeds' often championed for cropping systems during the 1970s is no longer a target, as it could only be achieved via massive increases in the use of agricultural inputs without any appreciable gains in productivity—and it has led to an explosion in the populations of weeds resistant to the herbicides used! The recent recognition of the positive effect of weeds on certain ecosystem services has led to the proposal of classifying weeds into three different types as a function of the nature of their positive effects on 'useful' biodiversity while having a neutral or moderately negative effect on the crop (Moonen & Bàrberi, 2008). These types are: 1) species having cultural value or which facilitate other organisms with an agricultural value; 2) species which participate in the provision of production services or regulation services such as pollination; and 3) species which can be used as indicators for agricultural management. However, this proposal has not resulted in the development of operational management strategies as weeds can at the same time have negative effects on crops and also be useful from the viewpoint of sheltering beneficial biodiversity. This has been shown for a large number of crop weeds in the United Kingdom (Storkey, 2006), which explains why farmers do not wish to maintain weeds anywhere other than in extensive agricultural systems. It is evident that progress remains to be made in the understanding of the trade-offs between the negative and the still largely unknown positive effects of weeds so as to be able to develop new methods for the management of their communities.

8.3.7 Perspectives

The application of the trait-based approach to weeds is very recent. This approach holds the promise of progress in a number of areas, in particular because it can be used to study the weed community as a whole and not only focus on some major species. First, it should allow a better understanding of the assembly of weed communities as a function of the choice of management practices and local environmental conditions, thus permitting predictions of the possible weed flora under novel conditions, especially in the case of innovative agricultural systems. To achieve this, progress will be required in the characterization of weed response traits to environmental gradients generated by agricultural practices, when it becomes possible to characterize these in a standardized manner. A second objective is to provide an evaluation of the impact of weeds on ecosystem properties and the trade-offs between these at the level of the cultivated field or the landscape. Attaining this second objective appears as a longer-term goal as it requires an understanding of the relationships between weeds, processes, and services for a large number of properties, and then linking this with the question of the trade-offs existing between these services. This question is essential and should be addressed in order to develop agricultural practices able to maintain local equilibria.

8.4 How do these examples help us develop innovative agricultural systems?

As shown above, a functional approach to diversity can be used to address applied ecological questions in the field of agriculture. Some of these questions are critical, given that agriculture is currently confronted with the need to reintroduce diversity into its systems and practices so as to achieve a less artificial functioning of agricultural areas (Robertson & Swinton, 2005). It allows an understanding of how complex plant communities are influenced by the interactions between agricultural practices and local conditions. In order to go further and become predictive, it is necessary to produce standardized methods for the quantification of management-induced environmental gradients, such as those that already exist for grasslands or in less human-modified systems (see Chapter 4), and then defining the response traits to those factors specific to agriculture. The functional approach also provides a method for approaching the question of ecosystem services and the trade-offs between them, ultimately leading in grasslands to the development of management tools and indicators. In the case of weeds, the state of progress is not as advanced. However, integrating all of the biotic interactions in which weeds are involved into a common framework, whether these be with cultivated plants or with herbivores or granivores which limit their expansion, should allow for the advancement of the development of new weed management systems (Gaba et al., 2014a). With the exception of tropical agro-forests whose functioning can be very close to that of natural forests (Malézieux et al., 2009), applications in other human-managed ecosystems are still in a very preliminary phase. Garnier and Navas (2012) give the example of temperate forest management, for which a framework comparable to that described in this chapter is currently being developed.

Another perspective concerns the contribution of the functional approach to the elaboration of new agricultural systems, established on the basis of the services they are required to provide (Gaba et al., 2015). To achieve this, processes of agro-ecosystems linked to the targeted services and the effects traits associated with these need to be identified. It will then be necessary to construct a cultivated plant community possessing both an adequate distribution of effect traits and also of response traits to the local environmental conditions (soil, climate, etc.), management practices (those agricultural techniques able to be used as a function of equipment and environmental conditions), and socio-economic conditions (seeds and agricultural inputs available, local regulations, etc.).

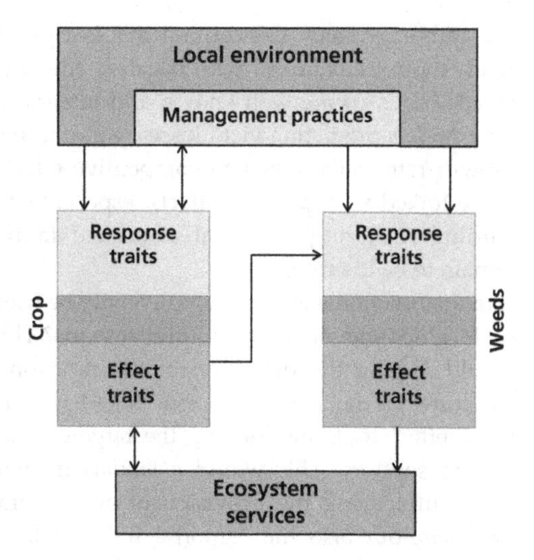

Figure 8.9 Conceptual diagram representing the relationships between the functional structure of a crop and the associated weed community. The functional structure of the crop is managed by the farmer on the basis of the required services, but it also varies in response to local environmental, socio-economic, and technical conditions. There exists therefore a double constraint on the crop's functional structure, marked by the direction of the arrows: the response traits are determined as a function of environmental filters imposed by the local environment and agricultural practices, while effect traits are determined as a function of the targeted services. Some crop traits can also affect the range of possible agricultural practices, which is indicated by a double arrow. The functional structure of the crop weed community depends on environmental factors and agricultural practices, but also on the effect traits of the crop itself, and influences the provision of services, the majority of which are different to those delivered by the crop. See the text for more details.

The diagram shown in Figure 8.9 illustrates the links occurring between the functional structures of the crop and the weed communities, and the services provided. It shows that the functional structure of the crop is actually controlled both by environmental filters—some of which are directly manipulated by farmers—and via the processes which provide the desired ecosystem services. The functional structure of the weed communities is dependent on environmental factors but also on the effect traits of the crop, as discussed in Sections 8.3.3 and 8.3.4 of this chapter.

Based on this diagram, once the relationships between the environment and the response traits are understood, which is underway, as well as those between effect traits and trade-offs between services, for which further work is required, it will be possible to develop crops with a mix of varieties or species to obtain an adequate functional structure in terms of distributions of response and effect traits. This strategy can also be applied to conceive species mixes growing in the same plot with differing phenologies or with different nutritional requirements (mixes of grasses and legumes—Bedoussac & Justes, 2010), allowing for the complementary use of local resources. It would also be possible to develop successions of species over the course of time, alternating cash crops and cover crops, which would allow for better soil management and more efficient control of pest species (Damour et al., 2014, 2015), as well as developing spatial mosaics at the level of the landscape to optimize certain regulatory functions such as pollination or the control of pest species (Gaba et al., 2014a).

8.5 Conclusions

The application of the trait-based approach to agricultural systems allows us to address two important issues: the characterization of complex gradients depending both on local environmental factors and management practices linked to human activities, and the construction of plant communities with a precise functional structure, responding to local conditions and providing particular ecosystem services while at the same time having minimal negative impacts. These issues are also fundamental in the case of the restoration of degraded ecosystems, and also for the establishment of planned urban biodiversity—two fields in which the trait-based approach has been little used so far. It is probable that these types of application will only become more important in the future.

8.6 Key points

1. The current context of a transition from intensive agriculture to more sustainable systems has prompted a vigorous debate about the role of biodiversity in agricultural systems. The functional approach to diversity has the potential to better understand the functioning of these systems, to develop new management tools, and to construct innovative cropping and grazing systems.

2. Permanent grasslands are the system for which the functional approach has been most often applied to date. The functioning of these systems is dependent on two principal types of environmental gradients, generated by the practices of defoliation and fertilization. Those traits which respond to management are essentially those linked with the utilization of resources, morphology and growth forms, and regeneration; they differ depending on the availability of nutrients and water, the type of defoliation (mowing or grazing, including the type of grazing animal and grazing practices), and the interactions between these factors. The functional structure of grasslands allows us to predict the major components of agricultural value and associated services: the quantity of biomass at peak production depends on vegetation height, and the date of this production peak varies with phenology and certain leaf traits, some of which also determine the digestibility of the produced biomass. Finally, the flexibility of grassland use increases with increasing divergence of flowering phenology and certain leaf traits. Grassland management tools designed to be used by farmers have been developed on the basis of these results, by regrouping grassland species on the basis of their functional and phenological strategies.
3. The functional approach allows for the characterization of the responses of crop weeds to cropping practices and/or their effects on the crop, while at the same time taking into account their positive environmental effects: erosion prevention, sheltering beneficial biodiversity, etc. In this context, the crop is considered as a particular environmental filter, which constrains the functional structure of weed communities by its direct impacts on the environment (resource use, modification of local temperature, etc.), but also indirectly via the associated cropping practices. This approach requires comparing the environmental gradients generated by cropping practices (sowing date, type of ploughing, herbicides, previous agricultural history, etc.), identifying the response traits linked with these gradients, some of which are still poorly known (e.g. response traits to herbicides), and also those effect traits which have an effect on the functioning of the agro-ecosystem. Concerning this last point, while those traits linked with negative effects of weeds on crops are well known and integrated into models predicting yield losses (e.g. size and growth rate traits linked to competitive effect), those linked with positive effects, especially regarding sheltering beneficial associated fauna, remain to be identified.
4. Can the functional approach to diversity be used to develop innovative agricultural systems? This would require the definition of the functional structure of the agro-ecosystem together with those effect traits modulating the targeted ecosystem services. This would *a minima* include those influencing the provision of agricultural products, but also the response traits to local environmental conditions that are modified by agricultural practices and constrained by the socio-economic context.

8.7 References

Agreil, C., Meuret, M., & Vincent, M. (2004). GRENOUILLE: une méthode pour gérer les ressources alimentaires pour des ovins sur milieux embroussaillés. *Fourrages, 180*, 467–481.

Al Haj Khaled, R., Duru, M., Decruyenaere, V., Jouany, C., & Cruz, P. (2006). Using leaf traits to rank native grasses according to their nutritive value. *Rangeland Ecology and Management, 59*, 648–654.

Al Haj Khaled R., Duru, M., Theau, J.-P., Plantureux, S., & Cruz, P. (2005). Variation in leaf traits through seasons and N-availability levels and its consequences for ranking grassland species. *Journal of Vegetation Science, 16*, 391–398.

Altieri, M. A. (1999). The ecological role of biodiversity in agroecosystems. *Agriculture, Ecosystems & Environment, 74*(1–3), 19–31. doi:10.1016/s0167-8809(99)00028-6

Andueza, D., Cruz, P., Farruggia, A., Baumont, R., Picard, F., & Michalet-Doreau, B. (2010). Nutritive value of two meadows and relationships with some vegetation traits. *Grass and Forage Science, 65*(3), 325–334. doi:10.1111/j.1365-2494.2010.00750.x

Ansquer, P. (2006). *Caractérisation agroécologique des végétations prairiales naturelles en réponse aux pratiques agricoles. Apports pour la construction d'outils de diagnostic.* Thèse de Spécialité, Institut National Polytechnique de Toulouse, Toulouse.

Ansquer, P., Duru, M., Theau, J. P., & Cruz, P. (2009a). Convergence in plant traits between species within grassland communities simplifies their monitoring. *Ecological Indicators*, *9*(5), 1020–1029. doi:10.1016/j.ecolind.2008.12.002

Ansquer, P., Duru, M., Theau, J. P., & Cruz, P. (2009b). Functional traits as indicators of fodder provision over a short time scale in species-rich grasslands. *Annals of Botany*, *103*(1), 117–126. doi:10.1093/aob/mcn215

Ansquer, P., Theau, J.-P., Cruz, P., Viegas, J., Al Haj Khaled, R., & Duru, M. (2004). Caractérisation de la diversité fonctionnelle des prairies à flore complexe: vers la construction d'outils de gestion. *Fourrages*, *179*, 353–368.

Aubertot, J.-N., Barbier, J.-M.,Carpentier, A., Gril, J.-J., Guichard, L., Lucas, P.,... Voltz, M. (eds) (2007). *Pesticides, agriculture et environnement. Réduire l'utilisation des pesticides et en limiter les impacts environnementaux*. Versailles: Quae.

Baker, H. G. (1965). Characteristics and modes of origin of weeds. In H. G. Baker & G. L. Stebbins (eds), *The Genetics of Colonizing Species* (pp. 147–168). New York & London: Academic Press.

Baker, H. G. (1974). The evolution of weeds. *Annual Review of Ecology and Systematics*, *5*, 1–24.

Bàrberi, P., Burgio, G., Dinelli, G., Moonen, A. C., Otto, S., Vazzana, C., & Zanin, G. (2010). Functional biodiversity in the agricultural landscape: relationships between weeds and arthropod fauna. *Weed Research*, *50*(5), 388–401. doi:10.1111/j.1365-3180.2010.00798.x

Bedoussac, L., & Justes, E. (2010). The efficiency of a durum wheat–winter pea intercrop to improve yield and wheat grain protein concentration depends on N availability during early growth. *Plant and Soil*, *330*(1–2), 19–35. doi:10.1007/s11104-009-0082-2

Bernard-Verdier, M., Navas, M.-L., Vellend, M., Violle, C., Fayolle, A., & Garnier, E. (2012). Community assembly along a soil depth gradient: contrasting patterns of plant trait convergence and divergence in a Mediterranean rangeland. *Journal of Ecology*, *100*(6), 1422–1433. doi:10.1111/1365-2745.12003

Booth, B. D., & Swanton, C. J. (2002). Assembly theory applied to weed communities. *Weed Science*, *50*(1), 2–13. doi:10.1614/0043-1745

Brooks, D. R., Storkey, J., Clark, S. J., Firbank, L. G., Petit, S., & Woiwod, I. P. (2012). Trophic links between functional groups of arable plants and beetles are stable at a national scale. *Journal of Animal Ecology*, *81*(1), 4–13. doi:10.1111/j.1365-2656.2011.01897.x

Brym, Z. T., Lake, J. K., Allen, D., & Ostling, A. (2011). Plant functional traits suggest novel ecological strategy for an invasive shrub in an understorey woody plant community. *Journal of Applied Ecology*, *48*(5), 1098–1106. doi:10.1111/j.1365-2664.2011.02049.x

Butler, S. J., Brooks, D., Feber, R. E., Storkey, J., Vickery, J. A., & Norris, K. (2009). A cross-taxonomic index for quantifying the health of farmland biodiversity. *Journal of Applied Ecology*, *46*(6), 1154–1162. doi:10.1111/j.1365-2664.2009.01709.x

Ceulemans, T., Merckx, R., Hens, M., & Honnay, O. (2011). A trait-based analysis of the role of phosphorus vs nitrogen enrichment in plant species loss across North-west European grasslands. *Journal of Applied Ecology*, *48*(5), 1155–1163. doi:10.1111/j.1365-2664.2011.02023.x

Chollet, S., Rambal, S., Fayolle, A., Hubert, D., Foulquié, D., & Garnier, E. (2014). Combined effects of climate, resource availability, and plant traits on biomass produced in a Mediterranean rangeland. *Ecology*, *95*(3), 737–748. doi:10.1890/13-0751.1

Ciais, P., Reichstein, M., Viovy, N., Granier, A., Ogee, J., Allard, V.,... Valentini, R. (2005). Europe-wide reduction in primary productivity caused by the heat and drought in 2003. *Nature*, *437*(7058), 529–533.

Cingolani, A. M., Cabido, M., Gurvich, D. E., Renison, D., & Díaz, S. (2007). Filtering processes in the assembly of plant communities: are species presence and abundance driven by the same traits? *Journal of Vegetation Science*, *18*, 911–920.

Colbach, N., Granger, S., Guyot, S. H. M., & Mézière, D. (2014). A trait-based approach to explain weed species response to agricultural practices in a simulation study with a cropping system model. *Agriculture, Ecosystems & Environment*, *183*, 197–204. doi: http://dx.doi.org/10.1016/j.agee.2013.11.013

Cruz, P., De Quadros, F. L. F., Theau, J. P., Frizzo, A., Jouany, C., Duru, M., & Carvalho, P. C. F. (2010). Leaf traits as functional descriptors of the intensity of continuous grazing in native grasslands in the south of Brazil. *Rangeland Ecology & Management*, *63*(3), 350–358.

Cruz, P., & Soussana, J. -F. (1997). The nitrogen requirement of major agricultural crops. Mixed crops. In G. Lemaire (ed.), *Diagnosis of the nitrogen status in crops* (pp. 131–144). Heidelberg: Springer-Verlag.

Damour, G., Dorel, M., Quoc, H. T., Meynard, C., & Risède, J. -M. (2014). A trait-based characterization of cover plants to assess their potential to provide a set of ecological services in banana cropping systems. *European Journal of Agronomy*, *52*, 218–228. doi:10.1016/j.eja.2013.09.004

Damour, G., Garnier, E., Navas, M.-L., Dorel, M., & Risède, J.-M. (2015). Using functional traits to assess the services provided by cover plants: A review of potentialities in banana cropping systems. *Advances in Agronomy*, *134*, 81–133. doi:10.1016/bs.agron.2015.06.004

de Bello, F., Lavorel, S., Díaz, S., Harrington, R., Cornelissen, J. H. C., Bardgett, R. D.,... Harrison, P. A. (2010). Towards an assessment of multiple ecosystem processes and services via functional traits. *Biodiversity and Conservation*, *19*, 2873–2893.

Deen, W., Cousens, R., Warringa, J., Bastiaans, L., Carberry, P., Rebel, K., . . . Wang, E. (2003). An evaluation of four crop: weed competition models using a common data set. *Weed Research, 43*(2), 116–129. doi:10.1046/j.1365-3180.2003.00323.x

Díaz, S., Hector, A., & Wardle, D. A. (2009). Biodiversity in forest carbon sequestration initiatives: not just a side benefit. *Current Opinion in Environmental Sustainability, 1*(1), 55–60. doi:10.1016/j.cosust.2009.08.001

Díaz, S., Lavorel, S., Chapin, F. S., III, Tecco, P. A., Gurvich, D. E., & Grigulis, K. (2007a). Functional diversity—at the crossroads between ecosystem functioning and environmental filters. In J. G. Canadell, D. E. Pataki, & L. F. Pitelka (eds), *Terrestrial Ecosystems in a Changing World* (pp. 81–91). Berlin: Springer-Verlag.

Díaz, S., Lavorel, S., McIntyre, S., Falczuk, V., Casanoves, F., Milchunas, D. G., . . . Campbell, B. D. (2007b). Plant trait responses to grazing—A global synthesis. *Global Change Biology, 13*, 313–341.

Doré, T., Makowski, D., Malézieux, E., Munier-Jolain, N., Tchamitchian, M., & Tittonell, P. (2011). Facing up to the paradigm of ecological intensification in agronomy: revisiting methods, concepts, and knowledge. *European Journal of Agronomy, 34*(4), 197–210. doi:10.1016/j.eja.2011.02.006

Doyle, C. J. (1997). A review of the use of models of weed control in Integrated Crop Protection. *Agriculture, Ecosystems & Environment, 64*(2), 165–172.

Dumont, B., Rook, A. J., Coran, C., & Röver, K. U. (2007). Effects of livestock breed and grazing intensity on biodiversity and production in grazing systems. 2. Diet selection. *Grass and Forage Science, 62*(2), 159–171. doi:10.1111/j.1365-2494.2007.00572.x

Duru, M., Balent, G., Gibon, A., Magda, D., Theau, J.-P., Cruz, P., & Jouany, C. (1998). Fonctionnement et dynamique des prairies permanentes. Exemple des Pyrénées centrales. *Fourrages, 153*, 97–113.

Duru, M., Cruz, P., Al Haj Khaled R., Ducourtieux, C., & Theau, J. -P. (2008). Relevance of plant functional types based on leaf dry matter content for assessing digestibility of native grass species and species-rich grassland communities in spring. *Agronomy Journal, 100*, 1622–1630.

Duru, M., Cruz, P., Ansquer, P., & Navas, M. L. (2014a). Standing herbage mass: an integrated indicator of management practices for examining how fertility and defoliation regime shape the functional structure of species-rich grasslands. *Ecological Indicators, 36*, 152–159. doi:10.1016/j.ecolind.2013.07.015

Duru, M., Cruz, P., Jouany, C., & Theau, J.-P. (2010a). Herb'type©: un nouvel outil pour évaluer les services de production fournis par les prairies permanentes. *INRA Productions Animales, 23*(4), 319–332.

Duru, M., Cruz, P., & Theau, J.-P. (2010b). A simplified method for characterising agronomic services provided by species-rich grasslands *Crop and Pasture Science, 61*, 420–433. doi:10.1071/CP09296

Duru, M., & Ducrocq, H. (1997). A nitrogen and phosphorus herbage nutrient index as a tool for assessing the effect of N and P supply on the dry matter yield of permanent pastures. *Nutrient Cycling in Agroecosystems, 47*, 59–69.

Duru, M., Jouany, C., Le Roux, X., Navas, M. L., & Cruz, P. (2014b). From a conceptual framework to an operational approach for managing grassland functional diversity to obtain targeted ecosystem services: case studies from French mountains. *Renewable Agriculture and Food Systems, 29*(03), 239–254. doi:10.1017/S1742170513000306

Duru, M., Lemaire, G., & Cruz, P. (1997). Grasslands. In G. Lemaire (ed.), *Diagnosis of the Nitrogen Status in Crops* (pp. 59–72). Berlin: Springer-Verlag.

Duru, M., Theau, J.-P., Jouany, C., & Cruz, P. (2011). Optimiser les services fourragers des prairies permanentes. Des outils pour caractériser et gérer la diversité floristiques. *FaçSade - Résultats des recherches du département INRA-SAD, 36*, 1–4.

Evju, M., Austrheim, G., Halvorsen, R., & Mysterud, A. (2009). Grazing responses in herbs in relation to herbivore selectivity and plant traits in an alpine ecosystem. *Oecologia, 161*, 77–85. doi:10.1007/s00442-009-1358-1

Fagan, K. C., Pywell, R. F., Bullock, J. M., & Marrs, R. H. (2008). Do restored calcareous grasslands on former arable fields resemble ancient targets? The effect of time, methods, and environment on outcomes. *Journal of Applied Ecology, 45*(4), 1293–1303.

Fayolle, A. (2008). *Structure des communautés de plantes herbacées sur les Grands Causses: stratégies fonctionnelles des espèces et interactions interspécifiques*. PhD, SupAgro Montpellier, Montpellier.

Franke, A. C., Lotz, L. A. P., Van der Burg, W. J., & Van Overbeek, L. (2009). The role of arable weed seeds for agroecosystem functioning. *Weed Research, 49*, 131–141.

Fried, G. (2007). *Variations spatiales et temporelles des communautés adventices des cultures annuelles en France*. PhD, Université de Bourgogne, Dijon.

Fried, G., Chauvel, B., & Reboud, X. (2009). A functional analysis of large-scale temporal shifts from 1970 to 2000 in weed assemblages of sunflower crops in France. *Journal of Vegetation Science, 20*(1), 49–58. doi:10.1111/j.1654-1103.2009.05284.x

Fried, G., Kazakou, E., & Gaba, S. (2012). Trajectories of weed communities explained by traits associated with species' response to management practices. *Agriculture, Ecosystems & Environment, 158*, 147–155. doi:10.1016/j.agee.2012.06.005

Fried, G., Norton, L. R., & Reboud, X. (2008). Environmental and management factors determining weed species composition and diversity in France. *Agriculture, Ecosystems & Environment, 128*, 68–76. doi:10.1016/j.agee.2008.05.003

Gaba, S., Bretagnolle, F., Rigaud, T., & Philippot, L. (2014a). Managing biotic interactions for ecological intensification of agroecosystems. *Frontiers in Ecology and Evolution, 2*. doi:10.3389/fevo.2014.00029

Gaba, S., Fried, G., Kazakou, E., Chauvel, B., & Navas, M.-L. (2014b). Agroecological weed control using a functional approach: a review of cropping systems diversity. *Agronomy for Sustainable Development, 34*(1), 103–119. doi:10.1007/s13593-013-0166-5

Gaba, S., Lescourret, F., Boudsocq, S., Enjalbert, J., Hinsinger, P., Journet, E.-P., ... Ozier-Lafontaine, H. (2015). Multiple cropping systems as drivers for providing multiple ecosystem services: from concepts to design. *Agronomy for Sustainable Development, 35*(2), 607–623. doi:10.1007/s13593-014-0272-z

Gardarin, A., Dürr, C., & Colbach, N. (2010a). Effects of seed depth and soil aggregates on the emergence of weeds with contrasting seed traits. *Weed Research, 50*(2), 91–101. doi:10.1111/j.1365-3180.2009.00757.x

Gardarin, A., Dürr, C., & Colbach, N. (2011). Prediction of germination rates of weed species: Relationships between germination speed parameters and species traits. *Ecological Modelling, 222*(3), 626–636. doi:10.1016/j.ecolmodel.2010.10.005

Gardarin, A., Dürr, C., Mannino, M. R., Busset, H., & Colbach, N. (2010b). Seed mortality in the soil is related to seed coat thickness. *Seed Science Research, 20*(4), 243–256. doi:10.1017/s0960258510000255

Gardarin, A., Garnier, E., Carrère, P., Cruz, P., Andueza, D., Bonis, A., ... Kazakou, E. (2014). Plant trait–digestibility relationships across management and climate gradients in permanent grasslands. *Journal of Applied Ecology, 51*, 1207–1217. doi:10.1111/1365-2664.12293

Garnier, E., Lavorel, S., Ansquer, P., Castro, H., Cruz, P., Dolezal, J., ... Zarovali, M. (2007). Assessing the effects of land use change on plant traits, communities, and ecosystem functioning in grasslands: a standardized methodology and lessons from an application to 11 European sites. *Annals of Botany, 99*, 967–985.

Garnier, E., & Navas, M.-L. (2011). Assessing the functional role of plant diversity in grasslands: a trait-based approach. In G. Lemaire, J. G. Hodgson, & A. Chabbi (eds), *Grassland Productivity and Ecosystem Services* (pp. 138–147). Wallingford, UK: CAB Int.

Garnier, E., & Navas, M.-L. (2012). A trait-based approach to comparative functional plant ecology: concepts, methods, and applications for agroecology. A review. *Agronomy for Sustainable Development, 32*, 365–399. doi:10.1007/s13593-011-0036-y

Ginane, C., Dumont, B., Baumont, R., Prache, S., Fleurance, G., & Farruggia, A. (2008). *Comprendre le comportement alimentaire des herbivores au pâturage: intérêts pour l'élevage et l'environnement*. Paper presented at the 15ème journées de Rencontres Recherches Ruminants (3R).

Gomez-Aparicio, L., Zavala, M. A., Bonet, F. J., & Zamora, R. (2009). Are pine plantations valid tools for restoring Mediterranean forests? An assessment along abiotic and biotic gradients. *Ecological Applications, 19*(8), 2124–2141. doi:10.1890/08-1656.1

Grime, J. P. (1974). Vegetation classification by reference to strategies. *Nature, 250*, 26–31.

Grime, J. P. (1998). Benefits of plant diversity to ecosystems: immediate, filter, and founder effects. *Journal of Ecology, 86*, 902–910.

Guérin, G., Moulin, C., & Tchakérian, E. (2009). Les apports de l'approche des systèmes pastoraux à la réflexion sur la gestion des ressources des zones herbagères. *Fourrages, 200*, 489–498.

Gunton, R. M., Petit, S., & Gaba, S. (2011). Functional traits relating arable weed communities to crop characteristics. *Journal of Vegetation Science, 22*(3), 541–550. doi:10.1111/j.1654-1103.2011.01273.x

Hawes, C., Begg, G. S., Squire, G. R., & Iannetta, P. P. M. (2005). Individuals as the basic accounting unit in studies of ecosystem function: functional diversity in shepherd's purse, *Capsella*. *Oikos, 109*(3), 521–534. doi:10.1111/j.0030-1299.2005.13853.x

Hawes, C., Haughton, A. J., Bohan, D. A., & Squire, G. R. (2009). Functional approaches for assessing plant and invertebrate abundance patterns in arable systems. *Basic and Applied Ecology, 10*(1), 34–42. doi:10.1016/j.baae.2007.11.007

Herault, B., Honnay, O., & Thoen, D. (2005). Evaluation of the ecological restoration potential of plant communities in Norway spruce plantations using a life-trait based approach. *Journal of Applied Ecology, 42*(3), 536–545. doi:10.1111/j.1365-2664.2005.01048.x

Hodgson, J. G., Montserrat-Martí, G., Cerabolini, B., Ceriani, R. M., Maestro-Martí, M., Peco, B., ... Villar-Salvador, P. (2005a). A functional method for classifying European grasslands for use in joint ecological and economic studies. *Basic and Applied Ecology, 6*, 119–131.

Hodgson, J. G., Montserrat-Martí, G., Tallowin, J., Thompson, K., Díaz, S., Cabido, M., ... Zak, M. R. (2005b). How much will it cost to save grassland diversity? *Biological Conservation, 122*, 263–273.

Huyghe, C. (2010). Usages émergents des surfaces prairiales et des espèces fourragères: les messages importants. *Fourrages*, (203), 213–219.

Jouany, C., Stroia, M. C., Farruggia, A., & Duru, M. (2002). Plant and soil indicators for P management in grassland. *Grassland Science in Europe, 7*, 698–699.

Kahmen, S., & Poschlod, P. (2004). Plant functional trait responses to grassland succession over 25 years. *Journal of Vegetation Science, 15*, 21–32.

Kahmen, S., Poschlod, P., & Schreider, K.-F. (2002). Conservation management of calcareous grasslands. Changes in plant species composition and response of functional traits during 25 years. *Biological Conservation, 104*, 319–328.

Kardol, P., & Wardle, D. A. (2010). How understanding above-ground–below-ground linkages can assist restoration ecology. *Trends in Ecology and Evolution, 25*(11), 670–679.

Kéfi, S., Rietkerk, M., Alados, C. L., Pueyo, Y., Papanastasis, V. P., ElAich, A., & de Ruiter, P. C. (2007). Spatial vegetation patterns and imminent desertification in Mediterranean arid ecosystems. *Nature, 449*, 213–218.

Kleyer, M. (1999). Distribution of plant functional types along gradients of disturbance intensity and resource supply in an agricultural landscape. *Journal of Vegetation Science, 10*, 697–708.

Klumpp, K., & Soussana, J.-F. (2009). Using functional traits to predict grassland ecosystem change: a mathematical test of the response-and-effect trait approach. *Global Change Biology, 15*, 2921–2934.

Koohafkan, P., Altieri, M. A., & Gimenez, E. H. (2012). Green Agriculture: foundations for biodiverse, resilient, and productive agricultural systems. *International Journal of Agricultural Sustainability, 10*(1), 61–75. doi:10.1080/14735903.2011.610206

Kropff, M. J., Weaver, S. E., & Smits, M. A. (1992). Use of ecophysiological models for crop-weed interference: relations amongst weed density, relative time of weed emergence, relative leaf area, and yield loss. *Weed Science, 40*, 296–301.

Lavorel, S., Díaz, S., Cornelissen, J. H. C., Garnier, E., Harrison, S. P., McIntyre, S.,... Urcelay, C. (2007). Plant functional types: are we getting any closer to the Holy Grail? In J. Canadell, D. Pataki, & L. Pitelka (eds), *Terrestrial Ecosystems in a Changing World* (pp. 149–164). Berlin: Springer-Verlag.

Lavorel, S., & Garnier, E. (2002). Predicting changes in community composition and ecosystem functioning from plant traits: revisiting the Holy Grail. *Functional Ecology, 16*, 545–556.

Lavorel, S., & Grigulis, K. (2012). How fundamental plant functional trait relationships scale-up to trade-offs and synergies in ecosystem services. *Journal of Ecology, 100*(1), 128–140. doi:10.1111/j.1365-2745.2011.01914.x

Lavorel, S., Grigulis, K., Lamarque, P., Colace, M.-P., Garden, D., Girel, J.,... Douzet, R. (2011). Using plant functional traits to understand the landscape distribution of multiple ecosystem services. *Journal of Ecology, 99*, 135–147.

Le Roux, X., Barbault, R., Baudry, J., Burel, F., Doussan, I., Garnier, E.,... Trommetter, M. (eds) (2009). *Agriculture et biodiversité. Valoriser les synergies*. Versailles: Quae.

Lemaire, G., & Gastal, F. (1997). N uptake and distribution in plant canopies. In G. Lemaire (ed.), *Diagnosis of the Nitrogen Status in Crops* (pp. 3–41). Heidelberg: Springer-Verlag.

Ling-Hwa, T., & Morishima, H. (1997). Genetic characterization of weedy rices and the inference on their origins. *Breeding Science, 47*(2), 153–160.

Lizée, M.-H., Mauffrey, J.-F., Tatoni, T., & Deschamps-Cottin, M. (2011). Monitoring urban environments on the basis of biological traits. *Ecological Indicators, 11*(2), 353–361. doi:10.1016/j.ecolind.2010.06.003

Lloyd, K. M., Pollock, M. L., Mason, N. W. H., & Lee, W. G. (2010). Leaf trait-palatability relationships differ between ungulate species: evidence from cafeteria experiments using native tussock grasses. *New Zealand Journal of Ecology, 34*(2), 219–226.

Lososova, Z., Chytry, M., Cimalova, S., Kropac, Z., Otypkova, Z., Pysek, P., & Tichy, L. (2004). Weed vegetation of arable land in Central Europe: gradients of diversity and species composition. *Journal of Vegetation Science, 15*, 415–422.

Lososova, Z., Chytry, M., & Kühn, I. (2008). Plant attributes determining the regional abundance of weeds on central European arable land. *Journal of Biogeography, 35*(1), 177–187. doi:10.1111/j.1365-2699.2007.01778.x

Lososova, Z., Chytry, M., Kühn, I., Hajek, O., Horakova, V., Pysek, P., & Tichy, L. (2006). Patterns of plant traits in annual vegetation of man-made habitats in central Europe. *Perspectives in Plant Ecology, Evolution and Systematics, 8*(2), 69–81. doi:10.1016/j.ppees.2006.07.001

Louault, F., Pillar, V. D., Aufrère, J., Garnier, E., & Soussana, J.-F. (2005). Plant traits and functional types in response to reduced disturbance in a semi-natural grassland. *Journal of Vegetation Science, 16*, 151–160.

Maillet-Mezeray, J., & Gril, J.-J. (2010). Zones tampons: état des connaissances techniques et mise en oeuvre. *Fourrages, 202*, 111–116.

Malézieux, E., Crozat, Y., Dupraz, C., Laurans, M., Makowski, D., Ozier-Lafontaine, H.,... Valantin-Morison, M. (2009). Mixing plant species in cropping systems: concepts, tools, and models. A review. *Agronomy for Sustainable Development, 29*, 43–62.

Mason, N. W. H., Peltzer, D. A., Richardson, S. J., Bellingham, P. J., & Allen, R. B. (2010). Stand development moderates effects of ungulate exclusion on foliar traits in the forests of New Zealand. *Journal of Ecology, 98*(6), 1422–1433. doi:10.1111/j.1365-2745.2010.01714.x

McDonald, A. J., Riha, S. J., & Ditommaso, A. (2010). Early season height differences as robust predictors of weed growth potential in maize: new avenues for adaptive management? *Weed Research, 50*(2), 110–119. doi:10.1111/j.1365-3180.2009.00759.x

McIntyre, S. (2008). The role of plant leaf attributes in linking land use to ecosystem function in temperate grassy vegetation. *Agriculture, Ecosystems & Environment, 128*(4), 251–258. doi:10.1016/j.agee.2008.06.015

McNaughton, S. J., Milchunas, D. G., & Frank, D. A. (1996). How can net primary productivity be measured in grazing ecosystems? *Ecology, 77*, 974–977.

Meiss, H., Mediene, S., Waldhardt, R., Caneill, J., & Munier-Jolain, N. (2010). Contrasting weed species composition in perennial alfalfas and six annual crops: implications for integrated weed management. *Agronomy for Sustainable Development, 30*(3), 657–666. doi:10.1051/agro/2009043

Meynard, J.-M., Dedieu, B., & Bos, A. P. B. (2012). Re-design and co-design of farming systems. An overview of methods and practices. In I. Darnhofer, D. Gibon, & B. Dedieu (eds), *Farming systems research into the 21st century: the new dynamic* (pp. 407–432). Dordrecht: Springer.

Mládek, J., Mládková, P., Hejcmanová, P., Dvorský, M., Pavlu, V., De Bello, F., . . . Pakeman, R. J. (2013). Plant trait assembly affects superiority of grazer's foraging strategies in species-rich grasslands. *PLoS ONE, 8*(7), e69800. doi:10.1371/journal.pone.0069800

Molénat, G., Foulquié, D., Autran, P., Bouix, J., Hubert, D., Jacquin, M., . . . Bibé, B. (2005). Pour un élevage ovin allaitant performant et durable sur parcours: un système expérimental sur le Causse du Larzac. *INRA Productions Animales, 18*, 323–338.

Moog, D., Kahmen, S., & Poschlod, P. (2005). Application of CSR- and LHS-strategies for the distinction of differently managed grasslands. *Basic and Applied Ecology, 6*(2), 133–143. doi:10.1016/j.baae.2005.01.005

Moonen, A.-C., & Bàrberi, P. (2008). Functional biodiversity: an agroecosystem approach. *Agriculture, Ecosystems & Environment, 127*(1–2), 7–21. doi:10.1016/j.agee.2008.02.013

Navas, M.-L. (1991). Using plant population biology in weed research: a strategy to improve weed management. *Weed Research, 31*(4), 171–179. doi:10.1111/j.1365–3180.1991.tb01756.x

Navas, M.-L. (2012). Trait-based approaches to unravelling the assembly of weed communities and their impact on agro-ecosystem functioning. *Weed Research, 52*(6), 479–488. doi:10.1111/j.1365-3180.2012.00941.x

Navas, M.-L., Friess, N., & Maillet, J. (1998). Influence of cucumber mosaic virus infection on the growth response of *Portulaca oleracea* (purslane) and *Stellaria media* (chickweed) to nitrogen availability. *New Phytologist, 139*(2), 301–309. doi:10.1046/j.1469-8137.1998.00197.x

Navas, M.-L., & Violle, C. (2009). Plant traits related to competition: how do they shape the functional diversity of communities? *Community Ecology, 10*(1), 131–137. doi:10.1556/ComEc.10.2009.1.15

Ordoñez, J. C., van Bodegom, P. M., Witte, J.-P.M., Wright, I. J., Reich, P. B., & Aerts, R. (2009). A global study of relationships between leaf traits, climate, and soil measures of nutrient fertility. *Global Ecology and Biogeography, 18*, 137–149.

Orwin, K. H., Buckland, S. M., Johnson, D., Turner, B. L., Smart, S., Oakley, S., & Bardgett, R. D. (2010). Linkages of plant traits to soil properties and the functioning of temperate grassland. *Journal of Ecology, 98*(5), 1074–1083. doi:10.1111/j.1365-2745.2010.01679.x

Pakeman, R. J. (2011). Multivariate identification of plant functional response and effect traits in an agricultural landscape. *Ecology, 92*(6), 1353–1365. doi:10.1890/10-1728.1

Pakeman, R. J., Lepš, J., Kleyer, M., Lavorel, S., Garnier, E., & the Vista Consortium (2009). Relative climatic, edaphic, and management controls of plant functional trait signatures. *Journal of Vegetation Science, 20*(1), 148–159. doi:10.1111/j.1654-1103.2009.05548.x

Pakeman, R. J., & Marriott, C. A. (2010). A functional assessment of the response of grassland vegetation to reduced grazing and abandonment. *Journal of Vegetation Science, 21*(4), 683–694. doi:10.1111/j.1654-1103.2010.01176.x

Pe'er, G., Dicks, L. V., Visconti, P., Arlettaz, R., Báldi, A., Benton, T. G., . . . Scott, A. V. (2014). EU agricultural reform fails on biodiversity. *Science, 344*(6188), 1090–1092. doi:10.1126/science.1253425

Pérez-Ramos, I. M., Roumet, C., Cruz, P., Blanchard, A., Autran, P., & Garnier, E. (2012). Evidence for a 'plant community economics spectrum' driven by nutrient and water limitations in a Mediterranean rangeland of southern France. *Journal of Ecology, 100*(6), 1315–1327. doi:10.1111/1365-2745.12000

Petit, S., Boursault, A., Le Guilloux, M., Munier-Jolain, N., & Reboud, X. (2011). Weeds in agricultural landscapes. A review. *Agronomy and Sustainable Development, 31*, 309–317. doi:10.1051/agro/2010020

Plantureux, S., Bellon, S., Burel, F., Chauvel, B., Dajoz, I., Guy, P., . . . Viaux, P. (2009). Prospective Agriculture Biodiversité (pp. 310): INRA Département Environnement Agronomie. http://prodinra.inra.fr/?locale=fr#!ConsultNotice:207204

Pontes, L. D. S., Soussana, J.-F., Louault, F., Andueza, D., & Carrère, P. (2007). Leaf traits affect the above-ground productivity and quality of pasture grasses. *Functional Ecology*, (21), 844–853.

Poorter, H., & Bergkotte, M. (1992). Chemical composition of 24 wild species differing in relative growth rate. *Plant, Cell and Environment, 15*, 221–229.

Pywell, R. F., Bullock, J. M., Roy, D. B., Warman, L., Walker, K. J., & Rothery, P. (2003). Plant traits as predictors of performance in ecological restoration. *Journal of Applied Ecology, 40*(1), 65–77. doi:10.1046/j.1365-2664.2003.00762.x

Quétier, F., Lavorel, S., Thuiller, W., & Davies, I. (2007a). Plant-trait-based modelling assessment of ecosystem-service sensitivity to land-use change. *Ecological Applications, 17*(8), 2377–2386.

Quétier, F., Thébault, A., & Lavorel, S. (2007b). Plant traits in a state and transition framework as markers of ecosystem response to land-use change. *Ecological Monographs, 77*, 33–52.

Reitalu, T., Johansson, L. J., Sykes, M. T., Hall, K., & Prentice, H. C. (2010). History matters: village distances, grazing, and grassland species diversity. *Journal of Applied Ecology, 47*(6), 1216–1224. doi:10.1111/j.1365-2664.2010.01875.x

Ricotta, C., & Moretti, M. (2011). CWM and Rao's quadratic diversity: a unified framework for functional ecology. *Oecologia, 167*(1), 181–188. doi:10.1007/s00442-011-1965-5

Robertson, G. P., & Swinton, S. M. (2005). Reconciling agricultural productivity and environmental integrity: a grand challenge for agriculture. *Frontiers in Ecology and the Environment, 3*(1), 38–46. doi:10.1890/1540-9295

Römermann, C., Bernhardt-Romermann, M., Kleyer, M., & Poschlod, P. (2009). Substitutes for grazing in semi-natural grasslands—do mowing or mulching represent valuable alternatives to maintain vegetation structure? *Journal of Vegetation Science, 20*(6), 1086–1098.

Rook, A. J., Dumont, B., Isselstein, J., Osoro, K., Wallis-DeVries, M. F., Parente, G., & MIlls, J. (2004). Matching type of livestock to desired biodiversity outcomes in pastures—a review. *Biological Conservation, 119*, 137–150.

Rossignol, N., Chadoeuf, J., Carrère, P., & Dumont, B. (2011). A hierarchical model for analysing the stability of vegetation patterns created by grazing in temperate pastures. *Applied Vegetation Science, 14*, 189–199.

Saatkamp, A., Römermann, C., & Dutoit, T. (2010). Plant functional traits show non-linear response to grazing. *Folia Geobotanica, 45*(3), 239–252. doi:10.1007/s12224-010-9069-2

Sandel, B., Goldstein, L. J., Kraft, N. J., Okie, J. G., Shuldman, M. I., Ackerly, D. D., ... Suding, K. N. (2010). Contrasting trait responses in plant communities to experimental and geographic variation in precipitation. *New Phytologist, 188*(2), 565–575. doi:10.1111/j.1469-8137.2010.03382.x

Sasaki, T., Okubo, S., Okayasu, T., Jamsran, U., Ohkuro, T., & Takeuchi, K. (2009). Two-phase functional redundancy in plant communities along a grazing gradient in Mongolian rangelands. *Ecology, 90*(9), 2598–2608. doi:10.1890/08-0144.1

Scherber, C., Heimann, J., Koeler, G., Mitschunas, N., & Weisser, W. W. (2010). Functional identity versus species richness: herbivory resistance in plant communities. *Oecologia, 163*(3), 707–717. doi:10.1007/s00442-010-1625-1

Sciama, D., Augusto, L., Dupouey, J.-L., Gonzalez, M., & Moares Domíngue, C. (2009). Floristic and ecological differences between recent and ancient forests growing on non-acidic soils. *Forest Ecology and Management, 258*, 600–608.

Sinclair, C., & Hoffmann, A. A. (2003). Monitoring salt stress in grapevines: are measures of plant trait variability useful? *Journal of Applied Ecology, 40*(5), 928–937. doi:10.1046/j.1365-2664.2003.00843.x

Smith, R. G. (2006). Timing of tillage is an important filter on the assembly of weed communities. *Weed Science, 54*(4), 705–712. doi:10.1614/WS-05-177R1.1

Smith, R. G., & Gross, K. L. (2007). Assembly of weed communities along a crop diversity gradient. *Journal of Applied Ecology, 44*(5), 1046–1056. doi:10.1111/j.1365-2664.2007.01335.x

Smith, R. G., Mortensen, D. A., & Ryan, M. R. (2009). A new hypothesis for the functional role of diversity in mediating resource pools and weed-crop competition in agroecosystems. *Weed Research, 50*, 37–48. doi:10.1111/j.1365-3180.2009.00745.x

Soussana, J.-F., & Duru, M. (2007). Grassland science in Europe facing new challenges: biodiversity and global environmental change. *CAB Reviews: Perspectives in Agriculture, Veterinary Science, Nutrition and Natural Resources, 2*(2), 1–11. doi:10.1079/PAVSNNR20071072

Stein, C., Unsicker, S. B., Kahmen, A., Wagner, M., Audorff, V., Auge, H., ... Weisser, W. W. (2010). Impact of invertebrate herbivory in grasslands depends on plant species diversity. *Ecology, 91*(6), 1639–1650.

Storkey, J. (2006). A functional group approach to the management of UK arable weeds to support biological diversity. *Weed Research, 46*(6), 513–522. doi:10.1111/j.1365-3180.2006.00528.x

Storkey, J., Brooks, D., Haughton, A. J., Hawes, C., Smith, B. M., & Holland, J. M. (2013). Using functional traits to quantify the value of plant communities to invertebrate ecosystem service providers in arable landscapes. *Journal of Ecology, 101*(1), 38–46. doi:10.1111/1365-2745.12020

Storkey, J., & Cussans, J. W. (2007). Reconciling the conservation of in-field biodiversity with crop production using a simulation model of weed growth and competition. *Agriculture, Ecosystems & Environment, 122*(2), 173–182.

Storkey, J., Moss, S. R., & Cussans, J. W. (2010). Using assembly theory to explain changes in a weed flora in response to agricultural intensification. *Weed Science, 58*(1), 39–46. doi:10.1614/ws-09-096.1

Tall, L., Caraco, N., & Maranger, R. (2011). Denitrification hot spots: dominant role of invasive macrophyte *Trapa natans* in removing nitrogen from a tidal river. *Ecological Applications*, *21*(8), 3104–3114.

Thornthwaite, C. W. (1948). An approach toward a rational classification of climate. *Geographical Review*, *38*, 55–94.

Van Calster, H., Vandenberghe, R., Ruysen, M., Verheyen, K., Hermy, M., & Decocq, G. (2008). Unexpectedly high 20th century floristic losses in a rural landscape in northern France. *Journal of Ecology*, *96*(5), 927–936. doi:10.1111/j.1365-2745.2008.01412.x

Volaire, F. (2008). Plant traits and functional types to characterise drought survival of pluri-specific perennial herbaceous swards in Mediterranean areas. *European Journal of Agronomy*, *29*(2–3), 116–124. doi:10.1016/j.eja.2008.04.008

Volaire, F., Norton, M. R., & Lelièvre, F. (2009). Summer drought survival strategies and sustainability of perennial temperate forage grasses in Mediterranean areas. *Crop Science*, *49*, 2386–2392.

von Gillhaussen, P., Rascher, U., Jablonowski, N. D., Plückers, C., Beierkuhnlein, C., & Temperton, V. M. (2014). Priority effects of time of arrival of plant functional groups override sowing interval or density effects: a grassland experiment. *PLoS ONE*, *9*(1), e86906. doi:10.1371/journal.pone.0086906

Walker, K. J., Preston, C. D., & Boon, C. R. (2009). Fifty years of change in an area of intensive agriculture: plant trait responses to habitat modification and conservation, Bedfordshire, England. *Biodiversity and Conservation*, *18*(13), 3597–3613.

Westoby, M. (1998). A leaf-height-seed (LHS) plant ecology strategy scheme. *Plant and Soil*, *199*, 213–227.

Wezel, A., Bellon, S., Doré, T., Francis, C., Vallod, D., & David, C. (2009). Agroecology as a science, a movement, and a practice. A review. *Agronomy for Sustainable Development*, *29*, 503–515. doi:10.1051/agro/2009004

White, P. S., & Pickett, S. T. A. (1985). Natural disturbance and patch dynamics, an introduction. In S. T. A. Pickett & P. S. White (eds), *The Ecology of Natural Disturbance and Patch Dynamics* (pp. 3–13). New York: Academic Press.

Zhang, W., Ricketts, T. H., Kremen, C., Carney, K., & Swinton, S. M. (2007). Ecosystem services and dis-services to agriculture. *Ecological Economics*, *64*(2), 253–260. doi: http://dx.doi.org/10.1016/j.ecolecon.2007.02.024

CHAPTER 9

Managing functional diversity data

9.1 Introduction

The previous chapters have illustrated the large variety of data used in the field of plant functional diversity: characteristics of organisms, environmental factors (resources, disturbance regimes, temperature, etc.), community structure, and ecosystem services and properties, all of which span different spatial and temporal scales. As has been shown throughout this book, combining these heterogeneous types of data is necessary to answer pending ecological questions relating to functional diversity. Additionally, the intense research activity on this theme over the last three decades has led to the production of numerous large data sets. This progress is common to many aspects of ecology, which is transforming before our eyes into a 'data-intensive science' (Kelling et al., 2009; Michener & Jones, 2012), that is to say, a discipline intensively generating and utilizing large amounts of data. The issues raised, on the one hand, by the heterogeneity of ecological data, and on the other hand, by the need to treat large volumes of data, have led to the emergence of '*ecoinformatics**' (In this chapter, definitions for terms in italics followed by an asterisk are given in Table 9.1), defined as 'a field of research and development focused on the interface between ecology, computer science, and information technology' (Jones et al., 2006). The objective of ecoinformatics is to allow scientists 'to generate new knowledge through innovative tools and approaches for discovering, managing, integrating, analysing, visualizing, and preserving relevant biological, environmental, and socioeconomic data and information' (Michener & Jones, 2012).

Advances in ecoinformatics have been substantial over the last ten or so years, but these have not yet been widely adopted in ecological research or teaching. In this chapter, we will present a number of recent initiatives contributing to the effort to improve the management of plant functional diversity data focusing, as argued in Chapter 2, on the level of the organism. We begin with a brief summary of the current state of availability of plant trait data.

9.2 Availability of trait data

Over the last three decades, the development of diverse screening methods (Grime & Hunt, 1975; Hendry & Grime, 1993; Keddy, 1992), and the compilation of data from numerous different sources, has led to the development of large plant trait databases allowing numerous questions related to plant functioning and evolution to be addressed (Enquist & Niklas, 2002; Reich et al., 2006; Wright et al., 2004; cf. Chapter 3), ranging from investigating the relationships between traits and environmental factors at different scales of time and space (Ordoñez et al., 2009; Pakeman et al., 2008; Reich & Oleksyn, 2004; cf. Chapters 4 and 8), to the relationships between diversity, community structure, and ecosystem functioning (Garnier et al., 2007; Makkonen et al., 2012; Swenson et al., 2012; cf. Chapters 5 to 8).

While these data sets are of potentially great value to the scientific community, their compatibility, and their availability to researchers others than those who have compiled them, remain problematic, as they are often constructed using heterogeneous formats and terminology or are not necessarily publicly available (see e.g. Kattge et al., 2011b). A first phase of pooling such data consists of making available certain data sets by

Table 9.1 A concise (eco)informatics glossary

Term	Definition
Automated reasoning	Algorithms and software systems for automating the computation of logical inferences (typically refers to procedures for deductive reasoning)
Class	A class regroups an ensemble of objects having the same properties and behaviours (i.e. the evolution of properties over time)
Concept	An abstract, general, objective, and stable verbal representation of a real world object
Conceptualization	The action of elaborating a concept or an ensemble of communicable concepts
Controlled vocabulary	A list of terms and their definitions, established by a community of users. Controlled vocabularies can be considered as the language spoken by researchers within a scientific domain so as to clearly understand each other
Data integration	The ensemble of procedures aiming to facilitate the correspondence and combination of information derived from different sources, in a pertinent and useful manner
Data model	A data model organizes data elements and standardizes how the data elements relate to one another
Distributed (data) sources	Data sources located on different computer servers, or in different locations on the same server
Ecoinformatics	A field of research and development focused on the interface between ecology, computer science, and information technology
Entity	A real world, existing object, but representable only by a concept
Instance	A real world, existing object, with behaviour corresponding to the class to which it belongs: all of the instances of a class share the same characteristics
Interoperability	The capacity of a product or system to function with other existing or future products or systems, without restriction or requiring further modification
Mapping	Procedure allowing for the matching of corresponding data between two *data models** (in particular between data located in different sources and between the concepts used to represent them in ontologies)
Metadata	Information about data, which is necessary to understand and interpret these data
Ontology	A formal representation of the concepts of a domain of interest and the relationships between these concepts
Open access	The making available on line of digital content, which can itself be open access (for example a Creative Commons licence), or subject to various intellectual property restrictions
Semantic integration	The process of inter-relating information from diverse sources
Semantic web initiative	A broadly scoped effort led by the world wide web consortium (W3C) to enable software systems to easily find, analyse, share, and integrate web content. The W3C has created many technology specifications for extending web content, including via ontologies using the web ontology language (OWL)
Semantics	The study of meaning. It focuses on the relation between signifiers, like words, phrases, signs, and symbols, and what they stand for, their denotation
Standard	A published reference whose diffusion and utilization are widespread and recognized by a large proportion of those working in the domain
Thesaurus	Allows for the organization and structuring of a controlled vocabulary based on semantic rules of hierarchy, association, or equivalence
Workflow	The coordinated conceptualization and automated control of the ensemble of tasks to be accomplished by the various actors implicated in a given operational process

attaching them as annexes to the original publication (e.g. the wood density data set: Chave et al., 2009; or the Glopnet data set: Wright et al., 2004, cf. Chapter 3), or by placing them on dedicated sites (e.g. 'ecological archives' of the Ecological Society of America, with, for example, traits relative to responses to fire for Mediterranean basin species: Paula et al., 2009). In parallel, a number of initiatives have allowed the development of publicly accessible trait databases: amongst those which have been described in publications should be noted: EcoFlora (http://www.ecoflora.co.uk), concerning 3842 higher plant species from the British Isles (Fitter & Peat, 1994); CLO-PLA (CLOnal PLAnts: http://clopla.butbn.cas.cz), a database concerning the architecture of clonal plants from central Europe (Klimeš & Klimešová, 1999); BiolFlor (http://www.biolflor.de), concerning the traits and ecological preferences of 3660 species of the flora of Germany (Klotz et al., 2002); BASECO (http://www.baseco.imbe.cnrs.fr), concerning the traits of Mediterranean species (Gachet et al., 2005); and LEDA (http://www.leda-traitbase.org) which regroups life-history traits of the flora of northwest Europe (Kleyer et al., 2008). The TRY database (Kattge et al., 2011a) represents a particular case amongst plant trait databases as it regroups and *semantically integrates** (at least for most traits used) data sets collected by a large number of different contributors, including all of the before mentioned databases. New data are being continuously added since the launching of the initiative in 2007 with the aim to provide a global archive of plant trait data to support ecology and biodiversity science. In May 2015, the database contained about 5 600 000 trait records concerning more than 1000 traits and 100 000 species from a large variety of different biomes and geographic zones worldwide.

However, despite the large amounts of data included in these databases, they do not contain all the potentially available plant trait data. In addition, due to incomplete metadata and heterogeneous terminologies, it is often difficult to integrate data from different sources. This is why it is necessary to develop a global initiative to encourage the development of consistent and accepted standards to facilitate a systematic recovery, integration, storage, management, and public availability of plant trait data. The establishment of such an initiative would require particular emphasis on the development of an accepted standard for *metadata**, an agreed *controlled vocabulary** and *ontology** for plant traits to facilitate *interoperability** and seamless *data integration** of *distributed data sources**. Beyond this objective of integration, these efforts would result in substantial added-value for the data: they could indeed then be used in contexts different than those for which they were initially collected, or serve to address questions of a broader scope than would be possible if each data set remained in isolation. The following section presents the elements of an ecoinformatics initiative which fulfils the objectives of providing the necessary standards to facilitate integration, sharing, and reutilization of data. The implementation of this initiative is only in its beginnings for data concerning plant functional diversity. Some of the aspects presented here are of a generic nature, while others are more specifically linked to trait data.

9.3 Ecoinformatics and plant functional diversity

One of the general characteristics of ecological data, and that of functional diversity data in particular, is their considerable heterogeneity, which constitutes a major obstacle to both their integration and their reutilization (Jones et al., 2006; Reichman et al., 2011). This heterogeneity results notably from the manner in which ecological research is conducted: while there exist some coordinated studies carried out at large scales, the vast majority of ecological data is collected by researchers working in an independent manner, which leads to the existence of numerous small data sets, the form of which is often specific to the area being addressed, or to each research group or even individual scientist (Heidorn, 2008). Two types of heterogeneity are generally recognized. The first is of a syntactic nature, and results from the format in which the data are stored and the way in which they are managed: with the data being formatted as tables and annexes in reports and publications, as files in spreadsheets, as tables in relational databases, etc. The second is of a *semantic** nature, and corresponds to the diverse nature of ecological data, the multiplicity of terms, and at a larger scale the *concepts** used in ecology, which constitutes a major impediment towards the *interoperability** of existing data sets. The development of ecoinformatics has resulted in the realization that these obstacles should be removed, and data sharing and integration should become high priorities in ecology. In this, ecology is following the path followed by molecular biology some three decades ago, at the beginning of the development of bioinformatics (Stein, 2008).

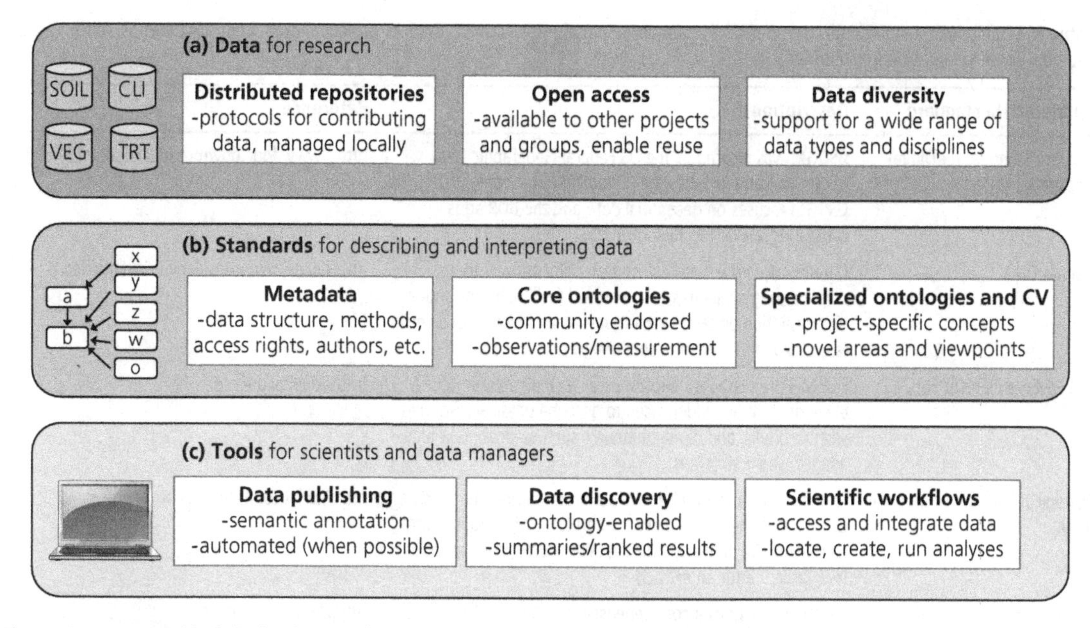

Figure 9.1 A proposed high-level architecture for ecological and environmental data management is shown consisting of three primary levels. Taken from Madin et al. (2008). Reproduced with permission from Elsevier.
(a) The data are stored in distributed data sources, symbolized by the cylinders labelled 'SOIL', 'CLI' (climate), 'VEG' (vegetation), 'TRT' (traits) in this example.
(b) These data are interfaced with a mediation system based on standard metadata, ontologies, and controlled vocabularies (CV).
(c) Software applications use community-endorsed controlled vocabularies and ontologies, and metadata standards from the middle level (b), to provide tools that are more effective for publishing, querying, integrating, and analysing data.
Ontologies are separated into framework ontologies and domain-specific ontologies, enabling contributions from multiple research groups, disciplines, and individuals. Cross-disciplinary data are maintained in local repositories, but made accessible to the broader research community through distributed systems based on shared, open protocols. Example repositories include the 'Long Term Ecological Research' network (LTER) in the United States (www.lternet.edu).

Figure 9.1, adapted from Madin et al. (2008), presents different elements of the problem to be solved: the upper part (in dark grey) gives some general characteristics concerning ecological data: they are often of heterogeneous nature (high data diversity), stored in dispersed archives, and for them to be made available for widespread use, they would need to be available under *open access**, which is currently the case for only a small fraction of existing data (Reichman et al., 2011). The lower part of the figure (in light grey) represents different types of usage of these data by scientists, managers, or stakeholders to answer scientific questions and/or propose, for example, environmental management plans. These uses require the establishment of tools to find and select data from diverse sources, the use of *workflows** to access and analyse this data, with the ultimate objective being to publish the results and conclusions stemming from the processes of data analyses in the form of reports, recommendations, scientific articles, etc. Madin et al. (2008) stipulate that this can only be done by incorporating an intermediate level between these two previous levels (represented in medium grey on Figure 9.1), which consists of defining the *standards** allowing the description and interpretation of the data: amongst these standards, the *metadata**, the *controlled vocabularies**, and the *ontologies** are key elements, and are described in the following paragraphs.

9.3.1. Ecological metadata

Metadata describe the 'who, what, when, where, and how' about every aspect of the data (Michener, 2006). They can be defined as information about data necessary to understand and interpret the data: the contents of the data set, their experimental context, their structure, and their accessibility

Table 9.2 Metadata standards (a) and metadata management tools (b) currently used in ecology. Taken from Michener & Jones (2012). Reproduced with permission from Elsevier.

(a) Metadata standard	Description	Reference
Content Standard for Digital Geospatial Metadata (CSDGM)	CSDGM was created by the US Federal Geographic Data Committee and includes the Biological Data Profile (BDP). The CSDGM focuses on geospatial data and the BDP adds categories relevant to biological data	http://www.fgdc.gov/metadata/csdgm/
Darwin Core	Darwin Core metadata include descriptors necessary for documenting museum specimens and facilitating the sharing of information pertaining to organisms and biological diversity (e.g., taxonomic classification, geographic location)	http://www.tdwg.org/activities/darwincore/
Dublin Core Element Set	Dublin Core metadata encompasses a small number of elements that are widely used to describe physical resources such as books, and digital materials such as video, text files, images, and web pages	http://dublincore.org/
Ecological Metadata Language (EML)	EML includes a comprehensive set of descriptors that can be used to document all elements of an array of ecological and environmental data and non-digital resources such as maps (see Table 9.3 for an extract)	http://knb.ecoinformatics.org /software/eml/
ISO 19115	ISO19115 includes a comprehensive set of more than 400 elements that describe geospatial data and services. ISO 19115 is a standard of the International Organization for Standardization (ISO)	http://www.iso.org/iso/
(b) Metadata tool	**Description**	**Reference**
MERMAid (Metadata Enterprise Resource Management Aid)	MERMAid is an online metadata entry and management tool that supports FGDC compliant metadata and the Biological Data Profile. The National Coastal Data Development Center (NCDDC) developed this US National Oceanic and Atmospheric Administration metadata tool	http://www.ncddc.noaa.gov/activities/mermaid/
Metavist	Metavist is a software tool for the metadata archivist, and is used to create FGDC compliant metadata. Metavist is a product of the US Forest Service and provides support for the Biological Data Profile	http://metavist.djames.net/
Morpho	Morpho is a comprehensive metadata management system that supports the creation and management of metadata that conform to EML, FGDC, and BDP standards. Morpho also interfaces with the Knowledge Network for Biocomplexity (KNB) Metacat server, which allows scientists to upload, download, store, query, and view public metadata and data	http://knb.ecoinformatics.org/morphoportal.jsp

(Michener et al., 1997). Numerous metadata standards are used in ecology (Michener & Jones, 2012; Table 9.2), which makes it somewhat difficult to choose the appropriate one.

The production of metadata in addition to directly studied biological data presents a large number of advantages (Michener, 2006; Scurlock et al., 2002). Amongst these, the most important are an increase in data longevity (Figure 9.2), facilitation of data re-utilization by those who collected the data, data sharing with other researchers, and the re-examination or reformatting of the data for different objectives and questions than those for which the data were initially collected. These points are even more important as ecological data should be able to be used independently of the person who initially collected them, and given that ecological research often requires a synthesis of information coming

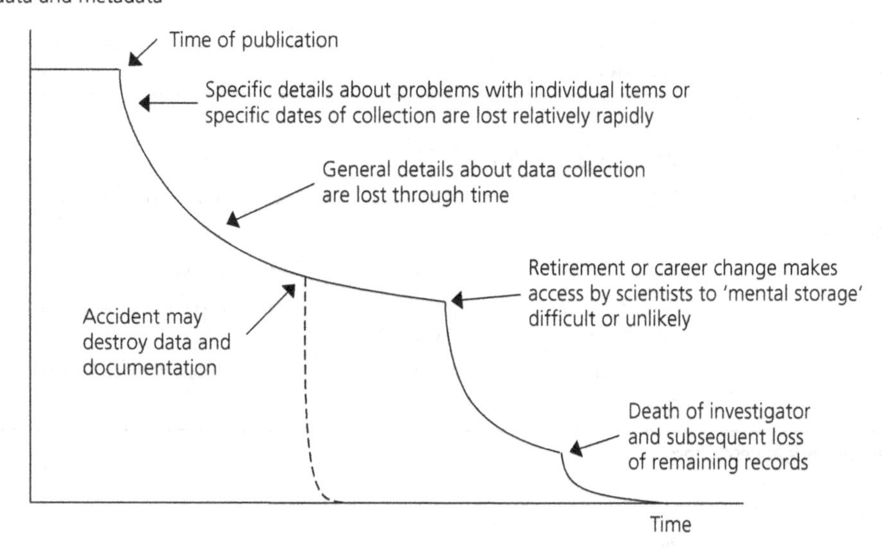

Figure 9.2 Example of the normal degradation in information content associated with data and metadata over time ('information entropy'). Accidents or changes in storage technology (dashed line) may eliminate access to remaining raw data and metadata at any time. Taken from Michener et al. (1997). Reproduced with permission from the Ecological Society of America.

from a large diversity of disciplines (Jones et al., 2006; Michener, 2006). A number of specifications can be used for the provision of metadata content so as to enrich ecological data, and different types of software have been developed to facilitate their entry (Table 9.2). Table 9.3 shows some specifications of the Ecological Metadata Language (EML: Michener et al., 1997), which is probably one of the most detailed standards developed for ecological data. However, as of today, there does not exist a unanimously adopted standard metadata format for ecology (Scurlock et al., 2002), but rather multiple, different standards, which are often not compatible. This might explain the fact that the use of metadata in ecology is not widespread.

Metadata specified for a general ecological context appear probably adapted to the field of research of the functional diversity of organisms. However, considerable work remains to be done to improve and generalize the use of metadata in ecological studies. The first task to be achieved towards a consolidated metadata standard for ecology would consist of defining a minimum set of descriptors, to accompany the collection and storage of data.

9.3.2. A controlled vocabulary and a thesaurus for plant traits

*Controlled vocabularies** and *thesauri** correspond to the second type of standards necessary for the description, discovery, and integration of data (Figure 9.1). A controlled vocabulary is tied to a specific field of knowledge, for which it will play the role of a reference. It reflects the concerted choice of a scientific community as to the designation of a list of key terms and the definitions of these terms. A thesaurus in itself allows for the organization and structuring of this list on the basis of hierarchic, associative, or equivalence semantic relationships. In contrast to metadata, this type of standard is highly specific to the research field concerned.

The necessity of developing such standards in the context of a trait-based approach to plant functional diversity can be illustrated by a simple example (Laporte et al., 2012). The trait 'plant height' is defined in the manual of methodological standards of Cornelissen et al. (2003, cf. Section 2.1.2 of Chapter 2) as 'the shortest distance between the upper limit of the photosynthetic tissues of a plant

Table 9.3 Standard EML (Ecological Metadata Language) ecological metadata descriptors and examples. The first two classes are extensively described, while for the three last ones, only a brief outline of the major descriptors is given. Adapted from Michener et al. (1997), with permission from the Ecological Society of America.

Descriptors	Examples
Class I : Data set descriptors	
A. Data set identity	Title or theme of data set
B. Data set identification code	Database accession numbers or site-specific codes used to uniquely identify data set
C. Data set description	
1. Originator(s)	Name(s) and address(es) of principal investigator(s) associated with data set
2. Abstract	Descriptive abstract summarizing research objectives, data contents (including temporal, spatial, and thematic domain), context, and potential uses of data set
D. Key words	Location (spatial scale), time period and sampling frequency (temporal scale), theme or contents (thematic scale)
Class II : Research origin descriptors	
A. 'Overall' project description	
1. Identity	Project title or theme
2. Originator(s)	Name(s) and address(es) of principal investigator(s) associated with project
3. Period of study	Date commenced, date terminated, or expected duration
4. Objectives	Scope and purpose of research programme
5. Abstract	Descriptive abstract summarizing broader scientific scope of 'overall' research project
6. Source(s) of funding	Grant and contract numbers, names and addresses of funding sources
B. 'Specific subproject' description	
1. Site description	Site type, geography, or climate
2. Experimental or sampling design	Description of statistical/sampling design
3. Research methods	Description of the methods
4. Project personnel	Principal and associated investigator(s), technicians, supervisors, students
Class III : Data set status and accessibility	
A. Status	
1. Latest update . . . 4. Data verification	
B. Accessibility	
1. Storage location and medium . . . 5. Costs associated with acquiring data	
Class IV : Data structural descriptors	
A. Data set file	
1. Identity . . . 7. Authentication procedures	
B. Variable information	
1. Variable identity . . . 5. Data format	
C. Data anomalies	Description of missing data, anomalous data, calibration errors, etc.

Table 9.3 *(continued)*

Descriptors	Examples
Class V : Supplemental descriptors	
A. Data acquisition	
1. Data forms or acquisition methods . . . 3. Data entry verification procedures	
B. Quality assurance/quality control procedures	Identification and treatment of outliers, description of quality assessments, calibration of reference standards, equipment performance results, etc.
C. Related materials	References and locations of maps, photographs, videos, GIS data layers, physical specimens, field notebooks, comments, etc.
D. Computer programs and data processing algorithms	Description or listing of any algorithms used in deriving, processing, or transforming data
E. Archiving	
1. Archival procedures . . . 2. Redundant archival sites	
F. Publications and results	Electronic reprints, lists of publications resulting from or related to the study, graphical/statistical data representations, etc.
G. History of data set usage	
1. Data request history . . . 3. Review history	

and the level of the soil'. In fact, this definition, which seems clear and precise, can also be applied to a number of other traits that could be found for example in the first version of the TRY database mentioned in Section 9.2, thus reflecting the richness and complexity of the terms used: 'maximum plant height', 'releasing height', 'canopy height observed', etc. (cf. Tables 3 and 4 in Kattge et al., 2011a). Are these different expressions synonymous, or do they each have different signification, despite being somewhat related? In the absence of precise and accepted standard terminologies, it is not possible to respond to this question satisfactorily. Standard terminologies relating to plant traits have been developed in particular by molecular biologists (Walls et al., 2012a; Table 9.4), but the coverage of traits used in ecology is relatively limited. For example, a request formulated on the server hosting the Plant Trait Ontology (TO: Table 9.4, queried on 25 May 2015), does not return any definitions for leaf dry matter content or wood density, two traits frequently used in ecology (cf. Chapter 3). There is consequently a strong need to develop a set of definitions and associated synonyms relevant to the specific field of plant functional diversity.

The recognition of this need has motivated the launch of an initiative whose aim is to construct a thesaurus of plant traits for ecology. One of the key ideas of this work has been to consider that, in order to be widely accepted and used, such a thesaurus should result from a collaborative construction (Laporte et al., 2012, and references therein). The tool ThesauForm–Traits, developed on this basis, has facilitated the work of a group of some fifteen experts in the field of plant traits, interacting via a web interface to define and annotate an initial list of 130 traits; a voting interface allows the determination of the definition(s) of traits, which can then be used as standards (see Pey et al., 2014, for an application to a trait-based approach of soil invertebrates). To construct an initial version of this thesaurus, the traits have been organized into different categories: for example, the category 'size' regroups plant height, plant mass, leaf length, etc., and the category 'chemical composition' regroups the concentration of mineral elements (nitrogen, phosphorus, calcium, etc.) of different organs. Table 9.5 presents an extract of the definitions contained in this thesaurus for some traits commonly used in comparative ecology (cf. Chapter 3). The objective is to progressively enrich this thesaurus (approximately 500 traits have been defined as of 30 May 2015), and to make them available via a web interface (Laporte et al., 2013), in the same manner as has been done for measurement protocols

Table 9.4 Terminological standards (a) and ecoinformatic resources (b) available for biology and plant ecology. Completed from Walls et al. (2012a), with permission from the Botanical Society of America.

	Domain and / or data type	References
(a) Terminology standard (abbreviation)		
Plant Ontology (PO)	Plant anatomical entities and plant structure developmental stages	(Ilic et al., 2007; Pujar et al., 2006) www.plantontology.org
Gene Ontology (GO)	Cellular components, biological processes, and molecular functions	(Gene Ontology Consortium, 2010) http://www.geneontology.org
Chemical Entities of Biological Interest (ChEBI)	Molecular entities that are natural products or are synthetic products used to intervene in the processes of living organisms	(de Matos et al., 2010; Degtyarenko et al., 2008) http://www.ebi.as.uk/chebi
Protein Ontology (PR)	Proteins based on evolutionary relatedness, protein forms produced from a given gene locus, and protein-containing complexes	(Bult et al., 2011; Natale et al., 2007) http://pir.georgetown.edu/pro/
Ontology for Biomedical Investigations (OBI)	Scientific investigations, including the protocols and instrumentation used, the material used, the data generated, and the types of analysis performed	(Brinkman et al., 2010) http://obi-ontology.org
Phenotypic Quality Ontology (PATO)	Phenotypic qualities (properties). This ontology can be used in conjunction with other ontologies such as anatomical ontologies to refer to phenotypes	(Mungall et al., 2010) http://obofoundry.org/wiki/index.php/PATO:Main_Page
Plant Trait Ontology (TO)	Phenotypic traits in plants; each trait is a distinguishable feature, characteristic, or quality of a plant	(Jaiswal et al., 2002) http://gramene.org/db/ontology/search?id=TO:0000387
Crop Ontology (CO)	An ontology dealing with traits and several structural and anatomic characteristics of 7 cultivated species with a specific terminology for each species	(Shrestha et al., 2010) http://cropontology.org/
Plant Infectious Disease Ontology (IDOPlant)	Plant infectious diseases, pathogens, and symptoms	(Walls et al., 2012b) http://purl.obolibrary.org/obo/idoplant.owl
Extensible Observation Ontology (OBOE)	A suite of ontologies for modelling and representing scientific observations	(Madin et al., 2007) https://semtools.ecoinformatics.org/oboe
Environment Ontology (EnvO)	Environmental features and habitats	http:/environmentontology.org/
Xeml Lab	A tool to develop experiments under controlled conditions, which generates metadata files containing information concerning genotypes, growth conditions, environmental disturbances, and the sampling strategy	(Hannemann et al., 2009) http://xeml.mpimp-golm.mpg.de/dnn/Resources/tabid/56/Default.aspx
NCBI Taxonomy	Biological taxa, based on the classification of the National Centre for Biotechnology Information	(Wheeler et al., 2007) http://obofoundry.org/cgi-bin/detail.cgi?id=ncbi_taxonomy
(b) Resources		
PROtocols, METHods, Explanations and Updated Standards Wiki (PrometheusWiki)	Collates protocols and methods for the domains of plant physiology, ecology, and environmental sciences	(Sack et al., 2010) http://www.publish.csiro.au/prometheuswiki
BioPortal	Source for finding, searching, and querying bio-ontologies	http://bioportal.bioontology.org
Ontology Lookup Service	Source for finding and searching bio-ontologies	(Côté et al., 2006) http://www.ebi.ac.uk/ontology-lookup
OntoBee	Source for finding, searching, and querying bio-ontologies	(Xiang et al., 2011) http://ontobee.org

Table 9.5 An extract from the thesaurus of plant traits, with the trait definitions retained after the voting procedure (see the text in Section 9.3.2 for more details) and the bibliographic reference from which the initial definition was taken. See Chapter 3 for the meaning of these traits. The terms and definitions for approximately 500 traits are available on line at the following address: http://top-thesaurus.org. Adapted from Laporte et al. (2012).

Terms	Definitions	Reference for the initial definition
Trait	Any morphological, physiological, or phenological heritable feature measurable at the level of an individual, from the cell to the entire organism, without reference to the environment or any other level of organization	Violle et al. (2007), modified as explained in Chapter 2
Specific leaf area	The ratio between the area of a leaf and its dry mass	Cornelissen et al. (2003)
Leaf mass per area	The ratio between the dry mass of a leaf and its area	Cornelissen et al. (2003)
Leaf life span	The period during which an individual leaf, or part of a leaf, remains alive and physiologically active	Cornelissen et al. (2003)
Specific root length	The ratio between the length of a root and its dry mass	Cornelissen et al. (2003)
Plant life form	A characteristic of the whole plant defined by the position of the perennating tissue relative to the ground. Perennating tissue refers to the embryonic (meristematic) tissue that remains inactive during a cold or dry season and resumes growth upon return of a favourable season *(see Box 2.1 in Chapter 2)*	Raunkiaer (1934)
Vegetative plant height	The shortest distance between the upper limit of the photosynthetic tissues of a plant and the ground level	Cornelissen et al. (2003)
Stem specific density	The ratio between the dry mass of a stem or a section thereof, and the corresponding volume	Cornelissen et al. (2003)
Bark thickness	The thickness of that part of a stem external to the wood, and which includes the vascular cambium	Cornelissen et al. (2003)
Seed mass	The dry mass of a seed *(see Box 2.1 in Chapter 2)*	Cornelissen et al. (2003)
Seed shape	The variance of diaspore length, width, and depth, after first transforming all values so that length (the largest dimension) is unity	Thompson et al. (1993)
Relative growth rate	The absolute growth rate of a whole plant, or a part of a plant, expressed relative to a measure of the whole plant or plant part being considered *(see Box 3.3 in Chapter 3)*	Evans (1972)
Leaf photosynthetic rate	The rate at which a leaf converts light energy into chemical energy	Hendry & Grime (1993)
Leaf dark respiration rate	The rate of basal metabolism of a leaf	Pérez-Harguindeguy et al. (2013)

in plant ecophysiology (the PrometheusWiki initiative: Sack et al., 2010; see Section 2.1.2 of Chapter 2). It is on the basis of this thesaurus that an ontology specific to the domain of plant functional diversity will be constructed, as explained in the following section.

9.3.3 Towards an ontology for functional diversity

*Ontologies**, which constitute the third type of standard discussed here, correspond to a key stage in the formalization of concepts within a given field of research. In information sciences, an ontology is an explicit, shared, and formal specification of a *conceptualization** (Gruber, 1995), and can be defined as a representation or formal classification of the concepts of the domain of interest (for example, plant traits) and the relationships between these concepts. Box 9.1 explains more precisely what an ontology is by using the example of concepts relating to the notion of a biological entity.

The formal structure of an ontology is based on rules of logic, which give the possibility, through the application of *automated reasoning**, to confront the data with interpretations and thus allow the emergence of new understandings; this is how ontologies allow for the identification of important

Box 9.1 What is an ontology?

Adapted from Madin et al. (2008).

Concepts and relationships

Ontologies are formal models that define concepts and their relationships within a scientific domain such as ecology. An ontological 'concept' denotes a collection of '*instances**' that share common characteristics. The backbone of ontologies is the 'is-a' relationship, which states that all instances of a sub-concept (i.e. subset) are also members of a super-concept and, therefore, inherit all characteristics of the super-concept (Box 9.1, Figure 1). For example, Tree would generally be defined as a sub-concept of Plant. There are other commonly used relationships that describe how concepts interact, including 'part-of' (or, conversely, 'has-part'), 'equivalence', and 'disjoint' relations. In a 'part-whole' (i.e. 'part-of' or 'has-part') relationship, the instances of one concept (e.g. Tree Branch) are components of instances of another concept (e.g. Tree, note 1:1). In an 'equivalence' relationship, two concepts denote the same set of instances (e.g. Animals and Metazoans), whereas in a 'disjoint' relationship, the instances of the two concepts are mutually exclusive (e.g. Plants and Animals). Relationships and multiplicities are inherited through an 'is-a' relationship; for example, instances of the Cohort concept have two or more Organism instances (noted as 2:n) as parts, because Cohort is a sub-concept of Population (Box 9.1, Figure 1).

Formal representation

Ontology modelling languages, such as the Web Ontology Language (OWL) for the *Semantic Web**, are based on a subfamily of mathematical logic called 'description logic'. The formal underpinnings of these languages offer advantages over less formal approaches, such as controlled vocabularies, thesauri, and concept maps. For example, ontology languages enable precise expressions of the meaning of a scientific assertion that can be checked for consistency and compared with other formal assertions. Through automated reasoning techniques, it is possible to automate the process of determining whether an ontology is internally consistent and to infer new relationships between concepts (beyond those explicitly given in the ontology). For example, in Figure 1, although Barnacles have Biological Parts (i.e. Barnacle 'is-an' Animal, Animal 'is-an' Organism, and Organism 'has member' Biological Part), and Tree Branches are Biological Parts (i.e. Tree Branch 'is-a' Biological Part), Barnacles cannot have Tree Branches because Animals are 'disjoint' from Plants. Although these relationship implications might be obvious to scientists, ontologies enable computers to deduce the implications of long chains of these formal assertions.

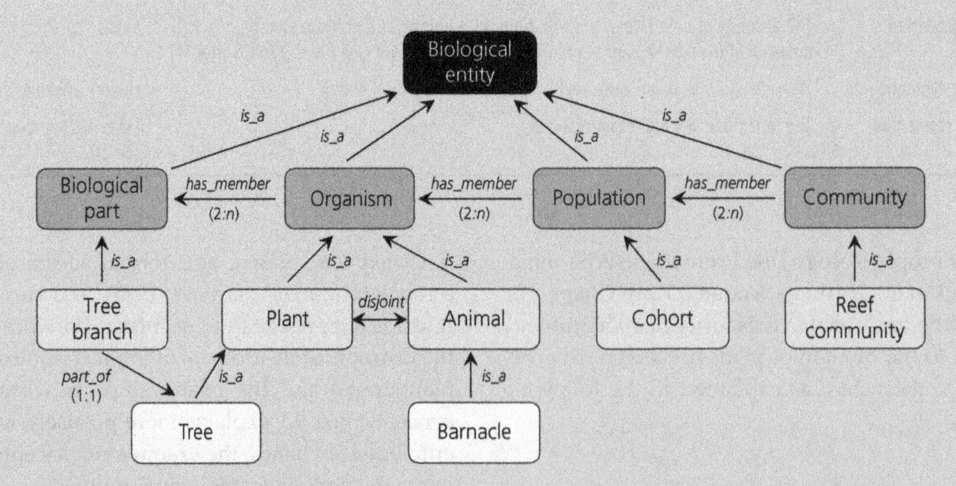

Box 9.1, Figure 1 An ontology fragment representing some Biological-Entity concepts and their relationships. In this graphical notation, rectangles denote concepts, arrows denote relationships, and multiplicities are given in parentheses. For example, any instance of Tree Branch is a part of one and only one (i.e. 1:1) instance of a Tree; but, conversely, an instance of Tree has at least two or more (i.e. 2:n) parts that are instances of Biological Part, because 'has-part' relationships and multiplicity are inherited from super-concepts. This ontology represents only one interpretation of the domain Biological Entity, where other interpretations can similarly be described and possibly interrelated using different ontologies. Taken from Madin et al. (2008). Reproduced with permission from Elsevier.

information that might otherwise remain hidden or implicit in a data set (Jones et al., 2006). For example, if we search for data on specific leaf area, these can be saved in different databases under the term 'specific leaf area' or under its inverse 'leaf mass per area'. The establishment of a thesaurus, such as described in Section 9.3.2, is essential in order to unequivocally define these two traits, and an ontology should be capable of determining that specific leaf area = 1/(leaf mass per area), from the explicit definitions of specific leaf area and leaf mass per area. Any query for specific leaf area or leaf mass per area sent to different databases, situated in different locations, via an ontological filter would *automatically* return the data stored under the two different terms.

Two major types of ontologies are used to formalize knowledge within a domain: ontologies called 'framework' and ontologies called 'domain specific'. The first describe very general concepts, and facilitate and guide the integration of information coming from on the one hand more specific vocabularies, and on the other hand domain-specific ontologies. Framework ontologies are developed to facilitate the interconnection of existing domain-specific ontologies, as well as providing a coherent framework for the construction of new ontologies (Madin et al., 2008). A number of framework ontologies have been developed for ecology. These are based on the formalization of the concepts of observations and measures, which are central to scientific method (Madin et al., 2008). Amongst these, OBOE (Extensible Observation Ontology: Madin et al., 2007) appears to be well adapted to represent plant trait data (see Kattge et al., 2011b, for the application of OBOE in the context of the TRY database). Within OBOE (Figure 9.3), an observation is made for an *entity** (biological organism, geographic location, environmental characteristic, etc.), and serves to group a set of measurements together to form a single 'observation event'. A measure gives a value for a characteristic of an observed entity (for example height during the observation of a tree). Measures are associated with measurement standards (measurement units), and can also be enriched with additional information, such as the protocols used in collecting the measure, the methods used, or the precision of the measure. An observation is carried out in a certain context, which implies a dependence on other observations.

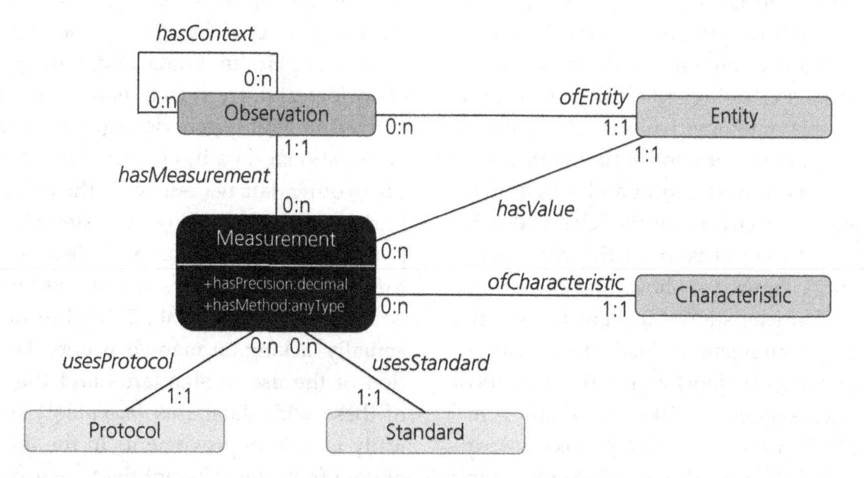

Figure 9.3 The main *classes** (rectangles) and properties (connections between the rectangles) of the observation and measurement ontology (OBOE) initially developed by Madin et al. (2007), and modified later by Saunders et al. (2011). An observation is made of an entity. A characteristic of an entity can be represented by a measure. Measures establish a relationship between the characteristics and a measurement standard via a value, and are obtained with a certain precision. Measures are carried out using a protocol in a certain place at a certain time. Observations can have multiple measures. Entities, characteristics, and measurement standards constitute entry points for domain-specific ontologies. Figure 9.4 gives an example of this type of application for the trait 'plant height'. The notations '1:1' and '0:n' are called multiplicities: they indicate how many objects within a given class can be linked to objects of another class. For example in the relationship 'of entity', an Observation will be linked to only one entity ('1:1'), while an entity could be linked to 0 or to n Observations ('0:n'). Reproduced with permission from the author.

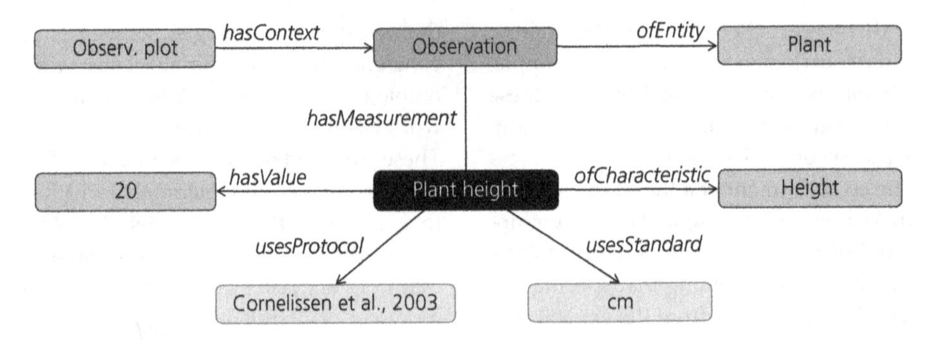

Figure 9.4 Example of the representation of a trait (plant height) in the OBOE formalism. In this formalism, a trait is a characteristic (height) of an entity (the plant), while the attribute is the value measured for the trait (20 cm) under the particular conditions of the observation.

Let's take the example of an observation carried out on a given tree, with a measure being made of its height. This observation has been done in a precise geographic location, which gives information about the 'context' (Figure 9.3). The geographic location provides important information not only for interpreting the height measurement of the observed tree, but also for comparing the measures made for any other tree in the same location. This is possible due to the establishment of a contextual relationship between the observations of trees and their geographic location.

The next stage consists of developing an ontology specific to the domain of ecology based on the use of the traits of organisms (the name 'PLATON' (PLAnt Trait ONtology for ecology) has been suggested for this ontology: Laporte, 2011), which would be compatible with the framework ontology OBOE. In this context, a trait and a trait value are represented respectively by the 'characteristics of an entity' and the 'measure of the characteristics of an entity'. Figure 9.4 shows the example of the height (a characteristic) of a plant (an entity), for which a measurement is made in a particular context (a particular plot) with a fixed protocol (that of Cornelissen et al., 2003) and unit (cm). This representation allows for the precise description of traits and their attributes, while remaining based on controlled vocabularies and/or ontologies that have been developed elsewhere: plant entities (tissues, organs, developmental stages, etc.) are, for example, defined in a very detailed manner in ontologies such as 'Plant Ontology', while some characteristics are defined in the 'Phenotypic Quality Ontology' (Table 9.4). The articulation between the concepts defined in these ontologies and those defined in the context of the trait thesaurus (Section 9.3.2) is a phase of work currently underway. The completion of this phase will allow for the finalization of PLATON, which represents a plant trait as the combination of plant organ (entity) and characteristics.

9.4 Conclusions

The approach to data management presented in this chapter constitutes a first step towards the long-term objective of improving the integration and sharing of plant trait data, and, more generally, that of ecological data. In fact, many of the aspects considered are highly generic (e.g. the representation of observational data in OBOE) and can certainly apply to other data necessary for the understanding of the functioning of ecological systems and their components. There exists for example a list of keywords for data relating to the structure and functioning of ecosystems (Porter et al., 2011), but definitions are actually lacking for many concepts. The generalization of the use of standards and the combination of these with databases (*mapping**) should specifically lead to improvements in the interoperability of data from the different fields of ecology (see e.g. Deans et al., 2015; Walls et al., 2014). This information revolution will allow for the greater accessibility and more efficient use of the large amounts of data collected during ecological studies through the development of information systems comparable to those currently used in molecular biology (e.g. GenBank: Benson et al., 2013; and EMBL database: Kanz et al., 2005, two comprehensive collections of nucleotide sequences), or for studies of taxonomic diversity (e.g. the Global Biodiversity Information Facility: Guralnick et al., 2007).

9.5 Key points

1. Over the course of the last three decades, the accumulation of data from differing sources has led to the construction of large plant trait databases and a first attempt to integrate the domain data. However, compatibility and data sharing between these different databases remains problematic and, despite the large number of records included, they do not contain all of the potentially available data.
2. One of the characteristics of functional diversity data (and more generally of ecological data) is their considerable heterogeneity, both syntactic (e.g. storage format and management approach) and semantic (e.g. the diverse nature of the data, and the multiplicity of terms and concepts used). This is why it is necessary to develop a global effort for the recovery, storage, management, semantic consolidation, and public availability of this data. The definition of standards—metadata, controlled vocabularies, thesauri, and ontologies—allowing for the description and interpretation of this data constitutes an important field of endeavour, the objective of which is to resolve this semantic heterogeneity, a key step to improving the interoperability of data sets and data discovery.
3. Metadata should allow the user of a data set to reply to questions of 'who, what, when, where, and how' concerning all of the different aspects of a data set. Despite their recognized importance, as yet there does not exist a standard format for metadata that is accepted within the field of ecology. A controlled vocabulary regroups a list of key terms relating to a particular field of knowledge as well as their associated definitions, while a thesaurus is used to organize and structure this list on the basis of the relationships between the terms. Such a thesaurus including approximately 500 terms widely used in comparative ecology is currently being finalized. Ultimately, this will serve as a basis for the development of an ontology—defined as a formal representation or classification of the concepts of a field of interest and the relationships between the concepts—specific to the field of plant functional diversity.
4. These different stages, combined with the mapping of existing databases, constitute the development of a global ecoinformatics approach which will allow for the greater accessibility and more efficient use of the large amounts of data collected during ecological studies, equivalent to that of bioinformatics as used for data stemming from genomic studies.

9.6 References

Benson, D. A., Cavanaugh, M., Clark, K., Karsch-Mizrachi, I., Lipman, D. J., Ostell, J., & Sayers, E. W. (2013). GenBank. *Nucleic Acids Research, 41*(Database issue), D36–D42. doi:10.1093/nar/gks1195

Brinkman, R., Courtot, M., Derom, D., Fostel, J., He, Y., Lord, P., ... the OBI consortium (2010). Modeling biomedical experimental processes with OBI. *Journal of Biomedical Semantics, 1*(Suppl 1), S7.

Bult, C. J., Drabkin, H. J., Evsikov, A., Natale, D., Arighi, C., Roberts, N., ... Wu, C. (2011). The representation of protein complexes in the Protein Ontology (PRO). *BMC Bioinformatics, 12*. doi:10.1186/1471-2105-12-371

Chave, J., Coomes, D., Jansen, S., Lewis, S. L., Swenson, N. G., & Zanne, A. E. (2009). Towards a worldwide wood economics spectrum. *Ecology Letters, 12*, 351–366.

Cornelissen, J. H. C., Lavorel, S., Garnier, E., Díaz, S., Buchmann, N., Gurvich, D. E., ... Poorter, H. (2003). A handbook of protocols for standardised and easy measurement of plant functional traits worldwide. *Australian Journal of Botany, 51*, 335–380.

Côté, R. G., Jones, P., Apweiler, R., & Hermjakob, H. (2006). The Ontology Lookup Service, a lightweight cross-platform tool for controlled vocabulary queries. *BMC Bioinformatics, 7*. doi:10.1186/1471-2105-7-97

de Matos, P., Alcantara, R., Dekker, A., Ennis, M., Hastings, J., Haug, K., ... Steinbeck, C. (2010). Chemical entities of biological interest: an update. *Nucleic Acids Research, 38*, D249–D254. doi:10.1093/nar/gkp886

Deans, A. R., Lewis, S. E., Huala, E., Anzaldo, S. S., Ashburner, M., Balhoff, J. P., ... Mabee, P. (2015). Finding our way through phenotypes. *PLoS Biol, 13*(1), e1002033. doi:10.1371/journal.pbio.1002033

Degtyarenko, K., De Matos, P., Ennis, M., Hastings, J., Zbinden, M., McNaught, A., ... Ashburner, M. (2008).

ChEBI: a database and ontology for chemical entities of biological interest. *Nucleic Acids Research, 36*, D344–D350. doi:10.1093/nar/gkm791

Enquist, B. J., & Niklas, K. J. (2002). Global allocation rules for patterns of biomass partitioning in seed plants. *Science, 295*, 1517–1520.

Evans, G. C. (1972). *The Quantitative Analysis of Plant Growth*. Oxford: Blackwell Scientific Publications.

Fitter, A. H., & Peat, H. J. (1994). The ecological flora database. *Journal of Ecology, 82*, 415–425.

Gachet, S., Véla, E., & Tatoni, T. (2005). BASECO: a floristic and ecological database of Mediterranean French flora. *Biodiversity and Conservation, 14*, 1023–1034.

Garnier, E., Lavorel, S., Ansquer, P., Castro, H., Cruz, P., Dolezal, J., . . . Zarovali, M. (2007). Assessing the effects of land use change on plant traits, communities, and ecosystem functioning in grasslands: a standardized methodology and lessons from an application to 11 European sites. *Annals of Botany, 99*, 967–985.

Gene Ontology Consortium. (2010). The Gene Ontology in 2010: extensions and refinements. *Nucleic Acids Research, 38*, D331–D335. doi:10.1093/nar/gkp1018

Grime, J. P., & Hunt, R. (1975). Relative growth-rate: its range and adaptive significance in a local flora. *Journal of Ecology, 63*(2), 393–422. doi: Stable URL: http://www.jstor.org/stable/2258728

Gruber, T. R. (1995). Toward principles for the design of ontologies used for knowledge sharing. *International Journal of Human-Computer Studies, 43*, 907–928.

Guralnick, R. P., Hill, A. P., & Lane, M. (2007). Towards a collaborative, global infrastructure for biodiversity assessment. *Ecology Letters, 10*, 663–672.

Hannemann, J., Poorter, H., Usadel, B., Blasing, O. E., Finck, A., Tardieu, F., . . . Gibon, Y. (2009). Xeml Lab: a tool that supports the design of experiments at a graphical interface and generates computer-readable metadata files, which capture information about genotypes, growth conditions, environmental perturbations, and sampling strategy. *Plant, Cell and Environment, 32*(9), 1185–1200. doi:10.1111/j.1365-3040.2009.01964.x

Heidorn, P. B. (2008). Shedding light on the dark data in the long tail of science. *Library Trends, 57*(2), 280–299.

Hendry, G. A. F., & Grime, J. P. (eds) (1993). *Methods in Comparative Plant Ecology*. London: Chapman & Hall.

Ilic, K., Kellogg, E. A., Jaiswal, P., Zapata, F., Stevens, P. F., Vincent, L. P., . . . Rhee, S. Y. (2007). The plant structure ontology, a unified vocabulary of anatomy and morphology of a flowering plant. *Plant Physiology, 143*(2), 587–599. doi:10.1104/pp.106.092825

Jaiswal, P., Ware, D., Ni, J., Chang, K., Zhao, W., Schmidt, S., . . . McCouch, S. R. (2002). Gramene: Development and integration of trait and gene ontologies for rice. *Comparative and Functional Genomics, 3*, 132–136.

Jones, M. B., Schildhauer, M. P., Reichman, O. J., & Bowers, S. (2006). The new bioinformatics: Integrating ecological data from the gene to the biosphere. *Annual Review of Ecology, Evolution and Systematics, 37*, 519–544.

Kanz, C., Aldebert, P., Althorpe, N., Baker, W., Baldwin, A., Bates, K., . . . Apweiler, R. (2005). The EMBL nucleotide sequence database. *Nucleic Acids Research, 33*(Database issue), D29–D33. doi:10.1093/nar/gki098

Kattge, J., Díaz, S., Lavorel, S., Prentice, I. C., Leadley, P., Bönisch, G., . . . Wirth, C. (2011a). TRY: a global database of plant traits. *Global Change Biology, 17*, 2905–2935. doi:10.1111/j.1365-2486.2011.02451.x

Kattge, J., Ogle, K., Bönisch, G., Díaz, S., Lavorel, S., Madin, J., . . . Wirth, C. (2011b). A generic structure for plant trait databases. *Methods in Ecology and Evolution, 2*(2), 202–213. doi:10.1111/j.2041-210X.2010.00067.x

Keddy, P. A. (1992). A pragmatic approach to functional ecology. *Functional Ecology, 6*, 621–626.

Kelling, S., Hochachka, W. M., Fink, D., Riedewald, M., Caruana, R., Ballard, G., & Hooker, G. (2009). Data-intensive science: A new paradigm for biodiversity studies. *BioScience, 59*(7), 613–620. doi:10.1525/bio.2009.59.7.12

Kleyer, M., Bekker, R. M., Knevel, I. C., Bakker, J. P., Thompson, K., Sonnenschein, M., . . . Peco, B. (2008). The LEDA Traitbase: a database of life-history traits of the Northwest European flora. *Journal of Ecology, 96*(6), 1266–1274. doi:10.1111/j.1365-2745.2008.01430.x

Klimeš, L., & Klimešová, J. (1999). CLO-PLA2—a database of clonal plants in Central Europe. *Plant Ecology, 141*, 9–19.

Klotz, S., Kühn, I., & Durka, W. (2002). BIOLFLOR: Eine Datenbank zu biologisch-ökologischen Merkmalen zur Flora von Deutschland. *Schriftenreihe für Vegetationskunde, 38*, 1–333.

Laporte, M.-A. (2011). *Définition de standards de données relatifs aux traits fonctionnels des végétaux pour l'étude de la biodiversité*. PhD, Université Montpellier II: Sciences et Techniques du Languedoc, Montpellier, France.

Laporte, M.-A., Garnier, E., & Mougenot, I. (2013). *A faceted search system for facilitating discovery-driven scientific activities: a use case from functional ecology*. Paper presented at the Proceedings of the first International Workshop on Semantics for Biodiversity (S4BioDiv 2013), Montpellier.

Laporte, M.-A., Mougenot, I., & Garnier, E. (2012). ThesauForm–Traits: a web-based collaborative tool to develop a thesaurus for plant functional diversity research. *Ecological Informatics, 11*, 34–44. doi: 10.1016/j.ecoinf.2012.04.004

Madin, J. S., Bowers, S., Schildhauer, M. P., & Jones, M. B. (2008). Advancing ecological research with ontologies. *Trends in Ecology and Evolution, 23*, 159–168.

Madin, J. S., Bowers, S., Schildhauer, M., Krivov, S., Pennington, D., & Villa, F. (2007). An ontology for describing and synthesizing ecological observation data. *Ecological Informatics, 2*, 279–296.

Makkonen, M., Berg, M. P., Handa, I. T., Hattenschwiler, S., van Ruijven, J., van Bodegom, P. M., & Aerts, R. (2012). Highly consistent effects of plant litter identity and functional traits on decomposition across a latitudinal gradient. *Ecology Letters, 15*(9), 1033–1041. doi:10.1111/j.1461-0248.2012.01826.x

Michener, W. K. (2006). Meta-information concepts for ecological data management. *Ecological Informatics, 1*, 3–7.

Michener, W. K., & Jones, M. B. (2012). Ecoinformatics: supporting ecology as a data-intensive science. *Trends in Ecology and Evolution, 27*(2), 85–93. doi:10.1016/j.tree.2011.11.016

Michener, W. K., Brunt, J. W., Helly, J. J., Kirchner, T. B., & Stafford, S. G. (1997). Nongeospatial metadata for the ecological sciences. *Ecological Applications, 7*, 330–342.

Mungall, C. J., Gkoutos, G. V., Smith, C. L., Haendel, M. A., Lewis, S. E., & Ashburner, M. (2010). Integrating phenotype ontologies across multiple species. *Genome Biology, 11*, R2.

Natale, D. A., Arighi, C. N., Barker, W. C., Blake, J., Chang, T. C., Hu, Z. Z., ... Wu, C. H. (2007). Framework for a protein ontology. *BMC Bioinformatics, 8*. doi:10.1186/1471-2105-8-s9-s1

Ordoñez, J. C., van Bodegom, P. M., Witte, J.-P. M., Wright, I. J., Reich, P. B., & Aerts, R. (2009). A global study of relationships between leaf traits, climate, and soil measures of nutrient fertility. *Global Ecology and Biogeography, 18*, 137–149.

Pakeman, R. J., Garnier, E., Lavorel, S., Ansquer, P., Castro, H., Cruz, P., ... Vile, D. (2008). Impact of abundance weighing on the response of seed traits to climate and land use change. *Journal of Ecology, 96*, 355–366.

Paula, S., Arianoutsou, M., Kazanis, D., Tavsanoglu, C., Lloret, F., Buhk, C., ... Pausas, J. G. (2009). Fire-related traits for plant species of the Mediterranean Basin. *Ecology, 90*, 1420.

Pérez-Harguindeguy, N., Díaz, S., Garnier, E., Lavorel, S., Poorter, H., Jaureguiberry, P., ... Cornelissen, J. H. C. (2013). New handbook for standardised measurement of plant functional traits worldwide. *Australian Journal of Botany, 61*, 167–234.

Pey, B., Laporte, M.-A., Nahmani, J., Auclerc, A., Capowiez, Y., Caro, G., ... Hedde, M. (2014). A thesaurus for soil invertebrate trait-based approaches. *PLoS ONE, 9* (10), e108985. doi:10.1371/journal.pone.0108985

Porter, J., O'Brien, M., Costa, D., Henshaw, D., Gries, C., Melendez, E., ... Laundre, J. (2011). *A controlled vocabulary for LTER data keywords*. Paper presented at the Environmental Information Management Conference 2011 (EIM 2011), Santa Barbara, CA.

Pujar, A., Jaiswal, P., Kellogg, E. A., Ilic, K., Vincent, L., Avraham, S., ... McCouch, S. R. (2006). Whole plant growth stage ontology for angiosperms and its application in plant biology. *Plant Physiology, 142*, 414–428.

Raunkiaer, C. (1934). *The Life Forms of Plants and Statistical Plant Geography* (English edition: H. Gilbert-Carter & A. Fausbøll, trans.). Oxford: Oxford University Press.

Reich, P. B., & Oleksyn, J. (2004). Global patterns of plant leaf N and P in relation to temperature and latitude. *Proc. Natl. Acad. Sci. USA, 101*, 11001–11006.

Reich, P. B., Tjoelker, M. G., Machado, J.-L., & Oleksyn, J. (2006). Universal scaling of respiratory metabolism, size, and nitrogen in plants. *Nature, 439*, 457–461.

Reichman, O. J., Jones, M. B., & Schildhauer, M. P. (2011). Challenges and opportunities of open data in ecology. *Science, 331*, 703–705.

Sack, L., Cornwell, W. K., Santiago, L. S., Barbour, M. M., Choat, B., Evans, J. R., ... Nicotra, A. (2010). A unique web resource for physiology, ecology, and the environmental sciences: PrometheusWiki. *Functional Plant Biology, 37*(8), 687–693. doi:10.1071/fp10097

Saunders, W., Bowers, S., & O'Brien, M. (2011). *Protégé extensions for scientist-oriented modeling of observation and measurement semantics*. Paper presented at the OWL: Experiences and Directions conference, 5–9 June 2011, San Francisco, CA.

Scurlock, J. M. O., Kanciruk, P., McCord, R. A., Olson, R. J., & Michener, W. K. (2002). Metadata. In H. A. Mooney & J. G. Canadell (eds), *The Earth system: biological and ecological dimensions of global environmental change* (pp. 409–411). New York: John Wiley & Sons.

Shrestha, R., Arnaud, E., Mauleon, R., Senger, M., Davenport, G. F., Hancock, D., ... McLaren, G. (2010). Multifunctional crop trait ontology for breeders' data: field book, annotation, data discovery, and semantic enrichment of the literature. *AoB PLANTS, 2010*, plq008. doi:10.1093/aobpla/plq008

Stein, L. D. (2008). Towards a cyberinfrastructure for the biological sciences: progress, visions, and challenges. *Nature Reviews Genetics, 9*, 678–688.

Swenson, N. G., Enquist, B. J., Pither, J., Kerkhoff, A. J., Boyle, B., Weiser, M. D., ... Nolting, K. M. (2012). The biogeography and filtering of woody plant functional diversity in North and South America. *Global Ecology and Biogeography, 21*(8), 798–808. doi:10.1111/j.1466-8238.2011.00727.x

Thompson, K., Band, S. R., & Hodgson, J. G. (1993). Seed size and shape predict persistence in soil. *Functional Ecology, 7*, 236–241.

Violle, C., Navas, M.-L., Vile, D., Kazakou, E., Fortunel, C., Hummel, I., & Garnier, E. (2007). Let the concept of trait be functional! *Oikos, 116*, 882–892.

Walls, R. L., Athreya, B., Cooper, L., Elser, J., Gandolfo, M. A., Jaiswal, P., ... Stevenson, D. W. (2012a). Ontologies as integrative tools for plant science. *American Journal of Botany, 99*(8), 1263–1275. doi:10.3732/ajb.1200222

Walls, R. L., Deck, J., Guralnick, R., Baskauf, S., Beaman, R., Blum, S., ... Wooley, J. (2014). Semantics in support of biodiversity knowledge discovery: an introduction to the biological collections ontology and related ontologies. *PLoS ONE, 9*(3), e89606. doi:10.1371/journal.pone.0089606

Walls, R., Smith, B., Elser, J., Goldfain, A., Stevenson, D. W., & Jaiswal, P. (2012b). *A plant disease extension of the Infectious Disease Ontology*. Paper presented at the 3rd International Conference on Biomedical Ontology (ICBO 2012), Graz, Austria.

Wheeler, D. L., Barrett, T., Benson, D. A., Bryant, S. H., Canese, K., Chetvernin, V., ... Yaschenko, E. (2007). Database resources of the National Center for Biotechnology Information. *Nucleic Acids Research, 35*, D5–D12. doi:10.1093/nar/gkl1031

Wright, I. J., Reich, P. B., Westoby, M., Ackerly, D. D., Baruch, Z., Bongers, F., ... Villar, R. (2004). The worldwide leaf economics spectrum. *Nature, 428*, 821–827.

Xiang, Z., Mungall, C. J., Ruttenberg, A., & He, Y. (2011). *Ontobee: a linked data server and browser for ontology terms*. Paper presented at the 2nd International Conference on Biomedical Ontology (26–30 July), Buffalo, NY, USA.

CHAPTER 10

Perspectives for functional diversity research

10.1 Introduction

This book has shown the utility of the trait-based comparative approach in addressing questions concerning the functioning and distribution of organisms, community structure, ecosystem properties, and the provision of ecosystem services in ecological and agricultural contexts. The characterization of the functional aspect of diversity using traits is the most recent incarnation of a tradition whose aim has been to bring together the field of evolutionary ecology and its emphasis on adaptation, and that of functional ecology which is focused on the fluxes of materials and energy in ecological systems. It has rejuvenated approaches to diversity in ecology and has led to advances beyond those possible when species are simply described by their taxonomic identity.

A number of events and factors have been decisive in advancing plant trait research over the last twenty years. First, the strong need to predict changes in vegetation in response to the different components of global change has initiated research into methods of simplifying the treatment of floristic complexity. Second, the development of a coherent ensemble of concepts and hypotheses for moving from the level of the individual to higher levels of organization, and also for explicitly taking into account the concept of ecosystem services, has proved to be particularly powerful and efficient. Third, the identification of a small number of easily measured traits pertinent to plant functioning and distribution has allowed their quantification for a large range of species from a wide range of situations from around the planet, thus providing the functional approach with a large capacity for generalization. Fourth, the development of protocols for standardized measurements and the realization of the necessity of data and information sharing between researchers working in different systems have been essential to the progress achieved so far.

The numerous advances achieved over the last few years have also revealed those research frontiers that will need to be pushed further in order to fully realize the potential of the approach. Some of the principal future research directions that we have identified are briefly discussed below. These are not exhaustive, and are also necessarily subjective as they emerge from our perception of the subject. They recapitulate the questions dealt with throughout this book and are presented so as to take into account the levels of organization of the ecological systems which structure them.

10.2 Basis and theory of the trait-based approach to diversity

If the conceptual basis of the trait-based approach to diversity is now well established (see Chapter 2, Figure 2.6), a formal theoretical framework articulating in a quantitative and mechanistic manner the response of organisms to environmental factors, the functional structure and community dynamics, and ecosystem properties remains to be developed. A number of efforts have been made in this direction (Enquist et al., 2015; Norberg et al., 2001; Savage et al., 2007; Webb et al., 2010), but they require testing and further precision before they become finalized. The Trait Drivers Theory (TDT) recently proposed by Enquist et al. (2015) includes key stages of the framework presented in Chapter 2, and gives a central role to distribution of traits in communities in understanding both the response of vegetation to the environment and the effect of community structure on ecosystem functioning. A major tenet of the TDT is that this understanding can be greatly improved by coupling the trait-based approach to diversity with the metabolic scaling theory.

One of the points to take into account in these theories concerns one of the foundations of the approach, relating to the nature of the entity involved in the interaction between organisms and the environment. As discussed in Chapter 4, the identification of response traits to environmental factors suggests that environmental filters act via certain combinations of the functions of organisms, depending on the factor involved. It is thus evident that it is the phenotype as a whole which is being filtered, and not just one or multiple functions. The question of the relationships between traits/functions and the performance of phenotypes has long been an issue in ecology (Ackerly & Monson, 2003; Craine et al., 2012), and is still currently highly debated. Advancing this issue requires the selection of the most pertinent traits in the causal chain linking functional traits, performance traits, and the fitness of individuals, as presented in Figure 2.2 (Chapter 2), and constitutes a central point in the theoretical framework proposed by Webb et al. (2010). Taking into account the phenotype in its wider sense is also dependent on integrating in a more direct manner the evolutionary components of its response to environmental factors.

Beyond theoretical aspects, the recent availability of trait (Kattge et al., 2011) and demographic (Salguero-Gómez et al., 2015) databases with large taxonomic coverage will allow one to better understand the relationships between plant functioning as assessed with trait-based approaches and the demography of populations (see Adler et al., 2014). Advances in this area depend on an improved dialogue between ecology and demography (Crone et al., 2011), and would be a step towards reconciling 'life-history' and 'ecological' perspectives to strategies as discussed in Chapter 4.

10.3 Functional diversity and intraspecific variability

The trait-based approach to ecology is partly based on the hypothesis that the variability in functioning—and thus in traits—within a species is lower than the variability found between species. The validity of this hypothesis has led to much debate over the last three decades, without the emergence of any consensus (Albert et al., 2011; Bradshaw, 1987; Garnier et al., 2001; Harper, 1982; Messier et al., 2010). Beyond the issue of the intraspecific trait variability per se (see Box 4.1), it is the scale of the observations and the level of precision at which the processes are studied which are central for this debate. To illustrate the problem, let us use the analogy of the parable of the blind men and the elephant used in Chapter 2 (Box 2.2): the fourth blind man, after having touched the elephant's knee and then descending along the leg, discovers at its extremity the presence of five fingers of very different sizes. Let us now ask the following question: given that we know that the elephant is an animal and that it can move, what does the size difference between the five fingers of the elephant tell us about the speed at which the elephant is able to run? Asking this question does not mean that we deny the existence of this difference, but rather that we are interested in its influence for the movement of the elephant.

Similarly, it is not the central importance of intraspecific variability in evolutionary processes which is debated in comparative ecological approaches, but rather the role that this variability can have for ecological processes. A formal framework allowing for the examination of the cases in which intraspecific trait variability might be important to take into account has been presented by Albert et al. (2011, Figure 4.7 in Chapter 4). This depends in particular on the relative importance of intraspecific variability vs interspecific variability in the pool of species being studied and the scales of space and time relevant to the studied processes. While there are cases for which taking intraspecific variability into account is obviously indispensable, there are also others for which it can probably be ignored (cf. Figure 4.7). Current knowledge does not allow us to determine this for all possible questions and situations. To allow us to do this, it is necessary to have at the same time sufficient data, standardized sampling protocols, and adequate indices to quantify intraspecific variation (Valladares et al., 2006), which represents a considerable body of work for the future.

Another aspect of the study of intraspecific variability concerns the quantification of response curves of different traits to environmental factors. As explained in Chapter 4, this response is specific to each trait × environmental factor combination, and has led to the development of the meta-phenomic approach (Poorter et al., 2010). An alternative approach has been proposed by Craine et al. (2012), which is based on the determination of response curves of the performance of whole organisms to selected environmental factors, rather than those of individual traits. This is progressively being made possible by combining screening experiments on thousands of species through large-scale phenotyping with the development of bioinformatic approaches. Craine et al. (2012) postulate that determining the response of whole organisms to water and light availability, and to soil nutrient concentrations, would allow for the explanation of species distributions in a large variety of habitats. The approaches proposed by Poorter et al. (2010) and Craine et al. (2012), strongly based on ecophysiology, should lead to improvements in our understanding of the responses of plants to environmental factors at a greater scale than possible until now.

10.4 Traits, functioning of organisms, and environmental factors

A number of research directions concerning the functional characterization of organisms appear promising. For plants, for which this research has been the most active over the last two decades, three important aspects can be put forward. The first one concerns the identification of the minimum number of independent axes of variation that adequately describes the functional variation among plants (Laughlin, 2014). This touches the second aspect, which relates to the coordination of structure and function among organs: in the context of the fast–slow continuum of plant functioning (discussed in Chapter 3), Reich (2014) has formulated the hypothesis that 'being fast for all resources at any one organ level (e.g. the leaf level) requires being fast for all resources at the other organ levels (e.g. stem and root levels)'. Whether this really holds is currently debated (Baraloto et al., 2010; Freschet et al., 2010), and requires further testing on a wide range of species. Answers to this question relate in particular to the third aspect, which concerns the functioning of roots. For obvious reasons due to methodological difficulties, research on roots is much less advanced than that focused on leaves. Some research axes are currently being intensively explored among which should be noted (Craine, 2009; McCormack et al., 2015; Roumet et al., 2006; and see Chapter 3): (1) the identification of traits linked to key root functions such as the acquisition of mineral elements, plant anchoring, rhizospheric activity, and decomposition; (2) studies of the relationships between root and shoot traits with a view of using the latter as proxies for estimating root functioning; and (3) standardization of the measurement of root traits to allow for large-scale comparisons.

While currently less developed than for plants, the trait-based approach to diversity has also led to substantial advances in addressing questions relative to other types of organisms. This is the case for invertebrates (Dennis et al., 2004; Moretti et al., 2009; Pey et al., 2014; Southwood, 1977), vertebrates (Blondel, 2003; Gaillard et al., 1989; Villéger et al., 2010), and microorganisms (Green et al., 2008; Rinaldo et al., 2002; Salles et al., 2009). Grime and Pierce (2012) further give a broad overview of how the concept of ecological strategies can be applied to animals. Despite a number of major initiatives for animals, such as the development of the FishBase database (http://www.fishbase.org/) for fishes, or the BETSI database (Pey et al., 2014) for soil macrofauna, the primary functions of interest as well as a consensually agreed list of traits associated with them remain to be defined in order to reveal the full potential of the approach for these organisms.

Finally, but certainly not the least importantly, better identification and quantification of environmental factors as they are perceived by organisms is essential; the current deficit in this field is probably one of the major obstacles to understanding the relationships between organisms and their environment (Austin, 2013; Garnier et al., 2007; Shipley, 2010), and is true for both natural and agricultural environments (Gaba et al., 2014b).

10.5 Community functional structure, ecosystem properties, and services

Understanding the mechanisms by which the traits of organisms, modulated by environmental factors and biotic interactions, affect the different components of the functional structure of communities (cf. Chapter 5) will require the combined efforts of theoretical and experimental approaches and modelling (Enquist et al., 2015; Webb et al., 2010), as well as improvements in methodological approaches to quantify this functional structure (de Bello et al., 2010a; Pavoine & Bonsall, 2011). In particular, the manner in which the different facets of diversity (taxonomic, phylogenetic, and functional) and its different levels (α, β, and γ) should be quantified is the matter of current debate. These difficulties need to be overcome in particular to identify the role of phylogeny in structuring communities (Cavender-Bares et al., 2009; Mouquet et al., 2012). Moving onwards, specific efforts to better understand the dynamics of communities (modifications of structure over the course of time), while essential for the prediction of effects on ecological systems, such as those brought about by global change have to be developed (Enquist et al., 2015; Webb et al., 2010). This requires notably the identification of linkages between functional traits, the regeneration niche of species (sensu Grubb, 1977), and, as discussed above, the demography of populations.

The scaling of community functional structure at large geographic scales is now made possible by combining trait databases with databases assembling floristic relevés (cf. Swenson et al., 2012), and constitutes the emerging field of functional biogeography (Violle et al., 2014). This field has the potential to assess ecosystem properties and services at an unprecedented scale (see Violle et al., 2015, for an application to permanent grasslands), provided that accurate relationships between traits and ecosystem properties are established.

In fact, the identification of traits and functional assemblages which have a role in determining ecosystem properties (primary and secondary production, degradation of organic matter and nutrient cycling, pools of mineral elements, resilience, etc.) and the services which they provide to human society remains in its early stages (see the recent review by Lavorel, 2013). The relative roles of dominance and functional complementarity for ecosystem properties and services are still not well understood (Lavorel, 2013, and see discussion in Chapter 6). The identification of traits linked to the combinations of processes and services involved in trade-offs is of prime importance, and the methodological issues concerning the quantification of the functional structure of communities mentioned above should also be considered here. It should also be kept in mind that only a limited number of processes related to services have been studied until now (cf. Chapter 7). The evaluation of services at a given site is an important challenge in order to respond to the demands of land managers, and requires the development of innovative approaches integrating ecological science with the economic and social sciences (Daily et al., 2009; Lamarque et al., 2011).

Another important point concerns the role of traits of primary producers in the organization of trophic networks (Grigulis et al., 2013; Grime & Pierce, 2012; Lavorel et al., 2013; Wardle et al., 2004). Wardle et al. (2004) proposed a framework of the propagation of the effects of plant traits towards herbivores and soil organisms, which remains to be experimentally tested, in particular by integrating recent advances in the area of community functional structure. The combined effect of the traits of organisms (plants, animals, decomposers, etc.) on the synergies and trade-offs between ecosystem properties and services also constitute a currently largely open field for future research (de Bello et al., 2010b; Lavorel, 2013; Lavorel & Grigulis, 2012). Due to the complex roles of different organisms in ecosystems, this work requires a multidisciplinary approach linking functional ecology, community ecology, soil science, and microbial ecology.

10.6 Perspectives in the field of agriculture

The potential applications of the trait-based approach based in the field of agriculture are wide ranging. Intensive agricultural systems, in which crop diversity is reduced to a few (or even only one) species, generally homogenous from a genetic point of view, and in which external inputs are used in large quantities, are being more and more criticized for their negative impacts on the environment. At the same time feeding a growing world population is becoming increasingly difficult (Aubertot et al., 2007; Doré et al., 2011; Giller et al., 1997; Griffon, 2010; Le Roux et al., 2009; Wezel et al., 2009). Reducing such impacts will require, for example, a reduction in the use of fertilizers and/or pesticides, and a greater use of natural regulatory systems in which a larger and more diverse range of organisms are involved (e.g. soil organisms for the control of soil fertility, and non-pest insects and/or microorganisms for the biological control of pest species: cf. Gaba et al., 2014a). It will also require introducing greater diversity into cultivated species (including more varieties), and growing them in mixtures, but also changes to agricultural practices and cropping systems (Weiner et al., 2010).

This transfer of ecological approaches towards agronomy is already occurring in circumstances where the level of realized or potential specific diversity is high. As has been shown for grassland systems (Chapter 8), tools based on traits can be useful for the management of complex systems, where mechanistic approaches or ecological modelling requiring considerable knowledge of all of the involved species are currently not possible. Trait-based approaches can also be used to select species or genotypes for specific purposes: for example, Damour et al. (2015) have recently assessed the potentialities of using the approach to select cover crops in banana cropping systems, while screening procedures comparable to those developed in comparative ecology have been implemented to select such relevant cover crops in different systems (Damour et al., 2014; Tribouillois et al., 2015). Malézieux et al. (2009) have further suggested that trait or functional group approaches should be used to characterize and model multi-species cropping systems such as agroforestry systems, upon which a large proportion of the world's farmers depend, particularly in tropical regions (Malézieux et al., 2009; Swift & Anderson, 1993). However, transferring these concepts to systems of low diversity has received little attention. A prerequisite is that the specificities of these cropping systems are taken into account: these are linked to the types and intensity of selective pressures (e.g. the nature and return interval of different disturbances such as ploughing and/or the application of pesticides), and the particular structure of communities that are strongly determined by the type of plant being cultivated. These selective forces should be identified at the same time as the responses of all organisms involved, either the cultivated plants themselves or other organisms coexisting with the crop (Gaba et al., 2014a). This approach has been applied only rarely in highly artificialized systems. Such an application would provide a rigorous test of the trait-based approach outside of the initial ecological context in which it was developed. Applicability in such systems would also be valuable for the development of 'innovative' cropping systems (see Figure 8.9 in Chapter 8), responding to new agricultural, societal, and environmental goals. These systems are based on a detailed understanding of the generic links existing between the structural components of cropping systems, their performance, and the environmental conditions necessary for their adequate functioning (Doré et al., 2011; Meynard et al., 2012). Future studies could focus on a comparative approach to agricultural systems based on the quantification of (1) the gradients generated by agricultural practices, (2) the response traits of organisms to these gradients, and (3) the effects of these responses on the desired ecosystem services.

10.7 Data management

As explained in Chapter 9, despite recent efforts to develop plant trait databases, a large amount of data concerning the traits of organisms remains dispersed and heterogeneous, both from the syntactic and semantic points of view. Recent advances in the area of semantics (cf. Chapter 9) represent the first step on the path to increasing interoperability, and efforts in this area need to be continued in the different fields concerned with ecology (Deans et al., 2015; Walls et al., 2014). However this only represents one aspect of a global approach to the management of ecological data.

In general, increasing scientific knowledge stems from the acquisition of data and their transformation into relevant information which is then integrated into a body of facts, principles, and scientific theories. In order for data to contribute efficiently to the advance of scientific knowledge, Strasser et al. (2011) and Michener and Jones (2012) have suggested that data should enter a life cycle involving eight stages, including (1) planning, (2) collecting, (3) assuring quality control, (4) describing, and (5) archiving, in particular for those projects requiring the collection of new data. Stages 6 (discovery) and 7 (integration) are more relevant to scientific synthesis and meta-analysis. The collection of new data and data synthesis leads necessarily to the last identified stage: that of analysis (8), which allows for the identification of new questions and approaches, and leads to the development of new experiments and/or original syntheses. These different stages are not necessarily exclusive, nor obligatorily sequential. From an operational point of view, this life cycle is accompanied by the establishment of a 'data management plan', which consists of implementing specific procedures for each of these stages (Michener & Jones, 2012). Some of these stages are still poorly integrated into current ecological practices, such as (for example) the systematic recording of the manner in which the data have been analysed, clarification of the manner in which the data can be diffused and reutilized, or planning for the long-term archiving of the data.

The major objective of ecoinformatics (Jones et al., 2006; Michener & Jones, 2012) is to allow the development of this type of global approach to data management, and this requires strong interactions between information sciences, computing, and the natural sciences. Its implementation requires overcoming a number of both technical and structural obstacles, as well as overcoming a major sociocultural limitation in that the sharing of data in the ecological community remains the exception rather than the rule (Reichman et al., 2011). It is to be hoped that the importance of the challenges ecology is facing at the beginning of the twenty-first century will be the trigger to overcome these obstacles and lead ecology into the age of information.

10.8 Conclusions

Understanding the variation in the degree of expression of the functions of organisms, and the resulting consequences at the different levels of organization of the living world, remains one of the major challenges facing ecology in the current context of global change induced by human activities. As stated by Westoby et al. (2002), 'A lot remains to be done, but there is a real hope that we are heading in the right direction.' Going in the right direction will require combining approaches using experimentation, observation, developing appropriate information technology, and systems modelling. However, even beyond these combinations, it will be collaborations between disciplines which will likely provide future major advances. Such collaborations are necessary both within the ecological and agricultural sciences (taxonomy, ecophysiology, biogeochemistry, environmental microbiology, demography, soil science, climatology, and macroecology), the human sciences (geography, sociology, ethnology, economics, and even psychology), and the statistical and information sciences, as well as among these disciplines. This initiative towards improved scientific synthesis should consider biodiversity in the general context of the functioning of ecosystems, whether these are little disturbed or heavily modified, and in interaction with human societies.

10.9 Key points

1. Research in the field of plant functional diversity over the last decades has led to advances relevant to the functioning of organisms, the structure of communities, and the determinants of ecosystem properties and services these deliver to human societies. What are the perspectives of this field?
2. A first aspect concerns the theory and bases of the approach, for which a formal theoretical framework is still lacking. The objective is to relate in a quantitative and mechanistic manner the responses of organisms to environmental factors, the functional structure and dynamics of communities, and ecosystem properties. A key point concerns the nature of the entity

to consider—trait, whole phenotype?—to assess interactions between organisms and their environment. A thorough evaluation of questions for which intraspecific variability of organism functioning needs to be taken into account is also required.

3. For plants, taking into account the different organs, including roots, in an integrated vision of plant functioning is another priority. Explicit consideration of the regeneration niche as well as the manner in which traits relate to population demographic parameters constitutes major challenges for future research. Improving the identification and quantification of environmental factors as they are perceived by organisms is also essential. For organisms other than plants (e.g. microorganisms, animals), a trait-based approach to functional diversity is being developed, but the principal functions of interest and a consensual baseline list of associated traits remain to be established before the approach can achieve its full potential.
4. Understanding the mechanisms by which environmental factors and biotic interactions determine the functional structure of communities, and the identification of the traits and assemblages which play a role in determining ecosystem properties and the services they provide to human societies, will require combining theoretical approaches, experimental approaches, and modelling. The potential for applications in an agricultural context, in which intensive management practices are currently being questioned, is wide ranging and remains largely unexplored.
5. Improving the management of functional diversity data—and more generally that of data from all of the fields of ecology—is a prerequisite if these data are to contribute efficiently to advances in scientific knowledge. The emerging field of ecoinformatics aims at removing the obstacles preventing their efficient use: data should enter a well-planned 'life cycle', from the planning of experiments up to their integration.
6. These various directions, which are not in any way exhaustive, should allow us to understand the causes of variation in the degree of expression of functions by organisms and their consequences at the different levels of organization of the living world. They constitute major challenges for ecology in the current context of global change induced by human activities.

10.10 References

Ackerly, D. D., & Monson, R. K. (2003). Waking the sleeping giant: the evolutionary foundations of plant functions. *International Journal of Plant Sciences, 164 (3 Suppl.)*, S1–S6.

Adler, P. B., Salguero-Gómez, R., Compagnoni, A., Hsu, J. S., Ray-Mukherjee, J., Mbeau-Ache, C., & Franco, M. (2014). Functional traits explain variation in plant life history strategies. *Proc. Natl. Acad. Sci. USA, 111*(2), 740–745. doi:10.1073/pnas.1315179111

Albert, C. H., Grassein, F., Schurr, F. M., Vieilledent, G., & Violle, C. (2011). When and how should intraspecific variability be considered in trait-based plant ecology? *Perspectives in Plant Ecology, Evolution and Systematics, 13*(3), 217–225. doi:10.1016/j.ppees.2011.04.003

Aubertot, J.-N., Barbier, J.-M., Carpentier, A., Gril, J.-J., Guichard, L., Lucas, P., ... Voltz, M. (eds). (2007). *Pesticides, agriculture et environnement. Réduire l'utilisation des pesticides et en limiter les impacts environnementaux.* Versailles: Quae.

Austin, M. P. (2013). Inconsistencies between theory and methodology: a recurrent problem in ordination studies. *Journal of Vegetation Science, 24*(2), 251–268. doi:10.1111/j.1654-1103.2012.01467.x

Baraloto, C., Paine, C. E. T., Poorter, L., Beauchene, J., Bonal, D., Domenach, A. M., ... Chave, J. (2010). Decoupled leaf and stem economics in rain forest trees. *Ecology Letters, 13*(11), 1338–1347. doi:10.1111/j.1461-0248.2010.01517.x

Blondel, J. (2003). Guilds or functional groups: does it matter? *Oikos, 100*, 223–231.

Bradshaw, A. D. (1987). Comparison – Its scope and limits. *New Phytologist, 106 (Suppl.)*, 3–21.

Cavender-Bares, J., Kozak, K. H., Fine, P. V. A., & Kembel, S. W. (2009). The merging of community ecology and phylogenetic biology. *Ecology Letters, 12*, 693–715.

Craine, J. M. (2009). *Resource Strategies of Wild Plants.* Princeton and Oxford: Princeton University Press.

Craine, J. M., Engelbrecht, B. M. J., Lusk, C. H., McDowell, N., & Poorter, H. (2012). Resource limitation, tolerance, and the future of ecological plant classification. *Frontiers in Plant Science, 3.* doi:10.3389/fpls.2012.00246

Crone, E. E., Menges, E. S., Ellis, M. M., Bell, T., Bierzychudek, P., Ehrlen, J., ... Williams, J. L. (2011). How do plant ecologists use matrix population models? *Ecology Letters*, *14*(1), 1–8. doi:10.1111/j.1461-0248.2010.01540.x

Daily, G. C., Polasky, S., Goldstein, J., Kareiva, P. M., Mooney, H. A., Pejchar, L., ... Shallenberger, R. (2009). Ecosystem services in decision making: time to deliver. *Frontiers in Ecology and the Environment*, *7*(1), 21–28. doi:10.1890/080025

Damour, G., Dorel, M., Quoc, H. T., Meynard, C., & Risède, J.-M. (2014). A trait-based characterization of cover plants to assess their potential to provide a set of ecological services in banana cropping systems. *European Journal of Agronomy*, *52*, 218–228. doi:10.1016/j.eja.2013.09.004

Damour, G., Garnier, E., Navas, M.-L., Dorel, M., & Risède, J.-M. (2015). Using functional traits to assess the services provided by cover plants: a review of potentialities in banana cropping systems. *Advances in Agronomy*, *134*, 81–133. doi:10.1016/bs.agron.2015.06.004

de Bello, F., Lavergne, S., Meynard, C. N., Leps, J., & Thuiller, W. (2010a). The partitioning of diversity: showing Theseus a way out of the labyrinth. *Journal of Vegetation Science*, *21*(5), 992–1000. doi:10.1111/j.1654-1103.2010.01195.x

de Bello, F., Lavorel, S., Díaz, S., Harrington, R., Cornelissen, J. H. C., Bardgett, R. D., ... Harrison, P. A. (2010b). Towards an assessment of multiple ecosystem processes and services via functional traits. *Biodiversity and Conservation*, *19*, 2873–2893.

Deans, A. R., Lewis, S. E., Huala, E., Anzaldo, S. S., Ashburner, M., Balhoff, J. P., ... Mabee, P. (2015). Finding our way through phenotypes. *PLoS Biol*, *13*(1), e1002033. doi:10.1371/journal.pbio.1002033

Dennis, R. L. H., Hodgson, J. G., Grenyer, R., Shreeve, T. G., & Roy, D. B. (2004). Host plants and butterfly biology. Do host-plant strategies drive butterfly status? *Ecological Entomology*, *29*, 12–26.

Doré, T., Makowski, D., Malézieux, E., Munier-Jolain, N., Tchamitchian, M., & Tittonell, P. (2011). Facing up to the paradigm of ecological intensification in agronomy: revisiting methods, concepts, and knowledge. *European Journal of Agronomy*, *34*(4), 197–210. doi:10.1016/j.eja.2011.02.006

Enquist, B. J., Norberg, J., Bonser, S. P., Violle, C., Webb, C. T., & Savage, V. M. (2015). Scaling from traits to ecosystems: developing a general Trait Driver Theory via integrating trait-based and metabolic scaling theories. *Advances in Ecological Research*, *52*, 249–318. doi: 10.1016/bs.aecr.2015.02.001

Freschet, G. T., Cornelissen, J. H. C., van Logtestijn, R. S. P., & Aerts, R. (2010). Evidence of the 'plant economics spectrum' in a subarctic flora. *Journal of Ecology*, *98*, 362–373.

Gaba, S., Bretagnolle, F., Rigaud, T., & Philippot, L. (2014a). Managing biotic interactions for ecological intensification of agroecosystem. *Frontiers in Ecology and Evolution*, *2*. doi:10.3389/fevo.2014.00029

Gaba, S., Fried, G., Kazakou, E., Chauvel, B., & Navas, M.-L. (2014b). Agroecological weed control using a functional approach: a review of cropping systems diversity. *Agronomy for Sustainable Development*, *34*(1), 103–119. doi:10.1007/s13593-013-0166-5

Gaillard, J. M., Pontier, D., Allaine, D., Lebreton, J. D., Trouvilliez, J., & Clobert, J. (1989). An analysis of demographic tactics in birds and mammals. *Oikos*, *56*(1), 59–76. doi:10.2307/3566088

Garnier, E., Laurent, G., Bellmann, A., Debain, S., Berthelier, P., Ducout, B., ... Navas, M.-L. (2001). Consistency of species ranking based on functional leaf traits. *New Phytologist*, *152*, 69–83.

Garnier, E., Lavorel, S., Ansquer, P., Castro, H., Cruz, P., Dolezal, J., ... Zarovali, M. (2007). Assessing the effects of land use change on plant traits, communities, and ecosystem functioning in grasslands: a standardized methodology and lessons from an application to 11 European sites. *Annals of Botany*, *99*, 967–985.

Giller, K. E., Beare, M. H., Lavelle, P., Izac, M. N., & Swift, M. J. (1997). Agricultural intensification, soil biodiversity, and agroecosystem function. *Applied Soil Ecology*, *6*, 3–16. doi:10.1016/S0929-1393(96)00149-7

Green, J. L., Bohannan, B. J. M., & Whitaker, R. J. (2008). Microbial biogeography: from taxonomy to traits. *Science*, *320*, 1039–1043.

Griffon, M. (2010). *Pour des agricultures écologiquement intensives*. La Tour d'Aigues: Editions de l'Aube.

Grigulis, K., Lavorel, S., Krainer, U., Legay, N., Baxendale, C., Dumont, M., ... Clément, J.-C. (2013). Relative contributions of plant traits and soil microbial properties to mountain grassland ecosystem services. *Journal of Ecology*, *101*(1), 47–57. doi:10.1111/1365-2745.12014

Grime, J. P., & Pierce, S. (2012). *The Evolutionary Strategies that Shape Ecosystems*. Oxford: Wiley-Blackwell.

Grubb, P. J. (1977). The maintenance of species-richness in plant communities: the importance of the regeneration niche. *Biological Reviews*, *52*, 107–145.

Harper, J. L. (1982). After description. In E. I. Newman (ed.), *The Plant Community as a Working Mechanism* (pp. 11–25). Oxford: Blackwell Scientific Publications.

Jones, M. B., Schildhauer, M. P., Reichman, O. J., & Bowers, S. (2006). The new bioinformatics: integrating ecological data from the gene to the biosphere. *Annual Review of Ecology, Evolution and Systematics*, *37*, 519–544.

Kattge, J., Díaz, S., Lavorel, S., Prentice, I. C., Leadley, P., Bönisch, G., ... Wirth, C. (2011). TRY: a global database of plant traits. *Global Change Biology* *17*, 2905–2935. doi:10.1111/j.1365-2486.2011.02451.x

Lamarque, P., Quetier, F., & Lavorel, S. (2011). The diversity of the ecosystem services concept and its implications for their assessment and management.

Comptes Rendus Biologies, *334*(5–6), 441–449. doi:10.1016/j.crvi.2010.11.007

Laughlin, D. C. (2014). The intrinsic dimensionality of plant traits and its relevance to community assembly. *Journal of Ecology*, *102*(1), 186–193. doi:10.1111/1365-2745.12187

Lavorel, S. (2013). Plant functional effects on ecosystem services. *Journal of Ecology*, *101*(1), 4–8. doi:10.1111/1365-2745.12031

Lavorel, S., & Grigulis, K. (2012). How fundamental plant functional trait relationships scale-up to trade-offs and synergies in ecosystem services. *Journal of Ecology*, *100*(1), 128–140. doi:10.1111/j.1365-2745.2011.01914.x

Lavorel, S., Storkey, J., Bardgett, R. D., de Bello, F., Berg, M. P., Le Roux, X., ... Harrington, R. (2013). A novel framework for linking functional diversity of plants with other trophic levels for the quantification of ecosystem services. *Journal of Vegetation Science*, *24*(5), 942–948. doi:10.1111/jvs.12083

Le Roux, X., Barbault, R., Baudry, J., Burel, F., Doussan, I., Garnier, E., ... Trommetter, M. (eds). (2009). *Agriculture et biodiversité. Valoriser les synergies*. Versailles: Quae.

Malézieux, E., Crozat, Y., Dupraz, C., Laurans, M., Makowski, D., Ozier-Lafontaine, H., ... Valantin-Morison, M. (2009). Mixing plant species in cropping systems: concepts, tools, and models. A review. *Agronomy for Sustainable Development*, *29*, 43–62.

McCormack, M. L., Dickie, I. A., Eissenstat, D. M., Fahey, T. J., Fernandez, C. W., Guo, D., ... Zadworny, M. (2015). Redefining fine roots improves understanding of below-ground contributions to terrestrial biosphere processes. *New Phytologist*. doi:10.1111/nph.13363

Messier, J., McGill, B. J., & Lechowicz, M. J. (2010). How do traits vary across ecological scales? A case for trait-based ecology. *Ecology Letters*, *13*, 838–848.

Meynard, J.-M., Dedieu, B., & Bos, A. P. B. (2012). Re-design and co-design of farming systems. An overview of methods and practices. In I. Darnhofer, D. Gibon, & B. Dedieu (eds), *Farming Systems Research into the 21st Century: The New Dynamic* (pp. 407–432). Dordrecht: Springer.

Michener, W. K., & Jones, M. B. (2012). Ecoinformatics: supporting ecology as a data-intensive science. *Trends in Ecology & Evolution*, *27*(2), 85–93. doi:10.1016/j.tree.2011.11.016

Moretti, M., de Bello, F., Roberts, S. P. M., & Potts, S. G. (2009). Taxonomical vs. functional responses of bee communities to fire in two contrasting climatic regions. *Journal of Animal Ecology*, *78*(1), 98–108. doi:10.1111/j.1365-2656.2008.01462.x

Mouquet, N., Devictor, V., Meynard, C. N., Munoz, F., Bersier, L. F., Chave, J., ... Thuiller, W. (2012). Ecophylogenetics: advances and perspectives. *Biological Reviews*, *87*(4), 769–785. doi:10.1111/j.1469-185X.2012.00224.x

Norberg, J., Swaney, D. P., Dushoff, J., Lin, J., Casagrandi, R., & Levin, S. A. (2001). Phenotypic diversity and ecosystem functioning in changing environments: A theoretical framework. *Proc. Natl. Acad. Sci. USA*, *98*(20), 11376–11381. doi:10.1073/pnas.171315998

Pavoine, S., & Bonsall, M. B. (2011). Measuring biodiversity to explain community assembly: a unified approach. *Biological Reviews*, *86*(4), 792–812. doi:10.1111/j.1469-185X.2010.00171.x

Pey, B., Nahmani, J., Auclerc, A., Capowiez, Y., Cluzeau, D., Cortet, J., ... Hedde, M. (2014). Current use of and future needs for soil invertebrate functional traits in community ecology. *Basic and Applied Ecology*, *15*, 194–206. doi:10.1016/j.baae.2014.03.007

Poorter, H., Niinemets, Ü., Walter, A., Fiorani, F., & Schurr, U. (2010). A method to construct dose-response curves for a wide range of environmental factors and plant traits by means of a meta-analysis of phenotypic data. *Journal of Experimental Botany*, *61*(8), 2043–2055. doi:10.1093/jxb/erp358

Reich, P. B. (2014). The world-wide 'fast–slow' plant economics spectrum: a traits manifesto. *Journal of Ecology*, *102*(2), 275–301. doi:10.1111/1365-2745.12211

Reichman, O. J., Jones, M. B., & Schildhauer, M. P. (2011). Challenges and opportunities of open data in ecology. *Science*, *331*, 703–705.

Rinaldo, A., Maritan, A., Cavender-Bares, K. K., & Chisholm, S. W. (2002). Cross-scale ecological dynamics and microbial size spectra in marine ecosystems. *Proceedings of the Royal Society of London. Series B: Biological Sciences*, *269*(1504), 2051–2059. doi:10.1098/rspb.2002.2102

Roumet, C., Urcelay, C., & Díaz, S. (2006). Suites of root traits differ between annual and perennial species growing in the field. *New Phytologist*, *170*, 357–358.

Salguero-Gómez, R., Jones, O. R., Archer, C. R., Buckley, Y. M., Che-Castaldo, J., Caswell, H., ... Vaupel, J. W. (2015). The *Compadre* Plant Matrix Database: an open online repository for plant demography. *Journal of Ecology*, *103*(1), 202–218. doi:10.1111/1365-2745.12334

Salles, J. F., Poly, F., Schmid, B., & Le Roux, X. (2009). Community niche predicts the functioning of denitrifying bacterial assemblages. *Ecology*, *90*(12), 3324–3332. doi:10.1890/09-0188.1

Savage, V. M., Webb, C. T., & Norberg, J. (2007). A general multi-trait-based framework for studying the effects of biodiversity on ecosystem functioning. *Journal of Theoretical Biology*, *247*(2), 213–229. doi:10.1016/j.jtbi.2007.03.007

Shipley, B. (2010). *From Plant Traits to Vegetation Structure. Chance and Selection in the Assembly of Ecological Communities*. Cambridge: Cambridge University Press.

Southwood, T. R. E. (1977). Habitat, templet for ecological strategies – Presidential address to the British Ecological

Society, 5 January 1977. *Journal of Animal Ecology*, *46*(2), 337–365.

Strasser, C., Cook, R., Michener, W. K., Budden, A., & Koskela, R. (2011). *DataONE – Promoting data stewardship through best practices*. Paper presented at the Environmental Information Management Conference 2011 (EIM 2011), Santa Barbara, CA.

Swenson, N. G., Enquist, B. J., Pither, J., Kerkhoff, A. J., Boyle, B., Weiser, M. D., ... Nolting, K. M. (2012). The biogeography and filtering of woody plant functional diversity in North and South America. *Global Ecology and Biogeography*, *21*(8), 798–808. doi:10.1111/j.1466-8238.2011.00727.x

Swift, M. J., & Anderson, J. M. (1993). Biodiversity and ecosystem function in agricultural systems. In E.-D. Schulze & H. A. Mooney (eds), *Biodiversity and Ecosystem Function* (pp. 15–41). Berlin: Springer-Verlag.

Tribouillois, H., Fort, F., Cruz, P., Charles, R., Flores, O., Garnier, E., & Justes, E. (2015). A functional characterisation of a wide range of cover crop species: growth and nitrogen acquisition rates, leaf traits, and ecological strategies. *PLoS ONE*, *10*(3), e0122156. doi:10.1371/journal.pone.0122156

Valladares, F., Sanchez-Gomez, D., & Zavala, M. A. (2006). Quantitative estimation of phenotypic plasticity: bridging the gap between the evolutionary concept and its ecological applications. *Journal of Ecology*, *94*(6), 1103–1116. doi:10.1111/j.1365-2745.2006.01176.x

Villéger, S., Miranda, J. R., Hernandez, D. F., & Mouillot, D. (2010). Contrasting changes in taxonomic vs. functional diversity of tropical fish communities after habitat degradation. *Ecological Applications*, *20*(6), 1512–1522. doi:10.1890/09-1310.1

Violle, C., Choler, P., Borgy, B., Garnier, E., Amiaud, B., Debarros, G., ... Viovy, N. (2015). Vegetation ecology meets ecosystem science: Permanent grasslands as a functional biogeography case study. *Science of The Total Environment*, *534*, 43–51. doi:10.1016/j.scitotenv.2015.03.141

Violle, C., Reich, P. B., Pacala, S. W., Enquist, B. J., & Kattge, J. (2014). The emergence and promise of functional biogeography. *Proc. Natl. Acad. Sci. USA*, *111*(38), 13690–13696. doi:10.1073/pnas.1415442111

Walls, R. L., Deck, J., Guralnick, R., Baskauf, S., Beaman, R., Blum, S., ... Wooley, J. (2014). Semantics in support of biodiversity knowledge discovery: an introduction to the biological collections ontology and related ontologies. *PLoS ONE*, *9*(3), e89606. doi:10.1371/journal.pone.0089606

Wardle, D. A., Bardgett, R. D., Klironomos, J. N., Setala, H., van der Putten, W. H., & Wall, D. H. (2004). Ecological linkages between aboveground and belowground biota. *Science*, *304*(5677), 1629–1633. doi:10.1126/science.1094875

Webb, C. O., Hoeting, J. A., Ames, G. M., Pyne, M. I., & Poff, N. L. (2010). A structured and dynamic framework to advance traits-based theory and prediction in ecology. *Ecology Letters*, *13*, 267–283.

Weiner, J., Andersen, S. B., Wille, W. K. M., Griepentrog, H. W., & Olsen, J. M. (2010). Evolutionary agroecology: the potential for cooperative, high density, weed-suppressing cereals. *Evolutionary Applications*, *3*(5–6), 473–479. doi:10.1111/j.1752-4571.2010.00144.x

Westoby, M., Falster, D. S., Moles, A. T., Vesk, P. A., & Wright, I. J. (2002). Plant ecological strategies: some leading dimensions of variation between species. *Annual Review of Ecology and Systematics*, *33*, 125–159.

Wezel, A., Bellon, S., Doré, T., Francis, C., Vallod, D., & David, C. (2009). Agroecology as a science, a movement, and a practice. A review. *Agronomy for Sustainable Development*, *29*, 503–515. doi:10.1051/agro/2009004

Index